Computer Vision for X-Ray Testing

Domingo Mery · Christian Pieringer

Computer Vision for X-Ray Testing

Imaging, Systems, Image Databases,
and Algorithms

Second Edition

 Springer

Domingo Mery
Department of Computer Science
Pontifical Catholic University of Chile
Macul, Santiago, Chile

Christian Pieringer
Independent Research Scientist and
Consultant—Artificial Intelligence
Santiago, Chile

ISBN 978-3-030-56771-2   ISBN 978-3-030-56769-9   (eBook)
https://doi.org/10.1007/978-3-030-56769-9

This Springer imprint is published by the registered company Springer Nature Switzerland AG
The registered company address is: Gewerbestrasse 11, 6330 Cham, Switzerland

To Ximena, Anais and Valeria,
who show me everyday
the X-rays of love

—Domingo

To María Isabel, Dante, and Laura,
thank you for accompanying me
through the adventures of this travel
called life

—Christian

Foreword to the Second Edition

For the second edition of this book, we had once again thought of asking Reinhard to write the foreword. We have no doubt that he would have agreed to do so with great pleasure and enthusiasm. However, he left us and said goodbye to this world in April. For this reason, in honor of him, we have preferred to leave this foreword blank, as a sign of how much we miss him.

Santiago, Chile   The Authors
June 2020

Foreword to the First Edition

The wavelengths of X-rays are far shorter than those of visible light and even shorter than those of ultraviolet light. Wilhelm Conrad Röntgen (1845–1923) was awarded the first Nobel prize in Physics in 1901 for his contributions to the detection of electromagnetic radiation, and to the generation of X-rays, which are a form of electromagnetic radiation. Radiographs are produced by having X-rays, emitted from a source, geometrically assumed to be a point in three-dimensional (3D) space, recorded on a screen. This screen might have a slightly curved surface, but we can also see it (via defined mapping) as an image plane.

X-ray technology provides a way to visualize the inside of visually opaque objects. Pixel intensities in recorded radiographs correspond basically to the density of matter, integrated along rays; those readers who are interested in a more accurate description may wish to look up the interaction of X-rays with matter by way of photo-absorption, Compton scattering, or Rayleigh scattering by reading the first chapter in this book.

X-ray technology aims at minimizing scattering, by having nearly perfect rays pass through the studied object. Thus, we have a very particular imaging modality: objects of study need to fit into a bounded space, defined as being between source and image plane, and pixel intensities have a meaning which differs from our commonly recorded digital images when using optical cameras.

When modeling an X-ray imaging system we can apply much of the projective geometry, mathematics in homogeneous spaces, or analogous parameter notations: we just need to be aware that we are looking 'backwards', from the image plane to the source (known as projection center), and no longer from the image plane into the potentially infinite space in front of an optical camera. Thus, it appears that the problem of understanding 3D objects is greatly simplified by simply studying a bounded space: using a finite number of source-plus-screen devices for recording this bounded space; applying photogrammetric methods for understanding multi-view recordings, and applying the proper interpretation (*e.g.*, basically density) to the corresponding pixel values. Thus, this very much follows a common scenario of a computer vision, while also including image preprocessing and segmentation, object detection, and classification. The book addresses all of these subjects in the particular context of X-ray testing based on computer vision.

The briefly sketched similarities between common (*i.e.*, optical-camera-based) computer vision and X-ray testing techniques might be a good motif to generate curiosity among people working in computer vision, in order to understand how their knowledge can contribute to, or benefit from, various methods of X-ray testing.

The book illustrates X-ray testing for an interesting range of applications. It also introduces a publically available software system and an extensive X-ray database. The book will undoubtedly contribute to the popularity of X-ray testing among those in the computer vision and image analysis community and may also serve as a textbook or as support material for undertaking related research.

Auckland   Reinhard Klette
April 2015

Preface to the Second Edition

The second edition of this book began to be written in the middle of 2019. On Wednesday afternoons we met in a Café on Zanelli Street in Providencia, near our homes. In the company of a good cappuccino and an avocado toast, we could review our progress and discuss what was best for our book. In the year 2020, during the COVID-19 times, our city went into lockdown for several weeks, and we had to move our meetings to video conferences... but with a good coffee that never left us. We were fully aware that computer vision had progressed a lot in the last few years, and therefore we wanted to be able to capture the most important advances in the book, without neglecting the classic bases. We enjoyed writing this book very much, and we hope that this will be reflected in the text, in the figures, in the images, and also in the codes written in Python in a friendly way. The first edition of the book [11] has been the only book that combines computer vision and X-ray testing. Today, a computer vision book that does not include deep learning cannot be conceived. We believe that this update is a must. In addition, the tendency today is to code computer vision algorithms in Python, for this reason this change is very relevant in the second edition of the book.

Scope

X-ray imaging has been developed not only for its use in medical imaging for human beings, but also for materials or objects, where the aim is to analyze— non-destructively—those inner parts that are undetectable to the naked eye. Thus, X-ray testing is used to determine if a test object deviates from a given set of specifications. Typical applications are analysis of food products, screening of baggage, inspection of automotive parts, and quality control of welds. In order to achieve efficient and effective X-ray testing, automated and semi-automated systems are being developed to execute this task. In this book, we present a general overview of computer vision methodologies that have been used in X-ray testing. In addition, some techniques that have been applied in certain relevant applications are presented: there are also some areas—like casting inspection—where automated systems are very effective, and other application areas—such as baggage screening—where human inspection is still used. There are certain application areas—like welds and cargo inspections—where the process is semi-automatic, and there is some research in areas—including food analysis—where processes are beginning to be characterized by the use of X-ray imaging. In this book, Python codes for image analysis and computer vision algorithms are presented with real X-ray images that are available in a public database created for testing and evaluation.

Organization

The book is organized as follows:

Chapter 1 (X-ray Testing): This chapter provides an introduction to the book. It illustrates principles about the physics of X-rays and describes X-ray testing and imaging systems, while also summarizing the most important issues on computer vision for X-ray testing.

Chapter 2 (Images for X-ray Testing): This chapter presents a description of the GDXray+ database, the dataset of more than 23,100 X-ray images used in this book to illustrate and test several computer vision methods. The database includes five groups of X-ray images: castings, welds, baggage, natural objects, and settings.

Chapter 3 (Geometry in X-ray Testing): This chapter presents a mathematical background of the monocular and multiple view geometry that is normally used in X-ray computer vision systems.

Chapter 4 (X-Ray Image Processing): This chapter covers the main techniques of image processing used in X-ray testing, such as image processing, image filtering, edge detection, image segmentation, and image restoration.

Chapter 5 (X-ray Image Representation): This chapter covers several topics that are used to represent an X-ray image (or a specific region of an X-ray image). This representation means that new features are extracted from the original image; this can provide us with more data than the raw information expressed as a matrix of gray values.

Chapter 6 (Classification in X-Ray Testing): This chapter covers known classifiers with several examples that can be easily modified in order to test different classification strategies. Additionally, the chapter covers how to estimate the accuracy of a classifier using hold-out, cross-validation, and leave-one-out approaches.

Chapter 7 (Deep Learning in X-ray Testing): This chapter covers deep learning strategies (convolutional neural networks, pre-trained models, transfer learning, generative adversarial networks, and detection methods) that can be used in X-ray testing. As in previous chapter, this chapter presents simple examples that can be easily modified to train and test different deep learning architectures.

Chapter 8 (Simulation in X-ray Testing): This chapter reviews some basic concepts of the simulation of X-ray images and presents simple geometric and imaging models that can be used in the simulation.

Chapter 9 (Applications in X-ray Testing): This section describes relevant applications for X-ray testing such as the inspection of castings and welds, baggage screening, quality control of natural products, and inspection of cargos and electronic circuits.

Who Is This Book for

This book covers an introduction to computer vision algorithms that can be used in X-ray testing problems such as defect detection, baggage screening, 3D recognition, quality control of food products, and inspection of cargos and electronic circuits, among others. This work may not be ideal for students of computer science or electrical engineering that want to obtain a deeper knowledge of computer vision (for which purpose there are many wonderful textbooks on image processing, pattern recognition, and computer vision[1]). Rather, it is a good starting point for undergraduate or graduate students who wish to learn basic computer vision and its application in problems of industrial radiology.[2] Thus, the aim of this book is to cover complex topics on computer vision in an easy and accessible way. For instance, we present complex topics (such as support vector machines, SIFT descriptors, and convolutional neural networks) in such a straightforward way that any student who does not have much knowledge of these fields, can still understand how they work without having to analyze complicated equations.

Hands on!

In this book there is a Python Library called pyxvis Library,[3] with around 150 functions for computer vision in X-ray testing. Each function has a 'help' with an example in order to show its use in X-ray testing. Additionally, the book gives several Python examples that can be followed by the reader. These examples use pyxvis Library. Moreover, there are more than 23,100 X-ray images on the GDXray+ database[4] that can be used to test different algorithms and codes. The available examples, Python Library and X-ray images can help people to learn more

---

[1]See for example [1, 2, 4–10, 14].

[2]Obviously, the algorithms outlined in this book can be used in similar applications such as glass inspection [3] or quality control of food products using optical images [12]—to name but a few.

[3]Available on-line on https://github.com/computervision-xray-testing/pyxvis, with all examples implemented in Google Colab.

[4]Available on-line on https://domingomery.ing.puc.cl/material/gdxray/.

about computer vision for X-ray testing. The reader can modify the codes and can create his/her own codes in order to develop new functions for X-ray testing. The reader does not need any advance knowledge of Python to read and understand this document, however, he/she must have familiarity with basic linear algebra, geometry, and general knowledge of programming. If the reader does not (want to) use Python, he/she can also understand the examples from a traditional perspective by way of analyzing the input and the output given in each example. For more online resources, such as papers, figures, and slides, the reader can visit the web-page of the present book at the following address: https://domingomery.ing.puc.cl/material/.

Santiago, Chile   Domingo Mery
2019–2020   Christian Pieringer

References

1. Bhuyan, M.K.: Computer Vision and Image Processing: Fundamentals and Applications. CRC Press (2019)
2. Bishop, C.: Pattern Recognition and Machine Learning. Springer (2006)
3. Carrasco, M., Pizarro, L., Mery, D.: Visual inspection of glass bottlenecks by multiple-view analysis. Int. J. Comput. Integr. Manuf. **23**(10), 925–941 (2010)
4. Duda, R., Hart, P., Stork, D.: Pattern Classification, 2 edn. John Wiley & Sons, Inc., New York (2001)
5. Faugeras, O., Luong, Q.T., Papadopoulo, T.: The Geometry of Multiple Images: The Laws That Govern the Formation of Multiple Images of a Scene and Some of Their Applications. The MIT Press, Cambridge MA, London (2001)
6. Forsyth, D.A., Ponce, J.: A modern approach. Computer Vision: A Modern Approach (2003)
7. Gonzalez, R., Woods, R.: Digital Image Processing, third edn. Pearson, Prentice Hall (2008)
8. Goodfellow, I., Bengio, Y., Courville, A.: Deep learning. MIT press (2016)
9. Hartley, R.I., Zisserman, A.: Multiple View Geometry in Computer Vision, Second edn. Cambridge University Press (2003)
10. Klette, R.: Concise Computer Vision: An Introduction into Theory and Algorithms. Springer Science & Business Media (2014)
11. Mery, D.: Computer vision for X-Ray testing. Switzerland: Springer International Publishing **10**, 978–3 (2015)
12. Mery, D., Pedreschi, F., Soto, A.: Automated design of a computer vision system for visual food quality evaluation. Food Bioprocess Technol. **6**(8), 2093–2108 (2013)
13. Pichara, K., Pieringer, C.: Advanced Computer Programming in Python. CreateSpace Independent Publishing Platform (2017)
14. Szeliski, R.: Computer vision: Algorithms and applications. Springer-Verlag New York Inc (2011)

Preface to the First Edition

This book has been written in many *spatiotemporal coordinates*. For instance, some equations and figures were performed during my Ph.D. at the Technical University of Berlin (1996–2000). During that period, but in Hamburg, I took several X-ray images—that have been used in this book—in YXLON X-ray International Labs. After completing my Ph.D., and during my work in Santiago, Chile as associate researcher at the University of Santiago of Chile (2001–2003) and faculty member at the Catholic University of Chile (2004–to date), I have written more than 40 journal papers on computer vision applied to X-ray testing. During this time, I have developed a Matlab Toolbox that has been used in my research projects and in my classes teaching image processing, pattern recognition, and computer vision for graduate and undergraduate students. Over the last few years, my graduate students have taken thousands of X-ray images in our X-ray Testing Lab at the Catholic University of Chile. Moreover, in my sabbatical year at the University of Notre Dame (2014–2015), I had the time and space to teach the computer vision course for students of computer sciences, electrical engineering and physics, and I have been able to bring together all those related papers, diagrams, and codes in this book.

The present work has been written not only in three different countries (Germany, Chile, and the United States) over the last 15 years, but also in many different small places that provided me with the time and peace to write a paragraph, a caption of a figure, a code, or whatever I could. For example, I remember a Café in Michigan City where I spent various hours last winter writing this book with a delicious cappuccino beside me; or my study room in Fisher Apartments on Notre Dame Campus, looking out the window at a squirrel holding a nut; or on a narrow tray table while taking an Inter-Regio train between Berlin and Hamburg, which was where I drew a diagram using a pen and probably a napkin; and of course, my delightful office at the Catholic University of Chile with its breathtaking view of the Andes Mountains.

This book has been put together on the basis of four main pillars that have been constructed over the last 15 years: the first pillar is the set of journal and conference papers that I have published. The second corresponds to the material used in my classes and the feedback received from students when I have been teaching image processing, pattern recognition, and computer vision. The third pillar is the Matlab Toolbox that I was able to develop during this time, and which has been tested in several experiments, classes and research projects, among others. The fourth pillar is the thousands of X-ray images that my research group has been taking in recent years at our Lab, and the X-ray images of die castings that I took in Hamburg. Over all this time, I have realized that this amount of work can all be brought together in a book that collects the most important contributions in computer vision used in X-ray testing.

Notre Dame and Santiago de Chile   Domingo Mery
2015

Acknowledgements

This book could not have been written without the steady support and understanding of my wife Ximena and my daughters Anais and Valeria. Consequently, this book is dedicated to my nuclear family. Special thanks to my sister and my mother, who left this world recently, and my father and brother for the confidence and support.

Deep thanks also go to Reinhard Klette (University of Auckland), wherever he is, who carefully read the first edition and gave me relevant and detailed suggestions as to improve the book. Finally, I would like to offer my gratitude to all my students, colleagues and co-authors that helped me in numerous discussions and suggestions, particularly the following: Analí Alfaro, Marco Arias, Carlos Arteta, Gonzalo Acuña, José Miguel Aguilera, Miguel Angel Berti, Sandipan Banerjee, Kevin Bowyer, Miguel Carrasco, Max Chacón, Aldo Cipriano, Patricio Cordero, Esteban Cortázar, Pedro Cortés, Adam Czajka, Dieter Filbert, Patrick Flynn, Rodrigo González, Sergio Hernández, Thomas Jaegger, Denis Hahn, Daniel Heihnsohn, Ana Hincapié, Aggelos Katsaggelos Valerie Kaftandjian, James Kapaldo, Germán Larraín, Tomás Larraín, Gabriel Leiva, Pei Li, Hans Lobel, Iván Lillo, Juan Carlos de la Llera, Daniel Maturana, Carlos Mena, Cristobal Moenne, Sebastián Montabone, Germán Mondragón, Diego Patiño, Mar Pérez-Sanagustin, Karim Pichara, Franco Pedreschi, Christian Pieringer, Luis Pizarro, Loreto Prieto, Vladimir Riffo, Daniel Saavedra, José Saavedra, Doris Sáez, Walter Sheirer, Alvaro Soto, Eric Svec, Romeu da Silva, Gabriel Tejeda, René Vidal, Esteban Villalobos, Yuning Zhao, Uwe Zscherpel, and Irene Zuccar.

—Domingo

To my wife, María Isabel, my daughter Laura and my son Dante, for understanding and support during the time dedicated to coding, studying, and listen to all my ideas and dreams. Thanks to my mother and my brothers, for their support and love. To my father, in his loving memory, who taught me always to keep the faith.

A sincere thanks to Domingo Mery for his wise guidance and my Ph.D. thesis and path as a researcher. And especially for inviting and trusting me to contribute to this book. To Karim Pichara, for his support and advice during my postdoctoral fellow, and to encourage me to take some challenges at the early stage of my academic and professional career.

Finally, I would like to express my gratitude to all my colleagues, co-authors, professors, teaching assistants, and students who, in one way or another, have helped to improve my work and skills through discussions and suggestions, but especially the following: Billy Peralta, Vladimir Riffo, Elwin van 't Wout, Carlos Sing-Long, Miguel Carrasco, and Márcio Catelan.

—Christian

This work was supported by the School of Engineering of the Catholic University of Chile, and Fondecyt Grants No. 1130934, 1161314, and 1191131 from CONICYT, Chile.

Contents

About the Authors

Domingo Mery was born in Santiago de Chile in 1965. He is Full Professor in the Department of Computer Science at UC. He received the Diploma (M.Sc.) degree in Electrical Engineering from the Technical University of Karlsruhe, Germany, in 1992, and the Ph.D. degree with distinction at the Technical University of Berlin, in 2000. He was a Research Scientist at the Institute for Measurement and Automation Technology at the Technical University of Berlin with the collaboration of YXLON X-Ray International. He was a recipient of a Scholarship from the Konrad-Adenauer-Foundation, and from a Scholarship from the German Academic Exchange Service (DAAD) for his Ph.D. work. He was Associate Research in 2001 at the Department of Computer Engineering at the University of Santiago, Chile. Now, he is a Full Professor at the Department of Computer Science at the Pontificia Universidad Católica de Chile (UC), Chile. He was Chair of the Computer Science Department in 2005–2009. He was an Associate Visiting Professor at the Computer Vision Research Lab of the University of Notre Dame in 2014–2015. He was Director of Research and Innovation of the School of Engineering at the UC in 2015–2018. He serves as Associate Editor of the IEEE Transactions on Information, Forensics and Security, the IEEE Transactions on Transactions on Biometrics, Behavior, and Identity Science, and the International Journal of Fuzzy Logic and Intelligent Systems. In addition, he is the Editor of I3 Journal for Research, Interdiscipline, and Innovation at the School of Engineering (UC). His research interests include image processing for fault detection in aluminum castings, X-ray imaging, real-time programming, and computer vision. He is author of more than 80 journal publications and more than 90 conference papers. He served as the Local Co-chair of ICCV2015 (Santiago de Chile). He was program general chair of the PSIVT2007, program chair PSIVT2009, and General Co-chair of PSIVT2011 (Pacific-Rim Symposium on Image and Video Technology), and 2007 Iberoamerican Congress on Pattern Recognition.

Awards in Non-destructive Testing:

- Ron Halmshaw Award for publishing the best paper during 2017 on industrial radiography in Insight.[5]
- IAPR[6] Best Paper Presentation at Pacific-Rim Symposium on Image and Video Technology (PSIVT2015), Auckland, New Zealand.
- John Grimwade Medal for publishing the best paper during 2013 in Insight.
- Ron Halmshaw Award for publishing the best paper during 2012 on industrial radiography in Insight.
- Ron Halmshaw Award for publishing the best paper during 2005 on industrial radiography in Insight.
- Best Paper Award. Panamerican Conference on Non-destructive Testing (PANNDT 2003).

Christian Pieringer was born in Valparaiso, Chile, in 1978. He received his Bachelor Science (B.Sc.) degree in Electronic Engineering from Pontificia Universidad Catolica de Valparaiso, Chile, in 2003; a Master in Engineering from Pontificia Universidad Catolica de Chile in 2010; and a Ph.D. in Computer Science by Pontificia Universidad Catolica de Chile. He received a Ph.D. grant from the National Commission of Technological and Scientific Research of the Chilean Government (CONICYT), and a postdoctoral fellowship from the PUC-Harvard seed funds program to develop interdisciplinary research on Machine Learning applied to Astronomy. He was a Project Engineer at the Institute of Mathematical and Computational Engineering (IMCE) leading data-driven solution used to industry level. He was a research professor at Universidad Tecnologica de Chile (INACAP) and lecturer in professional Diploma Certificates on Business Intelligence and Big Data for IoT in Pontificia Universidad Catolica de Chile. At present, he is the Chief Technological Officer (CTO) at Suncast, developing Artificial Intelligence based solutions for the optimization of the operation and maintenance in renewables energies. He is co-author of the textbook *Advanced Computer Programming in Python* [13].

Awards in Non-destructive Testing:

- Ron Halmshaw Award for publishing the best paper during 2017 on industrial radiography in Insight.

---

[5]Insight is the Journal of the British Institute of Non-destructive Testing.
[6]IAPR: International Association of Pattern Recognition.

Chapter 1
X-ray Testing

Abstract X-ray testing has been developed for the inspection of materials or objects, where the aim is to analyze—nondestructively—those inner parts that are undetectable to the naked eye. Thus, X-ray testing is used to determine if a test object deviates from a given set of specifications. Typical applications are the inspection of automotive parts, quality control of welds, baggage screening, analysis of food products, inspection of cargos, and quality control of electronic circuits. In order to achieve efficient and effective X-ray testing, automated and semi-automated systems based on computer vision algorithms are being developed to execute this task. In this book, we present a general overview of computer vision approaches that have been used in X-ray testing in the last decades. In this chapter, we offer an introduction to our book by covering relevant issues of X-ray testing.

Cover image: *X-ray images of woods (series* N0010 *colored with 'hot' colormap).*

© Springer Nature Switzerland AG 2021
D. Mery and C. Pieringer, *Computer Vision for X-Ray Testing*,
https://doi.org/10.1007/978-3-030-56769-9_1

1.1 Introduction

Since Röntgen discovered in 1895 [90] that X-rays can be used to identify inner structures, X-rays have been developed not only for their use in *medical imaging* for human beings, but also in *non-destructive testing* (NDT) for materials or objects, where the aim is to analyze (non-destructively) the inner parts that are undetectable to the naked eye [44]. NDT with X-rays, known as *X-ray testing*, is used in many applications such as the inspection of automotive parts, quality control of welds, baggage screening, analysis of food products, inspection of cargos, and quality control of electronic circuits among others. X-ray testing usually involves measurement of specific part features such as integrity or geometric dimensions in order to detect, recognize, or evaluate wanted (or unwanted) inner parts. Thus, X-ray testing is a form of NDT defined as a task that uses X-ray imaging to determine if a *test object* deviates from a given set of specifications, without changing or altering that object in any way.

The most widely used X-ray imaging systems employed in X-ray testing are Digital Radiography (DR) and Computed Tomography (CT) imaging.[1] On the one hand, DR emphasizes high throughput. It uses electronic sensors (instead of traditional radiographic film) to obtain a digital X-ray projection of the target object, consequently it is simple and quick. A flat amorphous silicon detector can be used as an image sensor in X-ray testing systems. In such detectors, and using a semiconductor, energy from the X-ray is converted directly into an electrical signal that can be digitalized into an X-ray digital image [91]. On the other hand, CT imaging provides a cross-sectional image of the target object so that each object is clearly separated from any others, however, CT imaging requires a considerable number of projections to reconstruct an accurate cross-sectional image, which is time consuming.

In order to achieve efficient and effective X-ray testing, automated and semi-automated systems are being developed to execute this task that can be difficult (e.g., recognition of very small defects), tedious (e.g., inspection of thousand of similar items) and sometimes dangerous (e.g., explosive detection in baggage screening). Compared to manual X-ray testing, automated systems offer the advantages of objectivity and reproducibility for every test. Fundamental disadvantages are, however, the complexity of their configuration, the inflexibility to any change in the evaluation process, and sometimes the inability to analyze intricate images, which is something that people can generally do well. Research and development is, however, ongoing into automated adaptive processes to accommodate modifications.

X-ray testing is one of the more accepted ways for examining an object without destroying it. The purpose of this non-destructive method is to detect or recognize certain parts of interest that are located inside a test object and are thus not detectable to the naked eye. A typical example is the inspection of castings [70, 73]. The

---

[1]Computed tomography is beyond the scope of this book due to space considerations, however, some simple examples and basic concepts are covered (see Sect. 1.6.5). For NDT applications using CT, the reader is referred to [18, 34].

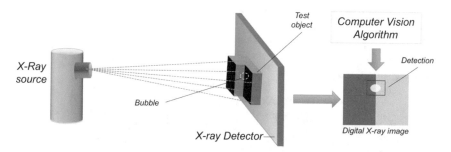

Fig. 1.1 Simple model of an X-ray computer vision system. In this example, a computer vision algorithm is used to detect a defect inside the test object automatically

material defects occurring in the casting process such as cavity, gas, inclusion, and sponge must be detected to satisfy the security requirements; consequently, it is necessary to check 100% of the parts.

The principle aspects of an X-ray testing system is illustrated in Fig. 1.1. Typically, it comprises the following steps:

- The test object is located in the desired position.
- The X-ray source generates X-rays which pass through the test object.
- The X-rays are detected and converted (e.g., by a flat panel or by an image intensifier and CCD-camera) in order to obtain a digital X-ray image.
- Computer vision algorithms are used to evaluate the X-ray image.

In last decades, flat detectors made of amorphous silicon have been widely used as image sensors in some industrial inspection systems [38, 83]. In these detectors, the energy from the X-ray is converted directly into an electrical signal by a semiconductor (without an image intensifier). However, using flat detectors is not always feasible because of their high cost compared to image intensifiers.

The properties of the X-rays that are used in X-ray testing are summarized in the following:

- X-rays can penetrate light blocking materials (e.g., metal) depending on a material's thickness;
- X-rays can be detected by photographic materials or electronic sensors;
- X-rays can spread a straight line; and
- X-rays can use many substances to stimulate fluorescence (fluoroscopy).

1.2 History

The discovery of X-rays by Röntgen in November 1895 [90] defines the beginning of the X-ray testing of metallic parts. A couple of days after the discovery of the 'X' radiation, he made radiographs of balance-weights in a closed box and a chamber of a shotgun (see Fig. 1.2). Röntgen observed that using X-rays, one can look not only into the inside of a human body, but also into metallic articles, if the strength and

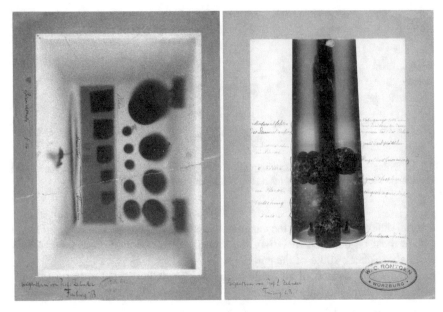

Fig. 1.2 X-ray image of balance-weights in a closed box and a shotgun taken by Wilhelm Conrad Röntgen in the summer of 1896. Courtesy of the Deutsches Röntgen-Museum in Würzburg

intensity of the X-rays are strong enough [62]. The potential use in the detection of hidden defects within armor-plates and machine parts was already envisioned at Yale University in 1896 [88].

The industrial use of X-rays began in Germany only two decades after their discovery. X-ray testing took place at that time with the help of radiographic films [92]. Radioscopy with fluorescent screens was developed only toward the end of the 1930s and at the beginning of the 1940s. In the following years, closed cabinets were already being used for X-ray testing of aluminum castings in the automobile industry [82].

In 1948 the image intensifier was developed, which converts X-rays into a visible light [106]. Image intensifier technology was originally developed as a low-light enhancer for military night-vision devices [41]. The introduction of the image intensifier led to considerable progress in the inspection technique, since otherwise the examiner would have to regard the X-ray image on a fluorescent screen. The brightness of the image was so small that the eyes needed a long time to adapt to the dark. Into the image amplifier an examiner could always look in the radiograph directly with the help of special optics. Image intensifiers, television equipment, and electrically controlled manipulators were developed further in the 1960s as radioscopic systems, which were widely used for casting and welding inspection in the 70s [88].

X-ray testing systems for baggage inspection were already developed in the 70s [25, 80] and the 80s [26, 58]. The object under test was scanned using fluoroscopy and the images were inspected on a a fluorescent screen.

Computer Tomography (CT) was developed in 1972 [16, 106]. With 2D-CT cross-sectional pictures of the object computed from its projections. These slices, which represent a reconstruction of the local distribution of the absorption coefficients of the object, are processed in order to find objects of interest in the test object. However, one disadvantage of the procedure is the high time requirement: for the reconstruction of meaningful slice images, both a minimum gate time per object position is necessary for a sufficient signal/noise ratio along with a minimum number of projections. For this reason, the use of computer tomography is so far limited in X-ray testing to the material development and research range, as well as to the examination of particularly important and expensive parts [49]. Later 3D-CT was developed, with which the whole object is reconstructed as voxels. State-of-the-art industrial computer tomography used this kind of CT [34, 110].

Approaches to the automatic image evaluation as well as image restoration were already used in the 80s with the help of the image processing techniques and CCD-cameras [82]. The first fully automatic X-ray testing systems were installed in the industry at the beginning of the 90s. One example can be found in the quality control of aluminum wheels performed by Alumetall Co. in Nuremberg, in which an automatic casting part recognition is also integrated using bar codes for the adjustment of the image analysis algorithms for different types of wheel [88].   At the end of the 90s, flat panel detectors from amorphous silicon were industrially used in some test systems [9, 50]. With these detectors the X-rays are converted by a semiconductor directly into electrical signals (without image intensifier). However, the X-ray testing with flat detectors was not always profitable due to their high costs (in the comparison to the image intensifier).

Before 9/11, X-ray testing of luggage mainly focused on capturing the images of their content: the reader can find in [74] an interesting analysis done in 1989 of several aircraft attacks in the world, and the existing technologies to detect the terrorists' threats based on Thermal-Neutron Activation (TNA), Fast-Neutron Activation (FNA), and dual-energy X-rays (used in medicine since early 70). In the 90s, Explosive Detection Systems (EDS) were developed based on X-ray imaging [75], and computed tomography through elastic scatter X-ray (comparing the structure of irradiated [99] advanced image analysis) to improve the detection performance. Nevertheless, the 9/11 attacks increased the security policies at airports, which also produced the interest of the scientific community for researching topics related to security using advanced computational techniques using pseudocoloring of X-ray images, for example, [96]. It has not been easy for X-ray baggage inspection to deal with low-density, often organic, materials (very important in baggage and food inspection). This is because typical X-ray inspection systems use conventional photon integration detectors that are unable to record the incoming X-ray energy. However, state-of-the-art multicolor detector technology could assist to overcome this problem. Thanks to recent advances in the development of photon counting detectors, multicolour X-ray imaging has become possible. Today, novel X-ray detectors have been developed. For example, detectors based on new semiconductors like CdTe or CZT [48, 101] that can count photons at high rates by discriminating different energy channels, in which image noise can be decreased, contrast can be enhanced and specific materials can be imaged; or wafer-scale CMOS flat panels with a pixel size of 100 μm × 100 μm in an array of 1220 × 12000 pixels [21].

In the last few decades, fully automatic and semi-automatic test systems have been used in many applications as we will cover in Chap. 9.

1.3 Physics of the X-rays

In general, X-rays are from same physical nature as visible light, radiowaves, microwaves, ultraviolet, or infrared. They are all electromagnetic waves, which spread at the speed of light, although with different wavelengths (see Table 1.1)

In the following, the formation of X-rays and their interaction with matter are explained. These principles of physics can be found in many textbooks (see, for example, [6, 64]).

1.3.1 Formation of X-rays

The formation of X-rays is performed in an X-ray tube in five steps as shown in Fig. 1.3:

1. A high DC voltage U is applied between cathode and anode.
2. The cathode is strongly heated by the voltage U_h, so that the kinetic energy of the heat is transferred to the mobile electrons in the cathode. The electrons are thus in a position to withdraw from the cathode.
3. The electrons emitted by the hot cathode are accelerated by high voltage U.
4. These high-energy electrons, which are called cathode rays, are incident on the anode.

Table 1.1 Electromagnetic spectrum [56]

Electromagnetic-waves ⟶	Radio-waves	Micro-waves	Infra-red	Visible light	Ultra-violet	X-rays	Gamma-rays
Wavelength in (m)	$1 \sim$ 10^4	$10^{-3} \sim$ 1	$7,7 \cdot 10^{-7} \sim$ 10^{-3}	$3,9 \cdot 10^{-7} \sim$ $7,7 \cdot 10^{-7}$	$10^{-8} \sim$ $3,9 \cdot 10^{-7}$	$10^{-12} \sim$ 10^{-8}	$10^{-14} \sim$ 10^{-12}
Frequency in (Hz)	$3 \cdot 10^8 \sim$ $3 \cdot 10^4$	$3 \cdot 10^{11} \sim$ $3 \cdot 10^8$	$3,9 \cdot 10^{14} \sim$ $3 \cdot 10^{11}$	$7,7 \cdot 10^{14} \sim$ $3,9 \cdot 10^{14}$	$3 \cdot 10^{16} \sim$ $7,7 \cdot 10^{14}$	$3 \cdot 10^{20} \sim$ $3 \cdot 10^{16}$	$3 \cdot 10^{22} \sim$ $3 \cdot 10^{20}$
Energy in (eV)	$1,2 \cdot 10^{-6} \sim$ $1,2 \cdot 10^{-10}$	$1,2 \cdot 10^{-3} \sim$ $1,2 \cdot 10^{-6}$	$1,6 \sim$ $1,2 \cdot 10^{-3}$	$3,2 \sim$ $1,6$	$1,2 \cdot 10^2 \sim$ $3,2$	$1,2 \cdot 10^6 \sim$ $1,2 \cdot 10^2$	$1,2 \cdot 10^8 \sim$ $1,2 \cdot 10^6$

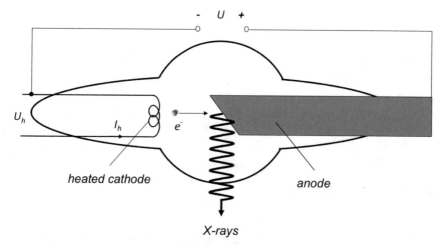

Fig. 1.3 Basic diagram of an X-ray tube

5. X-rays are produced when electrons of sufficiently high-energy incident on the anode are suddenly decelerated.

   There is a distinction   between discrete and continuous X-rays (commonly known as *Bremsstrahlung*).

Discrete X-rays
These result in transitions of electrons in the inner shells of an atom (see Fig. 1.4a). This happens when a highly accelerated electron e^- ① knocks an electron e_1^- from the atomic shell. Since both electrons leave the atom ②, a hole is formed (where e_1^- was) that is immediately filled by an outer electron (e.g., e_2^-) ③. In an atom, the electrons may be shown only on certain bands with a precisely specified energy level. The deeper the band is in the atom, the greater is the energy of that electron. When jumping from the electron to a lower band (in our example e_2^-) the energy difference between the two energy levels is emitted as electromagnetic radiation. Energy transitions in the region of the inner electron shells which have high binding energies lead to the emission of X-rays ④. Therefore, the spectrum of this radiation consists of lines at specific wavelengths or energies that are exclusively dependent on the nature of the atom (see Fig. 1.4c). These are called characteristic X-ray lines.

Continuous X-rays (Bremsstrahlung)
In addition to the discrete X-rays, there is a continuos radiation called Bremsstrahlung. This occurs when a highly accelerated electron approaches the domain of attraction of the atomic nucleus of the anode and are deflected due to the Coulomb force (Fig. 1.4b). There is no collision between nucleus and electron. Since the electron interacts with the Coulomb force, the direction and velocity of the electron are changed. In this deceleration, the electron loses some or all of the kinetic energy that is emitted in the form of X-rays to the outside. The closer the electron is to the nucleus, the greater is the deceleration and thus the energy of the Bremsstrahlung.

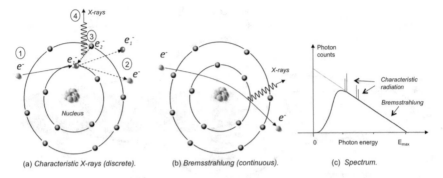

(a) Characteristic X-rays (discrete).   (b) Bremsstrahlung (continuous).   (c) Spectrum.

Fig. 1.4 X-ray formation and spectrum

As electrons can come close to the nucleus at any distance, this electromagnetic radiation has a continuous spectrum with an upper cut-off frequency E_{max} (see Fig. 1.4c). The maximum energy is obtained when an electron is completely decelerated, *i.e.*, when the kinetic energy of the electron ($E_{kin} = e \cdot U$) is converted entirely into photon energy ($E_{photo} = h \cdot \nu$), where e is the electric charge, U the anode voltage, h Planck's constant, and ν the frequency of the electromagnetic wave. The smallest possible X-ray wavelength becomes of $E_{kin} = E_{photo}(= E_{max})$ and $c = \lambda \nu$ with:

$$\lambda_{min} = \frac{h \cdot c}{E_{max}} = \frac{h \cdot c}{e \cdot U} \tag{1.1}$$

where c is the speed of light in vacuum. Changes to the heating of the cathode I_h (see Fig. 1.3) result in a proportional change of the energy flux density. An increase in the high-voltage U leads to the displacement of the maximum energy flux density to a higher energy.

1.3.2   Scattering and Absorption of X-rays

One aspect particularly important for X-ray testing is the attenuation of the intensity of X-rays when passing through matter. The attenuation is a function of X-ray energy and the material structure of the irradiated material (considerably in terms of density and thickness). The attenuation occurs by two processes: scattering and absorption. The scattering via classical scattering (Rayleigh scattering and Compton effect); and absorption through the photoelectric effect, pair production, and partly by the Compton effect. In the following, these are explained as interactions of X-rays with atoms.

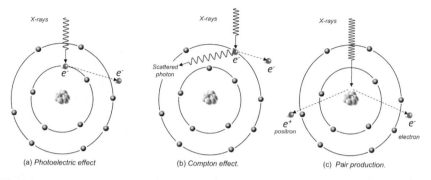

(a) Photoelectric effect   (b) Compton effect.   (c) Pair production.

Fig. 1.5 Interaction of X-rays with matter

Rayleigh Scattering

In this process, there is a scattering of X-rays from their original path, in which the radiation loses no energy. The lower the energy of the radiation, the more they are deflected from the original path of the rays.

Photo Effect

The photoelectric effect that occurs is likely to happen when the radiation energy just exceeds the binding energy of the electron. In the photoelectric effect, the energy of the incident photon is completely transferred to an electron, and mainly on one of the inner electron shells. The electron takes over the energy that the quantum of radiation it emits as kinetic energy and leaves the atomic union (Fig. 1.5a). This effect increases proportionally to $E^{-3}Z^5$, where E is the energy of the radiation and Z is the atomic number. The photoelectric effect plays a role in the small and medium energies of X-rays.

Compton Effect

In case the radiation energy is very much larger than the binding energy of the atomic electron, the X-ray radiation strikes out the electron from the atom. A portion of the energy of the X-ray radiation is transferred to the electron and converted into kinetic energy. The radiation is scattered and loses energy (see Fig. 1.5b). This results in a scattering due to the change of direction of the photons at the same time and absorption due to the energy loss. This effect is proportional to the atomic number of the atom Z and inversely proportional to the energy of the radiation to E.

Pair Production

In case the radiation energy is greater than 1.022 MeV and passes it straight into the proximity of the nucleus, the radiation can be turned into matter, producing an electron e^- and e^+ positron (see Fig. 1.5c), whose masses are $m_{e^-} = m_{e^+} = 511$ keV/c^2. The pair production is more frequent, the greater the quantum energy and the higher the atomic number of the irradiated material. In cases, where X-rays come from X-ray tubes there is no pair production, as the energy is always in the keV range.

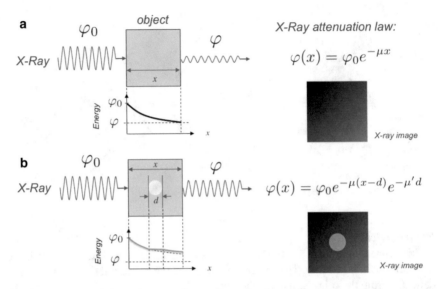

Fig. 1.6 X-ray image formation according to absorption law: **a** X-ray image of a homogenous object, and **b** X-ray image of an object with two different materials

Absorption and scattering can be described mathematically by the X-ray absorption law, which characterizes the intensity distribution of X-rays through matter:

$$\varphi(x) = \varphi_0 e^{-\mu x} \tag{1.2}$$

with φ_0 incident energy flux density, μ absorption coefficient, x thickness of the irradiated matter and φ energy flux density after passage through matter with the thickness of x (see Fig. 1.6a). The absorption coefficient μ depends on the incident photon energy and the density and atomic number of the irradiated material. It is composed of the coefficients of the classical dispersion σ_R, the photoelectric effect τ, the Compton effect σ_C, and the pair production χ:

$$\mu = \sigma_R + \tau + \sigma_C + \chi \tag{1.3}$$

Because of the continuous distribution of the energy of the Bremsstrahlung (see Fig. 1.4c) X-rays contain photons of different energies. In practice, therefore, the course of the absorption curve can only be determined empirically. In the case of aluminum, the course of the absorption coefficient in Fig. 1.7.

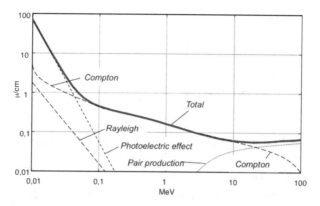

Fig. 1.7 Absorption coefficient for aluminum [56]

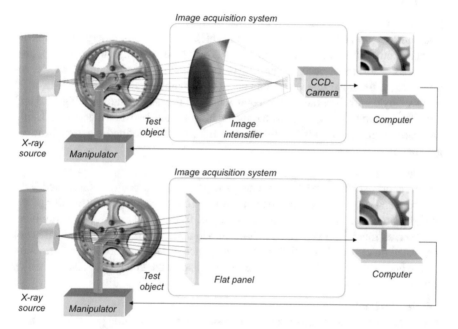

Fig. 1.8 X-ray testing systems. There are two kinds of image acquisition system: based on image intensifiers (top) and based on flat panels (bottom). In this example, an aluminum wheel is inspected using a manipulator

1.4 X-ray Testing System

The essential components of an automatic X-ray testing system (see Fig. 1.8), such as X-ray source, manipulator, image intensifier, and CCD-camera, are explained below.

1.4.1 X-ray Source

There are six requirements for an X-ray source [42]:

1. Adjustable quantum energy.
2. Possible large adjustable dose rate.
3. Intensity of the radiation as uniform as possible in the field of the object to be irradiated.
4. Smallest possible intensity of radiation outside the area to be irradiated.
5. Acceptable price.
6. Long life with constancy of features.

In this section, we describe the essential components of an X-ray source that fulfill the conditions mentioned. An explanation of the formation of X-rays can be found in Sect. 1.3.1.

Hot Cathode

The cathode is made of a filament from which the electrons emerge through the thermoelectric effect in the vacuum of the X-ray tube. Usually, tungsten (W), also known as wolfram, is used because of its high melting point (about 3380^0C). An influence of the dose rate (independent of the quantum energy of the X-rays) is achieved by controlling the electron emission over the heating current (Figs. 1.3 and 1.9). The quantum energy is adjusted by the high voltage between electrodes. Using an aperture that surrounds the filament, a thin, sharply defined electron beam is generated.

Anode

At the anode surface, the kinetic energy of the cathode beam is converted 99% into heat and only 1% into the desired X-rays. To reduce the geometric blur of the imaging process a small focal spot is required. In the focal spot of an X-ray tube, however, so much heat is created that the anode material may melt if the heat is not dissipated quickly and effectively. In order to increase the performance of an X-ray source and at the same time to reduce the focal spot, the anodes are constructed as follows:

Anode Material

The surface layer should be made of materials with a high melting point, high atomic number, and high thermal conductivity. The element tungsten (W) best meets the three criteria. In order to reduce the roughening during the operation, as well as to avoid cracking, it is alloyed with rhenium (Re).

Line Focus

To reduce the optical focus, the electron beam strikes the anode surface in the focal spot inclined by about $\alpha = 7^0 \sim 20^0$ from the vertical axis.

Rotating Anode

By rotating the anode the applied heat can be distributed over an entire ring without changing the size of the optical spot (see Fig. 1.9). The distribution of the high thermal load is better the larger the diameter of the ring and the higher the rotation speed.

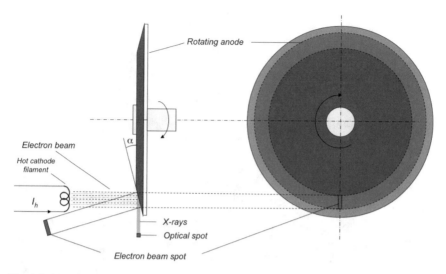

Fig. 1.9 Basic structure of an anode

Envelope
Given that between the electrodes the voltage is high voltage, anode, and cathode must be electrically isolated from each other. In addition, the tube envelope forms the vacuum vessel and the mechanical attachment of the tube components. Up until now, glass has been used for this purpose. However, in recent years envelopes made of metal and ceramics have been used.

1.4.2   Manipulator

A manipulator is a device that can be handled with the test objects in the desired manner without the operator using his/her hands to touch [93]. In an X-ray computer vision system, the task of the manipulator is the handling of the test objects as illustrated in Fig. 1.10. Due to the possibilities of movement, degrees of freedom of the manipulator, the test object can be brought into the desired position. For a manual inspection, the axes of a manipulator are moved by means of one or more joysticks. When an automatic inspection of this task is undertaken, it is handled by a Programmable Logic Controller (PLC) or an industrial computer.

A manipulator consists of sliding elements and rotary elements with which a translation or rotation of the object test can be performed. Previously, the manipulator moved the test object through the X-ray beam [54]. This solution resulted in a complicated mechanical construction with a high mechanical load, wear, and increased maintenance. Today it is possible to move the X-ray tube and the detector that is rigidly connected to it by a *C-arm manipulator*. These manipulators are much easier to control and are faster and cheaper [9, 50]. An example of such a manipulator is described in Sect. 3.3.4 (see Fig. 3.14).

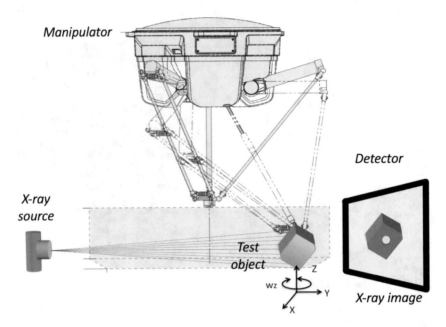

Fig. 1.10 X-ray system with a manipulator ABB-Picker: the manipulator can rotate and translate the test object to generate projections from different points of view

1.4.3 Image Intensifier

The X-ray image intensifier has two functions: *(i)* possible lossless conversion of X-ray projection information into a visible image and *(ii)* its brightness gain [41]. On the basis of the structure of an X-ray image intensifier shown in Fig. 1.11, the operation is explained. The X-ray radiation enters through an input screen into a vacuum tube. As the radiolucent input screen has to withstand the atmospheric pressure, it should not be too thin. Here metals are used with low atomic numbers that are trans-

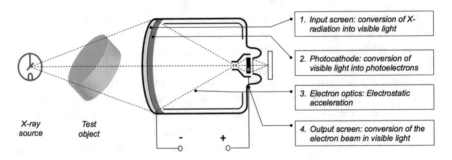

Fig. 1.11 Schematic illustration of the operation of an image intensifier

parent to X-rays, in which the absorption and scattering are relatively small. There-after, the radiation incident on the X-ray fluoroscopy screen, in which the conversion of X-radiation into visible light takes place. The X-rays are absorbed and about 2000 photons per X-ray quantum are triggered. The light strikes the photocathode and sets photoelectrons. These electrons are accelerated by approximately 25 kV, which are represented with reduced electron optics on an output phosphor screen. The output image of the image intensifier is then captured by a CCD-camera.

The disadvantage of the image intensifier is the geometric distortion due to the curvature of the input screen; details for this can be found in Sect. 3.3.2.

1.4.4   CCD-Camera

CCD-cameras use solid-state imaging sensors based on Charge-coupled device (CCD) arrays. In these imaging sensors, the active detector surface is divided into individual pixels in the CCD-sensor, while incident light is converted and trans-ported into an electrical charge. The principle of the charge transport is based on the charge transfer that takes place in the shift registers (Fig. 1.12).

The CCD-cameras are characterized by very good image geometry, high light-sensitivity and several megapixels for conventional cameras. In modern days, there are High Definition Television (HDTV) cameras up to 2,200,000 pixels. Further-more, a CCD-camera can achieve a resolution of 46 megapixels and the exposure time can be in a range between seconds and 1/8,000 s.

Due to the low sensitivity of the CCD-image sensor for direct X-ray radiation, the radiation must be converted into visible light. In an X-ray testing system with CCD-camera, this conversion happens in the image intensifier (see Sect. 1.4.3).

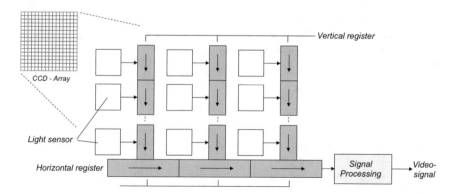

Fig. 1.12 Operation of a CCD-Array

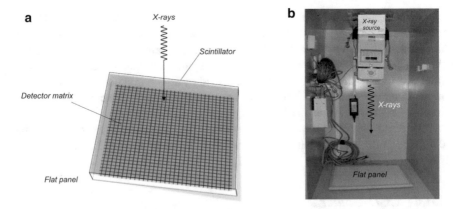

Fig. 1.13 Flatpanel: **a** Basic structure [46] und **b** Example: Canon, model CXDI-50G (resolution: 2208 × 2688 pixels and 4,096 grayscale image). In this example, the X-ray emitter tube is Poskom, model PXM-20BT

1.4.5   Flat Panel

A second possible image acquisition system is the flat panel detector based on amorphous silicon (a-Si), in which the X-ray, without going through an image intensifier with CCD-camera, is converted from a semiconductor directly into electrical signals (see Fig. 1.13). In this technology, a thin view of a-Si is deposited on a glass plate as a support. As in a CCD-chip, a pixel array with switching elements is generated in the silicon layer so that the charge which is stored in the individual pixels can be read out serially and electronically [16].

The advantages of this detector are larger image receiving surface, no geometric distortion, a high gray level resolution (12 ∼ 16 Bit/Pixel), that is very light and small. Due to the high gray level resolution and greater imaging surface less test positions are required for the inspection. The low weight allows for easier and faster mechanics [9, 50]. An flat detector is shown in Fig. 1.13.

1.4.6   Computer

In the context of X-ray testing, a computer system is typically used for the following tasks:

1. To control the image acquisition system.
2. To store acquired X-ray images.
3. To run computer vision algorithms that evaluate X-ray images.
4. To compute statistical analysis.

5. To display results.
6. To control the X-ray source.
7. To control the manipulator.

1.5 X-ray Imaging

In this section we present image formation, acquisition, and visualization.

1.5.1 X-ray Image Formation

In X-ray testing, X-ray radiation is passed through the test object, and a *detector* captures an X-ray image corresponding to the radiation intensity attenuated by the object.[2] According to the principle of *photoelectric absorption* (1.2): $\varphi = \varphi_0 \exp(-\mu x)$, where the transmitted intensity φ depends on the incident radiation intensity φ_0, the thickness x of the test object, and the energy dependent linear absorption coefficient μ associated with the material, as illustrated in Fig. 1.6.

In a photographic image, the surface of the object is registered. On the contrary, in an X-ray image, the inside of the object is captured. In order to illustrate the formation, we simulate the X-ray image of the object of Fig. 1.1 in several positions (in this example we use the approach outlined in Chap. 8). In this case, we have a homogenous test object with a spherical cavity inside. The result is shown in Fig. 1.14. In this example, we can observe, on the one hand, the absorption phenomenon. The thicker the object the more attenuated the X-rays. In our visualization, bright gray values are used for high output energy (low attenuation), and dark gray values for low-output energy (high attenuation). On the other hand, we can see the phenomenon of the summation of shadows, i.e., the output intensity of an image point corresponds to the summation of all the attenuations the X-ray encountered.

It is worth noting that if X-ray radiation passes through n different materials, with absorption coefficients μ_i and thickness x_i, for $i = 1, \ldots n$, the transmitted intensity φ can be expressed as

$$\varphi = \varphi_0 \exp\left(-\sum_{i=1}^{n} \mu_i x_i \right). \tag{1.4}$$

This explains the image generation of regions that are present within the test object, as shown in Figs. 1.6 and 1.14, where a gas bubble is clearly detectable. The contrast in the X-ray image between a flaw and a defect-free area of the object test is

---

[2]As explained in Sect. 1.3, X-rays can be *absorbed* or *scattered* by the test object. In this book we present only the first interaction because scattering is not commonly used for X-ray testing applications covered in this book. For an interesting application based on the X-ray scattering effect, the reader is referred to [108].

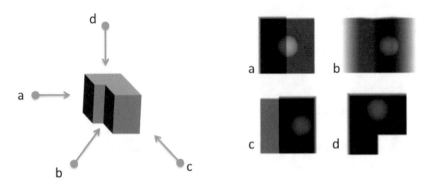

Fig. 1.14 Simulation of an X-ray image of object of Fig. 1.1 from four different points of view. Each arrow represents the orientation of the X-ray projection where the beginning corresponds to the X-ray source

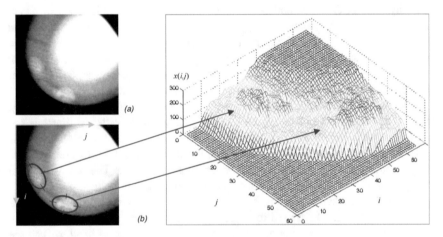

Fig. 1.15 Image formation process: **a** X-ray image of a wheel with two defects, **b** 3D plot of the gray values of the image

distinctive. In such X-ray images, we can see that the defects, like voids, cracks, or bubbles, show up as bright features. The reason is that the absorption in these areas is shorter. Hence, according to the principle of differential absorption, the detection of flaws can be achieved automatically using image processing techniques that are able to identify unexpected regions in a digital X-ray image. A real example is shown in Fig. 1.15 which clearly depicts two defects.

Another example is illustrated in Fig. 1.16a, where a backpack containing knives and a handgun is shown. However, X-ray images sometimes contain overlapped objects, making it extremely difficult to distinguish them properly, as shown in Fig. 1.16b where a handgun (superimposed onto a laptop) is almost impossible to detect.

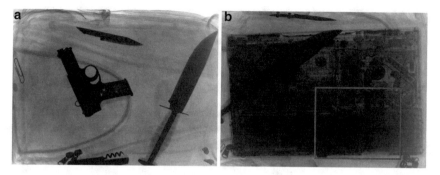

Fig. 1.16 X-ray images of a backpack. (Left) It is easy to recognize a handgun (and two knives). (Right) It is extremely difficult to detect the handgun (see red rectangle)

1.5.2   Image Acquisition

In X-ray examination, X-ray radiation is passed through the material under test, and a detector senses the radiation intensity attenuated by the material(s) of the test object. The spacial distribution of the attenuation coefficients of the elements of the object test define the X-ray information that is acquired by the sensor.

The X-ray image is usually captured with a CCD-camera (see Sect. 1.4.4) or a flat panel (see Sect. 1.4.5). The digitalized image is stored in a matrix. An example of a digitized X-ray image is illustrated in Fig. 1.17. The size of the image matrix corresponds to the resolution of the image. In this example the size is 286×384 picture elements, or *pixels*. Each pixel has a *gray value* associated. This value is between 0 and 255 for a scale of $2^8 = 256$ gray levels. Here, '0' means 100% black and a value of '255' corresponds to 100% white, as illustrated in Fig. 1.18. Typically, the digitized X-ray image is stored in a 2D matrix, e.g., $\mathbf{X}$, and its pixels are arranged in a grid manner. Thus, element $x(i, j)$ denotes the gray value of the ith row of the jth column, pixel (i, j), as shown in the matrix of Fig. 1.17.

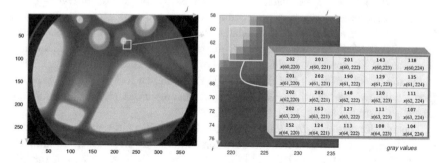

Fig. 1.17 Digital X-ray image

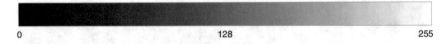

0   128   255

Fig. 1.18 256 gray level scale

The eye is only capable of resolving around 40 gray levels [20], however for computer vision applications, gray level resolution must be a minimum of 256 levels. In some applications, $2^{16} = 65,536$ gray levels are used [50], which allows one to evaluate both very dark and very bright regions in the same image.

1.5.3 X-ray Image Visualization

In many X-ray testing applications, it is necessary to display X-ray images. For example, when we present a result based on an X-ray image, or when a human evaluation of an X-ray image is required (e.g., baggage screening). In those cases, it is useful to have a suitable visualization of X-ray images.

A simple way to visualize an X-ray image is using a grayscale as shown in Fig. 1.17 that uses the grayscale of Fig. 1.18. Conventionally, X-ray images have been 'black and white' because of the gray nature of the radiographies and fluorescent screens. Usually, a common human eye can distinguish less than 50 gray values [20], however, a trained human eye is able to recognize up to 100 gray values[76].

Nowadays, it is possible to assign colors to grayscale images. With today's computing technology, especially with the ongoing advancements in displays, there is no reason to think that X-ray images must be visualized in grayscale only. In the seventeenth century, Newton said *indeed rays, properly expressed, are not colored* [3]. He was referring to light rays. Now, one can say that X-rays, *properly expressed*, are not gray... because they are not visible! We can just find a suitable way to visualize them. Thus, we can use the power of human vision that can distinguish thousands of colors [76].

In order to improve the visualization of an X-ray image, *pseudocoloring* can be used. In pseudocoloring, a gray value is converted into a color value. That is, we need a map function that relates the gray value x with a color value $(R(x), G(x), B(x))$ for red, green, and blue respectively if we use a RGB-based color map [35]. Some examples of the color maps are illustrated in Fig. 1.19 in which the transformations $(R(x), G(x), B(x))$ are shown for 'jet', 'hsv', 'parula', 'hot', 'rainbow', and 'sinmap' [35, 65, 76]. An example of a pseudocolored X-ray image is illustrated in Fig. 1.20.

The mentioned transformations correspond to linear mappings that can be loaded from a lookup table. In addition, there are some interesting algebraic or trigonometric transformations that can be used in pseudocoloring [1]. One of them is the 'sin transformation' generally defined as

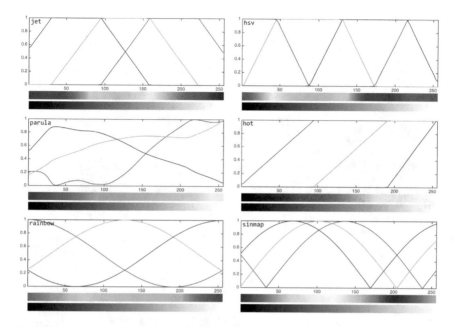

Fig. 1.19 Color maps used in pseudocoloring

$$R(x) = \mid a_R + k_R \sin(cos(\omega_R x) + \theta_R) \mid$$
$$G(x) = \mid a_G + k_G \sin(cos(\omega_G x) + \theta_G) \mid \qquad (1.5)$$
$$B(x) = \mid a_B + k_B \sin(cos(\omega_B x) + \theta_B) \mid,$$

where ω_C, θ_C, k_C, and a_C are frequency, phase, amplitude, and off-set for channel $C = R, G, B$. This color map is implemented in function sincolormap of pyxvis Library.[3] An example of a pseudocolored X-ray image is illustrated in Fig. 1.20 for 'rainbow' and 'sinmap'.

Python Example 1.1: In Fig. 1.20, we have an X-ray image of a pen case. In this example we show different visualizations of a small region of this image, namely the pencil sharpener. The example shows the classical grayscale representation, pseudocolors and a 3D representation:

Listing 1.1 : X-ray image representation.

```python
from pyxvis.io import gdxraydb
from pyxvis.io.visualization import show_xray_image, show_color_array,
    show_image_as_surface, dynamic_colormap

image_set = gdxraydb.Baggages()

# Input image
```

---

[3]pyxvis Library is an open source Python library that is used in all examples of this book (see Sect. 1.7.1).

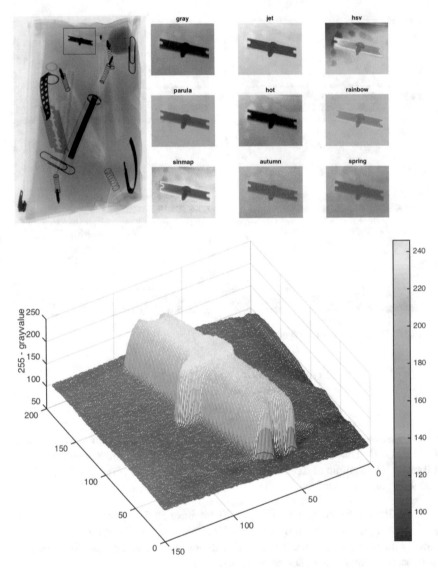

Fig. 1.20 Different visualizations of an X-ray image. [→ Example 1.1 🐍]

```
img = image_set.load_image(2, 4)

# Crop a region of interes within the image
roi = img[250:399, 340:529]

# Display the input image using customized color map
show_xray_image(img, color_map='gray')

# # Display the roi using various color maps
show_color_array(roi)
```

```
# Display the selected region into a 3D projection
show_image_as_surface(roi)
```

The output of this code is in Fig. 1.20. We can see the use of a color map for pseudocolor representations. A very interesting visualization is the 3D representation, where the z-axis corresponds to the gray value, in which the screw of the sharpener is clearly distinguishable. The output of this example is obtained using show_xray_image and show_color_array of pyxvis Library. The reader can experiment a different visualization using command dynamic_colormap of pyxvis Library, where a video of an X-ray image is presented. In this video, each frame is displayed using a different colormap that slightly varies from frame to frame.[4] □

1.5.4   Dual-Energy

In X-ray testing, dual-energy has been used successfully to provide information about the materials of the objects under test. Interesting applications can be found in baggage inspection and cargo inspection in the detection of organic or in-organic material [4, 7, 22, 52].

Coefficient μ in (1.2) can be modeled as $\mu/\rho = \alpha(Z, E)$, where ρ is the density of the material, and $\alpha(Z, E)$ is the mass attenuation coefficient that depends on the atomic number of the material Z, and the energy E of the X-ray photons. The absorption coefficient varies with energy (or wavelength) according to [23]:

$$\frac{\mu}{\rho} = \alpha(Z, E) = k\lambda^3 Z^3, \tag{1.6}$$

where k is a constant. Values for $\alpha(Z, E)$ are already measured and available in several tables (see [47]). In order to identify the material composition—typically for explosives or drug detection—the atomic number Z cannot be estimated using only one image, as a thin material with a high atomic number can have the same absorption as a thick material with a low atomic number [108]. For this purpose, a *dual-energy* system is used [86], where the object is irradiated with a High-energy level E_1 and a low-level energy E_2. In the first case, the absorbed energy depends mainly on the density of the material. In the second case, however, the absorbed energy depends primarily on the effective atomic number and the thickness of the material [97]. For two energies $i = 1, 2$, we obtain from (1.2) and (1.6):

$$\varphi_i/\varphi_0 = \exp(-\alpha(Z, E_i)\rho z), \tag{1.7}$$

Using dual-energy, it is possible to calculate the ratio:

---

[4]An example of a video generated with dynamic color is shown on https://youtu.be/Vsxff5CuTO0.

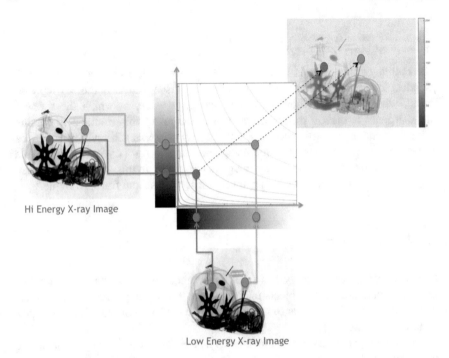

Fig. 1.21 Generation of a pseudocolor image using dual-energy. In this example the colors correspond to different materials. [$\rightarrow$ Example 1.2 🐍]

$$R = \frac{\ln(\varphi_2/\varphi_0)}{\ln(\varphi_1/\varphi_0)} = \frac{\alpha(Z, E_2)}{\alpha(Z, E_1)}, \tag{1.8}$$

where the term $-\rho z$ is canceled out, Z can be directly found using the known measurements $\alpha(Z, E)$ [43]. From both images, a new image is generated using a *fusion model*, usually a lookup-table that produces pseudocolor information [8, 31], as shown in Fig. 1.21.

🐍 Python Example 1.2: In Fig. 1.21 we have two X-ray images acquired from the same object at the same position but with different energies: the first one was taken at 5 mA and 70 kV and the second one at 5 mA and 100 kV. For an image generation of dual-energy, we can use the following Python code:

Listing 1.2 : Dual-energy.

```
from pyxvis.io import gdxraydb
from pyxvis.io.visualization import show_xray_image, show_color_array
from pyxvis.processing.images import dual_energy

import matplotlib.pylab as plt

# Select an images set and load images
image_set = gdxraydb.Baggages()
```

```
# Input images
img1 = image_set.load_image(60, 1)
img2 = image_set.load_image(60, 2)

# Display the input image using customized color map
show_xray_image([img1, img2], color_map='gray')

# Load LUT
lut = image_set.load_data(60, data_type='DualEnergyLUT')

# Compute dual energy image
energy_image = dual_energy(img1, img2, lut)

# Show results
plt.imshow(energy_image, cmap='viridis')
plt.axis('off')
plt.show()
```

The output of this code is in Fig. 1.21. We can see the use of a color map for pseudocolor representations. The output image is a grayscale image, however, the each gray value is displayed according to a 256 colors palette as shown in right bar. In this example we use dual_energy of pyxvis Library.   □

Some simple methods that deal with color X-ray images, based on dual-energy, have been developed to recognize objects in baggage inspection, see, for example, [22].

1.6 Computer Vision

Computer Vision is the science and technology of giving computers the ability to 'see' and 'understand' images taken by one or more cameras. The goal of computer vision is to study and develop algorithms for interpreting the visual world captured in images or videos. Typical topics of computer vision are detection and recognition, automated visual inspection, image stitching, image processing and analysis (enhancement, filtering, morphological operations, edge detection, and segmentation), video processing (optical flow and tracking), recognition of patterns, feature extraction and selection, local descriptors and classification algorithms, and finally, geometric vision topics such as projective geometry, camera geometric model, camera calibration, stereovision, and 3D reconstruction [11, 13, 27, 29, 30, 35, 36, 39, 51, 102].

In order to give an introduction to the topics of computer vision that have been used in X-ray testing and will be covered in this book, we follow Fig. 1.22 which illustrates an extended version of our simple model presented in Fig. 1.1.

In this general schema, X-ray images of a test object can be generated at different positions and different energy levels. Depending on the application, each block of this diagram can be (or not be) used. For example, there are applications such as weld inspection that uses a segmentation of a single mono-energetic X-ray image (black square), sometimes with pattern recognition approaches (red squares); applications like casting inspection that uses mono energetic multiple views where the

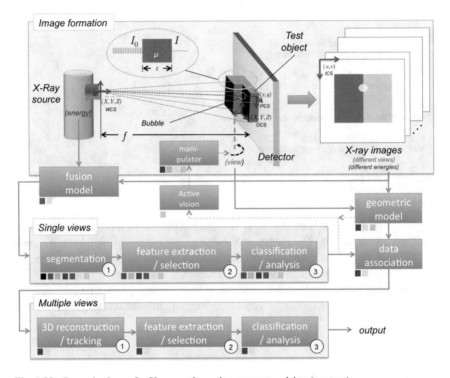

Fig. 1.22 General schema for X-ray testing using computer vision (see text)

decision is taken analyzing individual views (green squares) or corresponding multiple views (blue squares); applications including baggage screening that use dual-energy of single views (magenta squares) and multiple views (yellow squares); and finally, applications for cargo inspections that employ active vision where a next—best view is set according to the information of a single view (cyan squares). In each case, the blocks without the corresponding color square are not used.

1.6.1   Geometric Model

The X-ray image of a test object corresponds to a projection in perspective, where a 3D point of the test object is viewed as a pixel in the digital X-ray image, as illustrated in Fig. 1.22. A geometric model that describes this projection can be highly useful for 3D reconstruction and for data association between different views of the same object. Thus, 3D features or multiple view 2D features can be used to improve the diagnosis performed by using a single view.

As we will learn in Chap. 3, for the geometric model, four coordinate systems are used (see Fig. 1.22):

- OCS (X, Y, Z): Object Coordinate System, where a 3D point is defined using coordinates attached to the test object.
- WCS $(\bar{X}, \bar{Y}, \bar{Z})$: World Coordinate System, where the origin corresponds to the optical center (X-ray source) and the $\bar{Z}$-axis is perpendicular to the projection plane of the detector.
- PCS (x, y): Projection Coordinate System, where the 3D point is projected into the projection plane $\bar{Z} = f$, and the origin is the intersection of this plane with $\bar{Z}$-axis.
- ICS (u, v): Image Coordinate System, where a projected point is viewed in the image. In this case, (x, y)—axes are set to be parallel to (u, v)—axes.

The geometric model OCS $\rightarrow$ ICS, i.e., transformation $\mathbf{P} : (X, Y, Z) \rightarrow (u, v)$, can be expressed in homogeneous coordinates as [67]:

$$\lambda \begin{bmatrix} u \\ v \\ 1 \end{bmatrix} = \mathbf{P} \begin{bmatrix} X \\ Y \\ Z \\ 1 \end{bmatrix}, \tag{1.9}$$

where λ is a scale factor and $\mathbf{P}$ is a 3×4 matrix modeled as three transformations: (i) OCS $\rightarrow$ WCS, i.e., transformation $\mathbf{T}_1 : (X, Y, Z) \rightarrow (\bar{X}, \bar{Y}, \bar{Z})$, using a 3D rotation matrix $\mathbf{R}$, and 3D translation vector $\mathbf{t}$; (ii) WCS $\rightarrow$ (PCS), i.e., transformation $\mathbf{T}_2 : (\bar{X}, \bar{Y}, \bar{Z}) \rightarrow (x, y)$, using a perspective projection matrix that depends on focal distance f; and (iii) PCS $\rightarrow$ ICS, i.e., transformation $\mathbf{T}_3 : (x, y) \rightarrow (u, v)$, using scales factor α_x and α_y, and 2D translation vector (u_0, v_0). The three transformations OCS $\rightarrow$ WCS $\rightarrow$ PCS $\rightarrow$ ICS are expressed as

$$\mathbf{P} = \underbrace{\begin{bmatrix} \alpha_x & 0 & u_0 \\ 0 & \alpha_x & v_0 \\ 0 & 0 & 1 \end{bmatrix}}_{\mathbf{T}_3} \underbrace{\begin{bmatrix} f & 0 & 0 & 0 \\ 0 & f & 0 & 0 \\ 0 & 0 & 1 & 0 \end{bmatrix}}_{\mathbf{T}_2} \underbrace{\begin{bmatrix} \mathbf{R} & \mathbf{t} \\ \mathbf{0}^\mathsf{T} & 1 \end{bmatrix}}_{\mathbf{T}_1}. \tag{1.10}$$

The parameters included in matrix $\mathbf{P}$ can be estimated using a calibration approach [39].

In order to obtain multiple views of the object, n different projections of the test object can be achieved by rotating and translating it (for this task a manipulator can be used). For the p-th projection, for $p = 1 \ldots n$, the geometric model $\mathbf{P}_p$ used in (1.9) is computed from (1.10) including 3D rotation matrix $\mathbf{R}_p$ and 3D translation $\mathbf{t}_p$. Matrices $\mathbf{P}_p$ can be estimated using a calibration object projected in the n different positions [67] or using a bundle adjustment algorithm where the geometric model is obtained from the n X-ray images of the test object [69].

1.6.2 Single View Analysis

A computer vision system for single view analysis, as shown in Fig. 1.22, consists typically of the following steps: an X-ray image of the test object is taken and stored on a computer. The digital image is improved in order to enhance the details. The X-ray image of the parts of interest is found and isolated from the background of the scene. Significant features of the segmented parts are extracted. Selected features are classified or analyzed in order to determine if the test object deviates from a given set of specifications. Using a supervised pattern recognition methodology, the selection of the features and the training of the classifier are performed using representative images that are to be labeled by experts [27]. In this book, we will cover several techniques of image processing (Chap. 4), image representation (Chap. 5), and classification (Chap. 6) that have been in X-ray testing.

For the segmentation task, two general approaches can be used: a traditional image segmentation or a *sliding–window* approach. In the first case, image processing algorithms are used (e.g., histograms, edge detection, morphological operations, filtering, etc. [35]). Nevertheless, inherent limitations of traditional segmentation algorithms for complex tasks and increasing computational power have fostered the emergence of an alternative approach based on the so-called *sliding–window* paradigm. Sliding-window approaches have established themselves as state of the art in computer vision problems where a visually complex object must be separated from the background (see, for example, successful applications in face detection [105] and human detection [24]). In the sliding-window approach, a detection window is moved over an input image in both horizontal and vertical directions, and for each localization of the detection window, a classifier decides to which class the corresponding portion of the image belongs according to its features. Here, a set of candidate image areas are selected and all of them are fed to the subsequent parts of the image analysis algorithm. This resembles a brute force approach where the algorithm explores a large set of possible segmentations, and at the end the most suitable is selected by the classification steps. An example for weld inspection using sliding-windows can be found in Chap. 9.

1.6.3 Multiple View Analysis

It is well known that *A picture is worth a thousand words*, however, this is not always true if we have an *intricate* image as illustrated in Fig. 1.16b. In certain X-ray applications, e.g., baggage inspection, there are usually *intricate* X-ray images due to overlapping parts inside the test object, where each pixel corresponds to the attenuation of multiple parts, as expressed in (1.4).

In some cases, *active vision* can be used in order to adequate the viewpoint of the test object to obtain more suitable X-ray images to analyze. Therefore, an algorithm is designed for guiding the manipulator of the X-ray imaging system to poses where the detection performance should be higher [89] (see Fig. 1.22).

In other cases, multiple view analysis can be a powerful option for examining complex objects where uncertainty can lead to misinterpretation. Multiple view analysis offers advantages not only in 3D interpretation. Two or more images of the same object taken from different points of view can be used to confirm and improve the diagnosis undertaken by analyzing only one image. In the computer vision community, there are many important contributions in multiple view analysis (e.g., object class detection [100], motion segmentation [112], Simultaneous Localization And Mapping (SLAM) [53], 3D reconstruction [2], people tracking [28], breast cancer detection [103] and quality control [19]). In these fields, the use of multiple view information yields a significant improvement in performance.

Multiple view analysis in X-ray testing can be used to achieve two main goals: *(i)* analysis of 2D corresponding features across the multiple views, and *(ii)* analysis of 3D features obtained from a 3D reconstruction approach. In both cases, the attempt is made to gain relevant information about the test object. For instance, in order to validate a single view detection—filtering out false alarms—2D corresponding features can be analyzed [71]. On the other hand, if the geometric dimension of a inner part must be measured a 3D reconstruction needs to be performed [77].

As illustrated in Fig. 1.22, the input of the multiple view analysis is the *associated data*, i.e., corresponding points (or patches) across the multiple views. To this end, associated 2D cues are found using geometric constraints (e.g., epipolar geometry and multifocal tensors [39, 68]), and local scale-invariant descriptors across multiple views (e.g., like SIFT [63]).

Finally, 2D or 3D features of the associated data can be extracted and selected, and a classifier can be trained using the same pattern recognition methodology explained in Sect. 1.6.2.

Depending on the application, the output could be a measurement (e.g., the volume of the inspected inner part is $3.4\,cm^3$), a class (e.g., the test object is defective) or an interpretation (e.g., the baggage should be inspected by a human operator given that uncertainty is high).

1.6.4 Deep Learning

Originally, deep learning is inspired by ideas from neuroscience [40]. In recent years, deep learning has been successfully used in computer vision (see, for examples, in image and video recognition in [10, 59, 95]), and it has been established as the state of the art in many areas. The key idea of deep learning is to replace *handcrafted* features with features that are *learned* efficiently using a hierarchical feature extraction approach. There are several deep architectures such as deep neural networks, convolutional neural networks, energy-based models, Boltzmann machines, deep belief networks, and among others [10]. Convolutional Neural Networks (CNN), which were inspired by a biological model [60], is a very powerful method for image recognition [55]. In this book, we dedicate Chap. 7 to deep learning approaches that can be used in X-ray testing, namely convolutional neural networks, pre-trained models, transfer learning, generative adversarial networks, and detection methods.

1.6.5 Computed Tomography

Another method used in X-ray testing is Computed Tomography (CT) [18, 34], which produces a cross section of the object under test. The test object (or the X-ray source) can be rotated in order to obtain projections at different angles θ. As shown in Fig. 1.23, for each angle θ a new X-ray intensity profile $I(r, \theta)$ is obtained, where r is the distance to the origin of the object. According to the absorption's law (1.2) and a parallel-beam geometry, we obtain

$$I(r, \theta) = I_0 \exp \left(- \int_l \mu(x, y) ds \right) \tag{1.11}$$

in which (r, s) is a new coordinate system obtained by rotating (x, y) through θ with $x = r \cos \theta - s \sin \theta$ and $y = r \sin \theta + s \cos \theta$. Straight line l is the line of the X-ray beam from the X-ray source to the detector. Thus, the attenuation distribution $\mu(x, y)$ can be computed from all profiles $I(r, \theta)$.

In computed tomography, in general, a new function $P_\theta(r) = - \ln(I(r, \theta)/I_0)$ is used to calculate the object's cross-sectional plane from the measured projections [18]. The reconstruction of the object function $\mu(x, y)$ from it's projections presents a typical inverse problem [17]. A great number of algorithms are available, which can be classified into three groups:

1. Back-projection [57, 78]: This is the most basic method because it simply 'smears' each projection along the path of the X-rays. It allows for a crude reconstruction of the test object.
2. Projection-Slice theorem [20]: As illustrated in Fig. 1.23b, this theorem states that a one-dimensional Fourier transformation of a projection $P_\theta(r)$ at the angle θ is equal to the two-dimensional Fourier transformation of the object function along a straight line through the origin in Fourier coordinates at the angle θ [15, 84]. A projection $P_\theta(r)$ is obtained through parallel-beam geometry, e.g., by shifting the radiation emitter-detector arrangement radially after each measurement.[5] In practice, however, these ideal conditions cannot be realized. Only a limited number of projection measurements are available for reconstruction,

---

[5]Many reconstruction approaches assume parallel-beam geometry, whereas CT scanners usually employ fan-beam geometries. There are dedicated fan-beam algorithms (see, for example, [45]), however, there are methods that resample the fan-beam data in order to obtain an equivalent parallel-beam data (see, for example, [45, 79, 107]). Thus, traditional reconstruction approaches can be used.

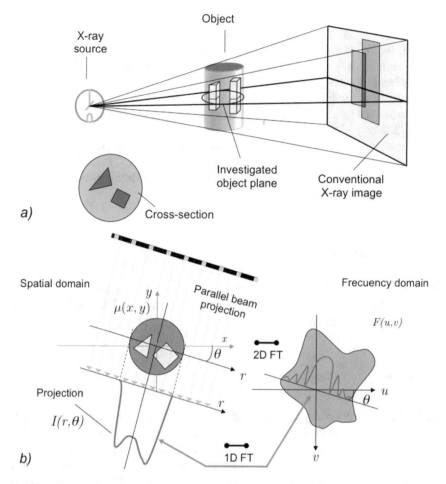

Fig. 1.23 Computed Tomography (CT) [72]: **a** Result of a CT reconstruction. **b** Projection slice theorem

and these are generated from a limited number of line integrals. As such, a two-dimensional function cannot be uniquely defined.

3. Filtered back-projection: In order to avoid the aforementioned problems, this method uses filters with low-pass characteristics. This has a negative impact, especially on high spatial resolution reconstructions, since great discontinuities in the measured values result from the object edges in the projections (highly absorptive material next to hollow spaces in the design). This leads to large artefacts, which can make image analysis impossible.

The aforementioned reconstruction problems have been addressed as an ill-posed problem [37]. There are diverse approaches for regularization and optimization algorithms that ensure their convergence. Some consider different a-priori information using probabilistic models [14] and geometric models. For example, some models have established a region of interest [61, 111] considering limited angles [32, 85, 98] or sparse representations [12, 32, 98], restricting the scope to a binary construction [5, 87], reconstructing faults in homogeneous material [5], or preserving borders [104, 109], to name just a few. The work of Retraint et al. [87] merits special interest because the authors conducted a binary reconstruction of the 3D image from just three X-ray projections (not necessarily orthogonal) using an Ising model [33].

Computed tomography provides cross-sectional images of the target object so that each object is clearly separated from any others, however, CT imaging requires a considerable number of projections to reconstruct an accurate cross-sectional image, which is time consuming.

1.7   Code and Data

The book provides supporting material at an associated website,[6] including a database of X-ray images and a Python Library, called pyxvis Library, for use with the book's many examples.

1.7.1   Pyxvis Library

In this book, we use many commands of pyxvis Library, i.e., an open source Python[7] library that we developed for X-ray testing with computer vision.[8] pyxvis Library contains more than 150 functions for image processing, projective geometry, multiple view analysis, feature extraction, feature transformation, feature analysis, feature selection, classification, convolutional neural networks, pre-trained models, transfer learning, generative adversarial networks, performance evaluation, and simulation (see Fig. 1.24).

Python Example 1.3: Each function of pyxvis Library has a 'help' with one or more examples. For example, this is the help for command dual_energy:

Listing 1.3 : Help of command dual_energy of pyxvis Library.

```
from pyxvis.io import gdxraydb
from pyxvis.processing.images import dual_energy

# Retrieve the function documentation
help(dual_energy)
```

---

[6]See https://domingomery.ing.puc.cl/material/.

[7]There are many textbooks that can be used to learn Python, see, for example, [66, 81, 94].

[8]pyxvis Library is available on https://github.com/computervision-xray-testing/pyxvis, with all experiments implemented in Google Colab.

Library	Group	Sub-Group	Description
pxvis	+ io	• data	- *data manipulation, read and write files, etc.*
		• gdxraydb	- *access to GDXray database*
		• misc	- *miscellaneous functions*
		• plots	- *plot of ellipses, space features, confusion matrices, ROC curves, etc.*
		• visualization	- *display images*
	+ geometry	• projective	- *projective geometry functions*
		• epipolar	- *multiple-view functions*
	+ processing	• images	- *image processing*
		• segmentation	- *segmentation of images*
		• helpers/kfunctions	- *kernal functions*
	+ features	• extraction	- *feature extraction*
		• seleccion	- *feature selection*
	+ learning	• classifiers	- *classification algorithms*
		• evaluation	- *hold-out, cross-validation, precision/recall*
		• cnn	- *convolutional neural networks*
		• pretrained	- *pre-trained models*
		• transfer	- *transfer learning*
		• gan	- *generative adversarial networks*
	+ simulation	• xsim	- *simulation of X-ray images*

Fig. 1.24 pyxvis Library: developed Python library for this book

The output of Example 1.3 is the following:

```
Help on function dual_energy in module images:

dual_energy(img_1, img_2, lut)
    Allows for dual-energy image computation.

    Args:
        img_1 (numpy array): first image
        img_2 (numpy array): second image
        lut (numpy array): the lookup table use during computation

    Raises:
        TypeError: invalid input type for images

    Returns:
        _dual_energy (numpy array): the dual-energy image. |
```

The reader can check the correct use of dual_energy in Example 1.2.

A quick reference for pyxvis Library and all of the examples of this book can be found in our repository (see footnote 8).

1.7.2 GDXray+ Database

We developed an X-ray database that contains more than 23,100 X-ray images.[9] The database is described in detail in Chap. 2 and Appendix A. The database includes five groups of X-ray images: castings, welds, baggage, natural objects, and settings. Each group has several series, and each series several X-ray images.

Most of the series are annotated or labeled. In those cases, the coordinates of the bounding boxes of the objects of interest or the labels of the images are available in standard text files. The size of GDXray+ is 4.5 GB.

1.8 General Methodology for X-ray Testing

In computer vision for X-ray testing, we identify three main areas:

- 1. **X-ray energies**: there is enough research evidence to show that multi-energy X-ray testing must be used when material characterization is required (e.g., to detect organic products). In other cases, such as inspection of castings, mono-energetic X-ray imaging is enough.
- 2. **X-ray multi-views**: the performance of the examination of a complex object can be better when analyzing multi-views (because a single view could present an unrecognizable pose). In othercases, such as inspection of welds, a single view is enough.
- 3. **X-ray computer vision**: there is a plethora of computer vision algorithms that can address many recognition/detection/inspection problems. There are cases (e.g., size of a fruit) in which a simple algorithm is enough, whereas in other applications (e.g., baggage inspection), more complex algorithms are required.

This taxonomy is called '3X-Strategy', as illustrated in Fig. 1.25. Each solution corresponds to a point in the 3X-space, which is defined as a combination of X-ray energies ($\mathbb{X}_1$), X-ray multi-views ($\mathbb{X}_2$) and X-ray computer vision algorithms ($\mathbb{X}_3$).

In X-ray testing, three main factors can have an impact on the solution:

1. The type of X-ray image, which depends on the X-ray energies used in the image acquisition process.
2. The point of view, that means the occlusion, which depends on whether or not other objects are superimposed over the target object, and the pose, which is related to the rotation of the object.
3. The image complexity, which depends on the number of objects present and how they are placed in the bag.

These factors have been addressed using a 3X-strategy: it is clear that certain objects of interest require more than one X-ray energy, more than one view, and

---

[9]GDXray+ is available on https://domingomery.ing.puc.cl/material/gdxray/.

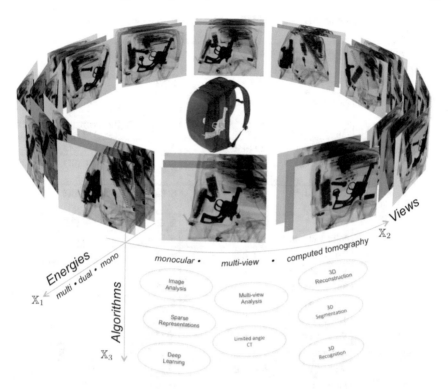

Fig. 1.25 3X-strategy: In 3X-space, a baggage inspection solution is defined as a combination of X-ray energies ($\mathbb{X}_1$), X-ray multi-views ($\mathbb{X}_2$) and X-ray computer vision algorithms ($\mathbb{X}_3$)

more than a simple algorithm. Thus, for X-ray testing there is a general methodology that can be understood as an ad-hoc combination of $\mathbb{X}_1$ for energies, $\mathbb{X}_2$ for views and $\mathbb{X}_3$ for algorithms that can be used. Table 1.2 provides possible 3X-combinations for certain applications. For example in baggage inspection, if we want to identify a flammable liquid in an uncluttered bag (i.e., low-image complexity) we need at least dual-energy, possibly only one view, and a simple computer vision algorithm. However, if we want to detect a handgun in a cluttered bag (i.e., high image complexity), we need several views, possibly a computed tomography, and a complex computer vision algorithm. If we want to detect a metallic handgun, dual-energy will be required. A 3X-strategy is to be designed for each kind of object to be detected.

Table 1.2 Information on possible combination of energies, views and algorithms for categories of objects

Application area	Object	$\mathbb{X}_1^*$ Energies 1 2 3	$\mathbb{X}_2^*$ Views 1 2 3	$\mathbb{X}_3^*$ Algorithms 1 2 3
Baggage inspection	Aerosols	□ ⊞ ⊡	⊞ ⊡ ⊡	⊞ ⊡ □
	Alcohol	□ ⊞ ⊡	⊞ ⊡ ⊡	⊞ ⊡ □
	Ammunition	□ ⊞ ⊡	⊞ ⊡ ⊡	⊞ ⊡ □
	Flammable liquids	□ ⊞ ⊡	⊞ ⊡ ⊡	⊞ ⊡ □
	Fruits and vegetables	□ ⊞ ⊡	⊞ ⊞ ⊡	⊞ ⊡ □
	Guns	⊞ ⊡ □	⊞ ⊞ ⊡	⊞ ⊞ ⊡
	Milk and honey	□ ⊞ ⊡	⊞ ⊡ ⊡	⊞ ⊡ □
	Pepper spray	□ ⊞ ⊡	⊞ ⊡ ⊡	⊞ ⊞ ⊡
	Seeds and grains	□ ⊞ ⊡	⊞ ⊞ ⊡	⊞ ⊡ □
	Sharp objects	⊞ ⊡ □	⊞ ⊞ ⊡	⊞ ⊞ ⊡
	Stun guns	⊞ ⊡ □	⊞ ⊞ ⊡	⊞ ⊞ ⊡
	Toxic substances	⊞ ⊡ □	⊞ ⊡ ⊡	⊞ ⊞ ⊡
	Woods and barks	□ ⊞ ⊡	⊞ ⊞ ⊡	⊞ ⊡ □
Quality control	Automotive parts	⊞ □ □	⊞ ⊞ ⊡	⊞ ⊞ ⊡
	Welds	⊞ □ □	⊞ □ □	⊞ ⊞ ⊡
	Food	⊞ □ □	⊞ □ □	⊞ ⊞ ⊡
	Electronic circuits	⊞ □ □	⊞ □ □	⊞ ⊞ ⊡
Cargo inspection	People	⊞ □ □	⊞ ⊡ □	⊞ ⊡ □
	Explosives	□ ⊞ ⊡	⊞ ⊡ ⊡	⊞ ⊡ □
	Fruits and vegetables	□ ⊞ ⊡	⊞ ⊡ ⊡	⊞ ⊡ □
	*1	Mono	Mono	Simple
	2	Dual	Multi	Medium
	3	Multi	CT	Complex

□ not used , ⊞ used, ⊡ probably used

1.9   Summary

In this book, we present a general overview of computer vision approaches that have been used in X-ray testing. In this chapter, we gave an introduction to our book by covering relevant issues of X-ray testing.

X-ray testing has been developed for the inspection of materials or objects, where the aim is to analyze—nondestructively—those inner parts that are undetectable to the naked eye. Thus, X-ray testing is used to determine if a test object deviates from a given set of specifications.

Typical applications are

- Inspection of automotive parts,
- Quality control of welds,
- Baggage screening,
- Analysis of food products,
- Inspection of cargo,
- Quality control of electronic circuits.

In order to achieve efficient and effective X-ray testing, automated and semi-automated systems based on computer vision algorithms are being developed to execute this task.

We gave an introduction to some physic and geometric principles related to computer vision. Following this, an overview of single and multiple view analysis, deep learning, and computed tomography was presented. Finally, we introduce a general methodology for computer vision for X-ray testing.

References

1. Abidi, B.R., Zheng, Y., Gribok, A.V., Abidi, M.A.: Improving weapon detection in single energy X-ray images through pseudocoloring. IEEE Trans. Syst., Man, Cybern., Part C: Appl. Rev. **36**(6), 784–796 (2006)
2. Agarwal, S., Snavely, N., Simon, I., Seitz, S., Szeliski, R.: Building Rome in a day. In: IEEE 12th International Conference on Computer Vision (ICCV2009), pp. 72–79 (2009)
3. Agoston, G.A.: The concept of color. Color Theory and Its Application in Art and Design, pp. 5–10. Springer, Berlin (1987)
4. Akcay, S., Breckon, T.: Towards automatic threat detection: a survey of advances of deep learning within X-ray security imaging (2020). arXiv:2001.01293
5. Allain, M., Idier, J.: Efficient binary reconstruction for non destructive evaluation using gammagraphy. Inverse Prob. **4**(23), 1371–1393 (2007)
6. Als-Nielsen, J., McMorrow, D.: Elements of Modern X-Ray Physics, 2nd edn. Willey, Hoboken (2011)
7. Baştan, M., Byeon, W., Breuel, T.M.: Object recognition in multi-view dual x-ray images. In: British Machine Vision Conference BMVC (2013)
8. Baştan, M., Yousefi, M.R., Breuel, T.M.: Visual words on baggage X-ray images. Computer Analysis of Images and Patterns, pp. 360–368. Springer, Berlin (2011)
9. Bavendiek, K., Krause, A., Beyer, A.: Durchsatzerhöhung in der industriellen Röntgenprüfung – Eine Kombination aus innovativem Prüfablauf und optimierter Bildauswertung. In: DGZfP Jahrestagung, vol. Berichtsband 63.1, pp. 301–306. Deutsche Gesellschaft für Zerstörungsfreie Prüfung e.V., Bamberg (1998)
10. Bengio, Y., Courville, A., Vincent, P.: Representation learning: a review and new perspectives. IEEE Trans. Pattern Anal. Mach. Intell. **35**(8), 1798–1828 (2013)
11. Bhuyan, M.K.: Computer Vision and Image Processing: Fundamentals and Applications. CRC Press, Boca Raton (2019)
12. Bian, J., Siewerdsen, J., Han, X., Sidky, E., Prince, J., C., P., Pan, X.: Evaluation of sparse-view reconstruction from flat-panel-detector cone-beam CT. Phys. Med. Biol. **55**(22), 6575–6599 (2010)
13. Bishop, C.: Pattern Recognition and Machine Learning. Springer, Berlin (2006)

14. Bouman, C., Sauer, K.: A unified approach to statistical tomography using coordinate descent optimization. IEEE Trans. Image Process., 480–492 (1996)
15. Bracewell, R.N.: Strip integration in radio astronomy. Aust. J. Phys. **9**(2), 198–217 (1956)
16. Bunke, J.: Computertomographie. In: Ewen, K. (ed.) Moderne Bildgebung: Physik, Gerätetechnik, Bildbearbeitung und -kommunikation, Strahlenschutz, Qualitätskontrolle, pp. 153–170. Georg Thieme Verlag, Stuttgart, New York (1998)
17. Buzug, T.: Computed Tomography. Springer, Berlin (2008)
18. Carmignato, S., Dewulf, W., Leach, R.: Industrial X-Ray Computed Tomography. Springer, Berlin (2018)
19. Carrasco, M., Pizarro, L., Mery, D.: Visual inspection of glass bottlenecks by multiple-view analysis. Int. J. Comput. Integr. Manuf. **23**(10), 925–941 (2010)
20. Castleman, K.: Digital Image Processing. Prentice-Hall, Englewood Cliffs (1996)
21. Cha, B.K., Jeon, S., Seo, C.W.: X-ray performance of a wafer-scale cmos flat panel imager for applications in medical imaging and nondestructive testing. Nucl. Instrum. Methods Phys. Res., Sect. A **831**, 404–409 (2016)
22. Chouai, M., Merah, M., Sancho-GÓmez, J.L., Mimi, M.: A machine learning color-based segmentation for object detection within dual X-ray baggage images. In: Proceedings of the 3rd International Conference on Networking, Information Systems & Security, pp. 1–11 (2020)
23. Cullity, B.D., Stock, S.R.: Elements of X-Ray Diffraction. Pearson, London (2001)
24. Dalal, N., Triggs, B.: Histograms of oriented gradients for human detection. In: Conference on Computer Vision and Pattern Recognition (CVPR2005), vol. 1, pp. 886–893 (2005)
25. Dennhoven, M., Kunze, C., Kuehn, R.: Baggage inspection device (1977). US Patent 4,047,035
26. Donges, G., Dietrich, R.: Baggage inspection system (1988). US Patent 4,759,047
27. Duda, R., Hart, P., Stork, D.: Pattern Classification, 2nd edn. Wiley, New York (2001)
28. Eshel, R., Moses, Y.: Tracking in a dense crowd using multiple cameras. Int. J. Comput. Vision **88**, 129–43 (2010)
29. Faugeras, O., Luong, Q.T., Papadopoulo, T.: The Geometry of Multiple Images: The Laws that Govern the Formation of Multiple Images of a Scene and Some of their Applications. The MIT Press, Cambridge (2001)
30. Forsyth, D.A., Ponce, J.: A modern approach. A Modern Approach, Computer Vision (2003)
31. Franzel, T., Schmidt, U., Roth, S.: Object detection in multi-view X-Ray images. Pattern Recognit., 144–154 (2012)
32. Frikel, J.: Sparse regularization in limited angle tomography. Appl. Comput. Harmon. Anal. **1**(34), 117–141 (2013)
33. Gallavotti, G.: Statistical mechanics. Texts and Monographs in Physics. Springer, Berlin (1999)
34. Goebbels, J.: Computed tomography. Handbook of Technical Diagnostics, pp. 249–258. Springer, Berlin (2013)
35. Gonzalez, R., Woods, R.: Digital Image Processing, 3rd edn. Prentice Hall, Pearson (2008)
36. Goodfellow, I., Bengio, Y., Courville, A.: Deep Learning. MIT Press, Cambridge (2016)
37. Hadamard, J.: Lectures on the Cauchy Problem in Linear Partial Differential Equations. Yale University Press, New Haven (1923)
38. Hanke, R., Fuchs, T., Uhlmann, N.: X-ray based methods for non-destructive testing and material characterization. Nucl. Instrum. Methods Phys. Res., Sect. A **591**(1), 14–18 (2008)
39. Hartley, R.I., Zisserman, A.: Multiple View Geometry in Computer Vision, 2nd edn. Cambridge University Press, Cambridge (2003)
40. Hassabis, D., Kumaran, D., Summerfield, C., Botvinick, M.: Neuroscience-inspired artificial intelligence. Neuron **95**(2), 245–258 (2017)
41. Heinzerling, J.: Bildverstärker-Fernseh-Kette. In: Ewen, K. (ed.) Moderne Bildgebung: Physik, Gerätetechnik, Bildbearbeitung und -kommunikation, Strahlenschutz, Qualitätskontrolle, pp. 115–126. Georg Thieme Verlag, Stuttgart, New York (1998)

42. Heinzerling, J.: Röntgenstrahler. In: Ewen, K. (ed.) Moderne Bildgebung: Physik, Gerätetechnik, Bildbearbeitung und -kommunikation, Strahlenschutz, Qualitätskontrolle, pp. 77–85. Georg Thieme Verlag, Stuttgart, New York (1998)

43. Heitz, G., Chechik, G.: Object separation in X-ray image sets. In: IEEE Conference on Computer Vision and Pattern Recognition (CVPR-2010), pp. 2093–2100 (2010)

44. Hellier, C.: Handbook of Nondestructive Evaluation, 2nd edn. McGraw Hill, New York (2013)

45. Herman, G., Lung, H.: Reconstruction from divergent beams: a comparison of algorithms with and without rebinning. Comput. Biol. Med. **10**(2), 131–139 (1980)

46. Horbaschek, H.: Technologie und Einsatz von Festkörperdetektoren in der Röntgentechnik (1998). Vortrag der Firma Siemens Pforchheim in der 9. Sitzung des Unterausschusses Bildverarbeitung in der Durchstrhlungprüfung (UA BDS) der Deutschen Gesellschaft für Zerstörungsfreie Prüfung e.V. (DGZfP), Ahrensburg

47. Hubbell, J., Seltzer, S.: Tables of X-Ray mass attenuation coefficients and mass energy-absorption coefficients from 1 keV to 20 MeV for elements Z = 1 to 92 and 48 additional substances of dosimetric interest (1996). http://www.nist.gov/pml/data/xraycoef/index.cfm

48. Iniewski, K.: CZT sensors for computed tomography: from crystal growth to image quality. J. Instrum. **11**(12), C12,034 (2016)

49. Jaeger, T.: Optimierungsansätze zur Lösung des limited data problem in der Computertomographie. Verlag Dr. Köster, Berlin (1997)

50. Jaeger, T., Heike, U., Bavendiek, K.: Experiences with an amorphous silicon array detector in an ADR application. In: International Computerized Tomography for Industrial Applications and Image Processing in Radiology, DGZfP Proceedings BB 67-CD, pp. 111–114. Berlin (1999)

51. Klette, R.: Concise computer vision: an introduction into theory and algorithms. Springer Science & Business Media (2014)

52. Kolkoori, S., Wrobel, N., Deresch, A., Redmer, B., Ewert, U.: Dual high-energy X-ray digital radiography for material discrimination in cargo containers. In: 11th European Conference on Non-Destructive Testing (ECNDT 2014), 6–10 Oct 2014, Prague, Czech Republic (2014)

53. Konolige, K., Agrawal, M.: FrameSLAM: from bundle adjustment to realtime visual mapping. IEEE Trans. Rob. **24**(5), 1066–1077 (2008)

54. Kosanetzky, J.M., Krüger, R.: Philips MU231: Räderprüfanlage. Technischer Bericht, Philips Industrial X-ray GmbH, Hamburg (1997)

55. Krizhevsky, A., Sutskever, I., Hinton, G.E.: ImageNet classification with deep convolutional neural networks. NIPS, pp. 1106–1114 (2012)

56. Kuchling, H.: Taschenbuch der Physik, 12th edn. Harri Deutsch, Thun-Frankfurt, Main (1989)

57. Kuhl, D., Edwards, R.: Image separation radioisotope scanning. Radiology **80**(4), 653–662 (1963)

58. Kunze, C., Dennhoven, M.: Inspection system for baggage (1980). US Patent 4,216,499

59. LeCun, Y., Bengio, Y., Hinton, G.: Deep learning. Nature **521**(7553), 436–444 (2015)

60. LeCun, Y., Bottou, L., Bengio, Y.: Gradient-based learning applied to document recognition. In: Proceedings of the Third International Conference on Research in Air Transportation (1998)

61. Lehr, C., Liedtke, C.: 3D reconstruction of volume defects from few X-ray. Computer analysis of images and patterns, pp. 257–284. Springer, Berlin (1999)

62. Lossau, N.: Röntgen: Eine Entdeckung verändert unser Leben, 1 edn. Köln, vgs (1995)

63. Lowe, D.: Distinctive image features from scale-invariant keypoints. Int. J. Comput. Vision **60**(2), 91–110 (2004)

64. Martz, H.E., Logan, C.M., Schneberk, D.J., Shull, P.J.: X-Ray Imaging: Fundamentals, Industrial Techniques and Applications. CRC Press, Boca Raton (2016)

65. MathWorks: Image Processing Toolbox for Use with MATLAB: User's Guide. The MathWorks Inc. (2014)

66. Matthes, E.: Python crash course: a hands-on, project-based introduction to programming. No Starch Press (2015)
67. Mery, D.: Explicit geometric model of a radioscopic imaging system. NDT & E Int. **36**(8), 587–599 (2003)
68. Mery, D.: Exploiting multiple view geometry in X-ray testing: part I, theory. Mater. Eval. **61**(11), 1226–1233 (2003)
69. Mery, D.: Automated detection in complex objects using a tracking algorithm in multiple X-ray views. In: Proceedings of the 8th IEEE Workshop on Object Tracking and Classification Beyond the Visible Spectrum (OTCBVS 2011), in Conjunction with CVPR 2011, Colorado Springs, pp. 41–48 (2011)
70. Mery, D.: Aluminum casting inspection using deep learning: a method based on convolutional neural networks. J. Nondestr. Eval. **39**(1), 12 (2020)
71. Mery, D., Filbert, D.: Automated flaw detection in aluminum castings based on the tracking of potential defects in a radioscopic image sequence. IEEE Trans. Robot. Autom. **18**(6), 890–901 (2002)
72. Mery, D., Filbert, D., Jaeger, T.: Image processing for fault detection in aluminum castings. In: MacKenzie, D., Totten, G. (eds.) Analytical Characterization of Aluminum and Its Alloys. Marcel Dekker, New York (2003)
73. Mery, D., Jaeger, T., Filbert, D.: A review of methods for automated recognition of casting defects. Insight **44**(7), 428–436 (2002)
74. Murphy, E.: A rising war on terrorists. IEEE Spectr. **26**(11), 33–36 (1989)
75. Murray, N., Riordan, K.: Evaluation of automatic explosive detection systems. In: 29th Annual 1995 International Carnahan Conference on Security Technology, 1995. Proceedings. Institute of Electrical and Electronics Engineers, pp. 175 –179 (1995). https://doi.org/10.1109/CCST.1995.524908
76. Neri, E., Caramella, D., Bartolozzi, C.: Image processing in radiology. In: Baert, A.L, Knauth, M., Sartor, K (eds.) Medical Radiology. Diagnostic Imaging. Springer, Berlin (2008)
77. Noble, A., Gupta, R., Mundy, J., Schmitz, A., Hartley, R.: High precision X-ray stereo for automated 3D CAD-based inspection. IEEE Trans. Robot. Autom. **14**(2), 292–302 (1998)
78. Oldendorf, W.: Isolated flying spot detection of radiodensity discontinuities-displaying the internal structural pattern of a complex object. IRE Trans. Biomed. Electron. **8**(1), 68–72 (1961)
79. Peters, T., Lewitt, R.: Computed tomography with fan beam geometry. J. Comput. Assist. Tomogr. **1**(4), 429–436 (1977)
80. Peugeot, R.S.: X-ray baggage inspection system (1975). US Patent 3,919,467
81. Pichara, K., Pieringer, C.: Advanced Computer Programming in Python. CreateSpace Independent Publishing Platform (2017)
82. Purschke, M.: Radioskopie – Die Prüftechnik der Zukunft? In: DGZfP Jahrestagung, vol. Berichtsband 68.1, pp. 77–84. Deutsche Gesellschaft für Zerstörungsfreie Prüfung e.V., Celle (1999)
83. Purschke, M.: IQI-sensitivity and applications of flat panel detectors and X-ray image intensifiers - a comparison. Insight **44**(10), 628–630 (2002)
84. Radon, J.: Über die Bestimmung von Funktionen durch ihre Integrale längs gewisser Mannigfaltigkeiten. Ber. Sächs. Akad. Wiss. Math. Phys. Kl. **69**, 262–277 (1917)
85. Rantala, M., Vanska, S., Jarvenpaa, S., Kalke, M., Lassas, M., Moberg, J., Siltanen, S.: Wavelet-based reconstruction for limited-angle. IEEE Trans. Med. Imaging **25**(2), 210–217 (2006)
86. Rebuffel, V., Dinten, J.M.: Dual-energy X-ray imaging: benefits and limits. Insight-Non-Destr. Test. Cond. Monit. **49**(10), 589–594 (2007)
87. Retraint, F., Peyrin, F., Dinten, J.: Three-dimensional regularized binary image reconstruction from three two-dimensional projections using a randomized ICM algorithm. Int. J. Imaging Syst. Technol. **9**, 135–146 (1998)
88. Richter, H.U.: Chronik der Zerstörungsfreien Materialprüfung, 1st edn. DGZfP, Verlag für Schweißen und verwendete Verfahren, DVS-Verlag GmbH, Berlin, Deutsche Gesellschaft für Zerstörungsfreie Prüfung (1999)

89. Riffo, V., Mery, D.: Active X-ray testing of complex objects. Insight **54**(1), 28–35 (2012)
90. Röntgen, W.: Eine neue Art von Strahlen: I Mitteilung. In: Sitzungsbericht der Würzburger Physikal.-Medicin. Gesellschaft. Verlag und Druck der Stahel'schen K. Hof- und Universitäts- Buch- und Kunsthandlung, Würzburg (1895)
91. Rowlands, J.: The physics of computed radiography. Phys. Med. Biol. **47**(23), R123 (2002)
92. Schaefer, M.: 100 Jahre Röntgenprüftechnik - Prüfsysteme früher und heute. In: DGZfP Jahrestagung, pp. 13–26. Deutsche Gesellschaft für Zerstörungsfreie Prüfung e.V., Aachen (1995)
93. Schwieger, R.: Stillegung, sicherer Einschluß und Abbau kerntechnischer Anlagen. Institut für Werkstoffkunde, Universität Hannover, Technischer Bericht (1999)
94. Shaw, Z.A.: Learn Python 3 the Hard Way: A Very Simple Introduction to the Terrifyingly Beautiful World of Computers and Code. Addison-Wesley Professional (2017)
95. Simonyan, K., Zisserman, A.: Very deep convolutional networks for large-scale image recognition (2014). ArXiv:abs/1409.1556
96. Singh, M., Singh, S.: Optimizing image enhancement for screening luggage at airports. In: Proceedings of the 2005 IEEE International Conference on Computational Intelligence for Homeland Security and Personal Safety, 2005. CIHSPS 2005, pp. 131–136 (2005). https://doi.org/10.1109/CIHSPS.2005.1500627
97. Singh, S., Singh, M.: Explosives detection systems (eds) for aviation security. Signal Process. **83**(1), 31–55 (2003)
98. Soussen, C., Idier, J.: Reconstruction of three-dimensional localized objects from limited angle X-ray projections: an approach based on sparsity and multigrid image representation. J. Electron. Imaging **17**(3) (2008)
99. Strecker, H.: Automatic detection of explosives in airline baggage using elastic X-ray scatter. Medicamundi **42**, 30–33 (1998)
100. Su, H., Sun, M., Fei-Fei, L., Savarese, S.: Learning a dense multi-view representation for detection, viewpoint classification and synthesis of object categories. In: International Conference on Computer Vision (ICCV2009) (2009)
101. Szeles, C., Soldner, S.A., Vydrin, S., Graves, J., Bale, D.S.: Cdznte semiconductor detectors for spectroscopic X-ray imaging. IEEE Trans. Nucl. Sci. **55**(1), 572–582 (2008)
102. Szeliski, R.: Computer Vision: Algorithms and Applications. Springer, New York Inc (2011)
103. Teubl, J., Bischof, H.: Comparison of Multiple View Strategies to Reduce False Positives in Breast Imaging. Digital Mammography, pp. 537–544 (2010)
104. Tian, Z., Jia, X., Yuan, K., Pan, T., Jiang, S.: Low-dose CT reconstruction via edge-preserving total variation regularization. Phys. Med. Biol. **56**(18), 5949–5967 (2011)
105. Viola, P., Jones, M.: Robust real-time object detection. Int. J. Comput. Vision **57**(2), 137–154 (2004)
106. Völkel: Grundlagen für den Prüfer mit Röntgen- und Gammastrahlung (Durchstrahlungsprüfung). Amt für Standarisierung, Meßwesen und Warenprüfung, Fachgebiet Zerstörungsfreie Werkstoffprüfung (1989)
107. Wang, L.: Cross-section reconstruction with a fan-beam scanning geometry. IEEE Trans. Comput. **100**(3), 264–268 (1977)
108. Wells, K., Bradley, D.: A review of X-ray explosives detection techniques for checked baggage. Applied Radiation and Isotopes (2012)
109. Yu, D., Fessler, J.: Edge-preserving tomographic reconstruction with nonlocal regularization. IEEE Trans. Med. Imaging **2**(21), 159–173 (2002)
110. Zabler, S., Maisl, M., Hornberger, P., Hiller, J., Fella, C., Hanke, R.: X-ray imaging and computed tomography for engineering applications. tm-Technisches Messen **1**(ahead-of-print) (2020)
111. Zhou, Y., Thibault, J., Bouman, C., Sauer, K., Hsieh, J.: Fast Model-Based X-ray CT Reconstruction Using Spatially Nonhomogeneous ICD Optimization. IEEE Trans. Image Process. **20**(1), 161–175 (2011)
112. Zografos, V., Nordberg, K., Ellis, L.: Sparse motion segmentation using multiple six-point consistencies. In: Proceedings of the Asian Conference on Computer Vision (ACCV2010) (2010)

Chapter 2
Images for X-ray Testing

Abstract In this chapter, we present the dataset that is used in this book to illustrate and test several methods. The database consists of 23,189 X-ray images. The images are organized in a public database called GDXray+ that can be used free of charge, but for research and educational purposes only. The database includes five groups of X-ray images: castings, welds, baggage, natural objects, and settings. Each group has several series, and each series several X-ray images. Most of the series are annotated or labeled. In such cases, the coordinates of the bounding boxes of the objects of interest or the labels of the images are available in standard text files. The size of GDXray+ is 4.5 GB and it can be downloaded from our website.

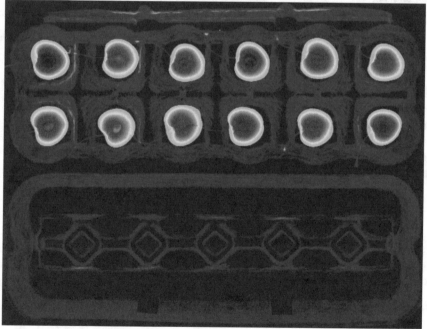

Cover image: *X-ray image of cherries in an egg crate (X-ray image* N0006_0027 *colored with 'jet' colormap).*

© Springer Nature Switzerland AG 2021
D. Mery and C. Pieringer, *Computer Vision for X-Ray Testing*,
https://doi.org/10.1007/978-3-030-56769-9_2

2.1   Introduction

Public databases of X-ray images can be found for medical imaging,[1] however, to the best knowledge of the author, up until now there have not been any public databases of digital X-ray images for X-ray testing for general purposes.[2]

As a service to the X-ray testing community, we collected more than 23,100 X-ray images for the development, testing, and evaluation of image analysis and computer vision algorithms. The images are organized in a public database called GDXray+.[3] In order to illustrate our database, a random selection of 120 X-ray images is shown in Fig. 2.1. The database includes five groups of X-ray images: castings, welds, baggage, natural objects, and settings. Each group has several series, and each series several X-ray images. Most of the series are annotated or labeled. In those cases, the coordinates of the bounding boxes of the objects of interest or the labels of the images are available. In Table 2.1, we can see some statistics. The size of GDXray+ is 4.54 GB, and it can be downloaded from our website (see Fig. 2.2).

In this chapter, we will view the structure of GDXray+ database, a description for each group (with some series examples), some examples of applications that have been published using images of GDXray+ and some examples in Python that can be used to manipulate the database. More details about GDXray+ are given in Appendix A

2.2   Structure of the Database

GDXray+ is available in a public repository. The repository contains 5 group folders one for each group: Castings, Welds, Baggage, Nature, and Settings. For each group, we define an initial: C, W, B, N, and S, respectively. As shown in Table 2.1, each group has several series. Each series is stored in an individual sub-folder of the corresponding group folder. The sub-folder name is Xssss, where X is the initial of the group and ssss is the number of the series. For example, the third series of group

---

[1] See, for example, a good collection in http://www.via.cornell.edu/databases/.

[2] There are some galleries of X-ray images available on the web with a few samples, see, for instance, http://www.vidisco.com/ndt_solutions/ndt_info_center/ndt_x_ray_gallery with approximately 50 X-ray images; and a very large dataset (more than 1 million images) for baggage inspection with no annotations [36].

[3] Available on https://domingomery.ing.puc.cl/material/gdxray/. Originally the name was GDXray [33]. The name comes from 'The Grima X-ray database' (Grima was the name of our Machine Intelligence Group at the Department of Computer Science of the Pontificia Universidad Católica de Chile). Now, we release an extended version of the dataset that we call GDXray+. The X-ray images included in GDXray+ can be used free of charge, but for research and educational purposes only. Redistribution and commercial use is prohibited. Any researcher reporting results which use this database should acknowledge the GDXray+ database by citing this chapter.

Fig. 2.1 Random X-ray images of GDXray+ database

Castings is stored in sub-folder C0003 of folder Castings. The X-ray images of a series are stored in file Xssss_nnnn.png. Again Xssss is the name of the series. The number nnnn corresponds to the number of the X-ray image of this series. For

Table 2.1 Statistics of GDXray+ database

Groups	Series	Images	Size (MB)
Castings	85	3768	664.8
Welds	4	98	209.5
Baggages	86	10863	3403.4
Nature	13	8290	191.9
Settings	8	170	73.1
Total	196	23189	4542.6

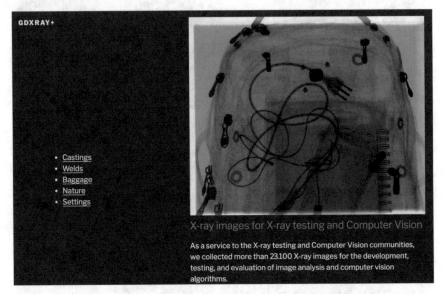

Fig. 2.2 Screenshot of GDXray+ website. The figure shows X-ray image of a backpack using pseudo coloring ('hot' colormap): B0083_0031.png

example, the fifth X-ray image of series C0003 is C0003_0005.png and is stored in folder Castings/C0003. The whole structure is summarized in Table 2.2. All X-ray images of GDXray+ are stored in 'png' (Portable Network Graphics)[4] format.

2.3   Castings

The group Castings contains 3,768 X-ray images arranged in 68 series. The X-ray images were taken mainly from automotive parts (aluminum wheels and knuckles). Some examples are illustrated in Figs. 2.3, 2.4 and 2.5. The details of each series are

---

[4]See http://www.libpng.org/pub/png/.

Table 2.2 Structure of GDXray+

Database	Groups	Series	X-ray images
GDXray →	Castings →	C0001 →	C0001_0001.png ... C0001_0072.png
		⋮	
		C0085 →	C0085_0001.png ... C0085_0686.png
	Welds	→ W0001 →	W0001_0001.png ... W0001_0010.png
		⋮	
		W0004 →	W0004_0001.png ... W0004_0010.png
	Baggage →	B0001 →	B0001_0001.png ... B0001_0014.png
		⋮	
		B0086 →	B0086_0001.png ... B0086_1000.png
	Nature	→ N0001 →	N0001_0001.png ... N0001_0013.png
		⋮	
		N0013 →	N0013_0001.png ... N0013_0006.png
	Settings →	S0001 →	S0001_0001.png ... S0001_0018.png
		⋮	
		S0008 →	S0008_0001.png ... S0008_0018.png

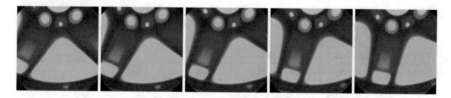

Fig. 2.3 Some X-ray images of an aluminum wheel (group Castings series C0001)

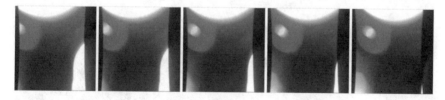

Fig. 2.4 Some X-ray images of a knuckle (group Castings series C0059)

given in Table A.2. Experiments on these data can be found in several publications as shown in Table 2.3. It is interesting to highlight that series C0001 (see Fig. 2.3) contains not only a sequence of 72 X-ray images taken from an aluminum wheel by rotating its central axis in 5^0, but also annotations of bounding boxes of the ground truth of 226 small defects and the calibration matrix of each image that relates the 3D coordinates of the aluminum wheel with 2D coordinates of the X-ray image.

Table 2.3　Applications of series Castings

Series	Application	References
C0001	Detection of defects in multiple views	[3, 16, 22, 26, 39, 40]
	Estimation of epipolar geometry with distortion	[24]
	Calibration of X-ray imaging system with image intensifiers	[26]
	Simulation of casting defects	[26]
	Detection defects using deep learning (CNN) and classic features	[5, 6, 19, 20, 35]
C0002	Experiments on detection of defects in single views	[7, 14, 23, 41]
C0008	Simulation of casting defects	[11]
C0010	Detection defects using deep learning (CNN) and classic features	[19, 20, 35]
C0015	Detection defects using deep learning (CNN) and classic features	[5, 6, 19, 20, 35]
C0017	Simulation of casting defects	[13, 28]
C0021	Detection defects using deep learning (CNN) and classic features	[19, 20, 35]
C0031	Detection defects using deep learning (CNN) and classic features	[5, 6, 19, 20, 35]
C0032	Experiments on detection of defects in multiple views	[16]
C0034	Detection defects using deep learning (CNN) and classic features	[5, 6, 19, 20, 35]
C0037	Simulation of casting defects	[13, 28]
C0054	Detection of casting on moving castings	[27]
C0055	Image restoration in blurred X-ray images	[25]
C0061	Detection defects using deep learning (CNN) and classic features	[19, 20, 35]

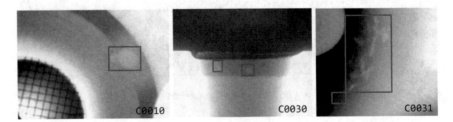

Fig. 2.5　Some annotated images showing bounding boxes of casting defects

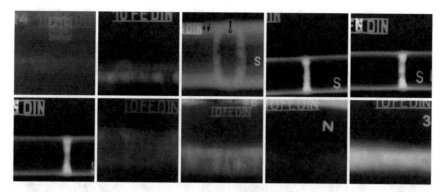

Fig. 2.6 Some X-ray images of group Welds series W0003. This series corresponds to the BAM database

Table 2.4 Applications of series Welds

Series	Application	References
W0001	Detection of defects in welds using classic methods	[2, 8, 17, 21, 49]
	Simulation of welding defects	[12, 17, 28]
	Detection of defects in welds using deep learning	[9, 10, 47, 48]
W0002	Evaluation of performance of detection algorithm	[2]
W0003	Detection of defects in welds using classic methods	[38, 46, 49]
	Detection of defects in welds using deep learning	[9, 10, 48]

2.4 Welds

The group Welds contains 98 images arranged in 4 series. The X-ray images were taken by the Federal Institute for Materials Research and Testing, Berlin (BAM).[5] Some examples are illustrated in Fig. 2.6. The details of each series are given in Table A.4. Experiments on these data can be found in several publications as shown in Table 2.4. It is interesting to highlight that series W0001 and W0002 (see Fig. 2.7) contains not only 10 X-ray images selected from the whole BAM database (series W0003), but also annotations of bounding boxes and the binary images of the ground truth of 641 defects.

---

[5]The X-ray images of series W0001 and W0003 are included in GDXray, thanks to the collaboration of the Institute for Materials Research and Testing (BAM), Berlin http://dir.bam.de/dir.html.

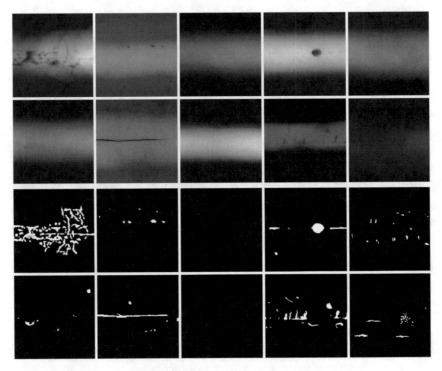

Fig. 2.7   Some images of group Welds series W0001 (X-ray images) and W0002 (ground truth)

2.5   Baggage

The group Baggage contains 10,863 X-ray images arranged in 86 series. The X-ray images were taken from different containers such as backpacks, pen cases, and wallets, etc. Some examples are illustrated in Figs. 2.8, 2.9, 2.10, 2.11, and 2.12. The details of each series are given in Table A.23. Experiments on these data can be found in several publications as shown in Table 2.5. It is interesting to highlight that series B0046, B0047, and B0048 (see, for example, Fig. 2.8) contains 600 X-ray images that can be used for automated detection of handguns, shuriken, and razor blades (bounding boxes for these objects of interest are available as well). In this case, the training can be performed using series B0049, B0050, and B0051 that includes X-ray images of individual handguns, shuriken, and razor blades, respectively, taken from different points of view as shown in Fig. 2.9.

Fig. 2.8 Some X-ray images of a bag containing handguns, *shuriken*, and razor blades (group Baggage series B0048)

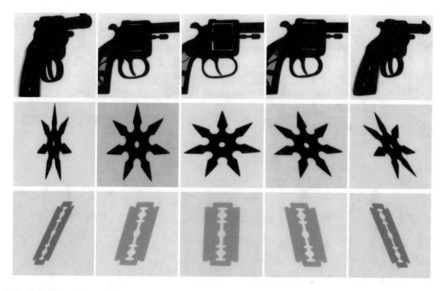

Fig. 2.9 Some X-ray images of handguns (series B0049), *shuriken* (series B0050), and razor baldes (series B0051) of group Baggage

Table 2.5 Applications of series Baggage

Series	Application	References
B0005	Experiments on detection of pins in multiple views	[16, 44]
	Detection of razor blades using active vision	[44]
B0007	Training of a classifier of razor blades	[44]
B0009-43	Experiments on detection of handguns	[4, 32]
B0045	Experiments on detection of objects in multiple views	[18, 34]
	Active vision	[44]
B0046-51	Simulation of threat objects	[29]
	Detection of threat objects using sparse representations	[29]
	Detection of threat objects using 3D reconstruction	[43]
	Detection of threat objects using active vision	[42]
	Detection of threat objects using deep learning	[1]
B0049-51	Detection of threat objects	[35, 45]
B0055	Experiments on detection of objects in sequences of four views	[18]
B0056	Experiments on detection of objects in sequences of six views	[18]
B0057	Experiments on detection of objects in sequences of eight views	[18]
B0058	Training of a classifier for clips, springs, and razor blades	[18, 34]
B0061-73	Detection of razor blades using active vision	[44]
B0078-82	Detection of threat objects	[35, 45]
B0083	Detection of threat objects	[45]

Fig. 2.10 A knife was rotated in 1^0 and by each position, an X-ray image was captured. In this figure, X-ray images at 0^0, 10^0, 20^0, ... 350^0 are illustrated (see series B00008 of group Baggage)

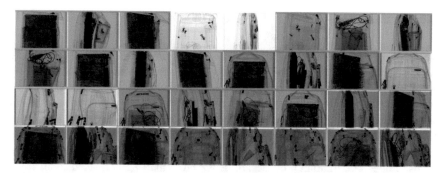

Fig. 2.11 Backpacks with no threat objects. They can be used to superimpose the isolated threat objects of Fig. 2.9 (see series B00083 of group Baggage). [→ Example 2.1 🐍]

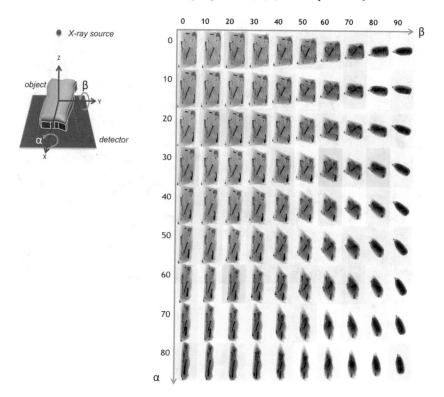

Fig. 2.12 X-ray images of a pen case from 90 different points of view. They are obtained by rotating α and β as shown in the left model (see series B00045 of group Baggage)

2.6   Natural Objects

The group Nature contains 8,290 X-ray images arranged in 13 series. The X-ray images were taken from different natural objects such as salmon filets, fruit, and wood pieces. Some examples are illustrated in Figs. 2.13, 2.14, and 2.15. The details of each series are given in Table A.1. Experiments on these data can be found in several publications as shown in Table 2.6. It is interesting to highlight that series N0012 and N0013 (see Fig. 2.16) contains not only 6 X-ray images of salmon filets, but also annotations of bounding boxes and the binary images of the ground truth of 73 fish bones. For training proposes, there are more than 7,500 labeled small crops (10 × 10 pixels), of regions of X-ray of salmon filets with and without fish bones in series N0003.

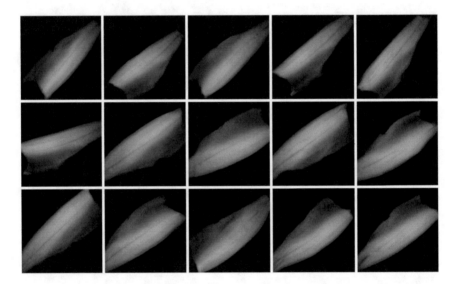

Fig. 2.13 Some X-ray images of salmon filets (group Nature series N0011)

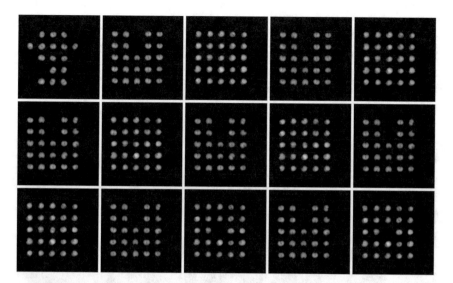

Fig. 2.14 Some X-ray images of cherries (group `Nature` series N0006)

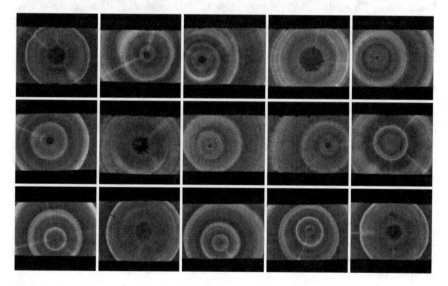

Fig. 2.15 Some X-ray images of wood (group `Nature` series N0010)

Table 2.6 Applications of series Nature

Series	Application	References
N0003	Automated design of a visual food quality system	[31]
N0003	Automated fish bone detection	[30]
N0008	Quality control of kiwis	[37]
N0011	Automated fish bone detection	[30]

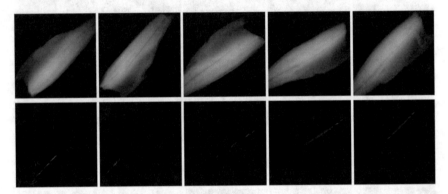

Fig. 2.16 Some images of group Nature series S0012 (X-ray images of salmon filets) and S0013 (ground truth for fish bones)

2.7 Settings

The group Settings contains 170 X-ray images arranged in 8 series. The X-ray images were taken from different calibration objects such checkerboards and 3D objects with regular patterns. Some examples are illustrated in Figs. 2.17 and 2.18. The details of each series are given in Table A.5. Experiments on these data can be found in several publications as shown in Table 2.7. It is interesting to highlight that series S0008 (see Fig. 2.17) contains not only 18 X-ray images of a copper checkerboard, but also the calibration matrix of each view. In addition, series S0007 can be used for modeling the distortion of an image intensifier. The coordinates of each hole of the calibration pattern in each view are available, and the coordinates of the 3D model are given as well.

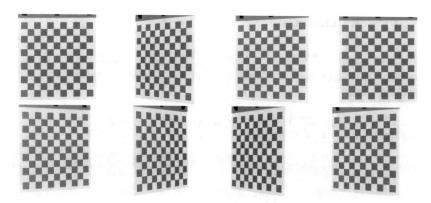

Fig. 2.17 Some X-ray images of a copper checkerboard used by calibration (group Settings series S0008)

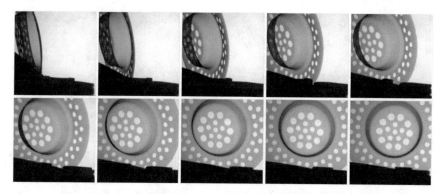

Fig. 2.18 Some X-ray images of circular pattern in different points of view used by calibration (group Settings series S0007)

Table 2.7 Applications of series Settings

Series	Application	References
S0001	Calibration of a multiple view X-ray imaging system for active vision	[44]
S0002	Distortion model of an image intensifier	[24, 26]
S0007	Explicit geometric model of a radioscopic imaging system	[15]

2.8   Python Commands

In order to manipulate GDXray+ database easily, some helpful Python functions were developed in pyxvis Library. In this section, we present a summary of them with some examples.

Python Example 2.1: In this example, we show how simple it is to display the X-ray images of a series of GDXray+.:

Listing 2.1 : Display of X-ray images of GDXray+.

```
from pyxvis.io import gdxraydb
from pyxvis.io.visualization import show_series

gdxraydb.xgdx_stats()

image_set = gdxraydb.Baggages()

image_set.describe()

print(image_set.get_dir(60))

try:
    series_dir = image_set.get_dir(83)
except ValueError as err:
    print(err)

show_series(image_set, 8, range(1, 352, 10), n=18, scale=0.2)
```

The output of this code is illustrated in Fig. 2.11.   □

Some pyxvis Library functions that can be used to manipulate GDXray+ are the following:

- gdx_browse of gui: This GUI function[6] is used to browse GDXray+ database. An example is illustrated in Fig. 2.19. An additional example using pseudo coloring is shown in Fig. 2.20;, the user can select one of 10 different color maps. This function can be used to display the bounding boxes of an X-ray image. For example, in X-ray image N0012_0004.png the bounding boxes of the ground truth are displayed in Fig. 2.19. Each bounding box is stored as a row in file ground_truth.txt of folder N0012 of GDXray+. The format of this file is as follows: one bounding box per row; the first number of the row is the number of the image of the series, and the next four values are the coordinates x_1, x_2, y_1, y_2 of a bounding box. Thus, the rectangle of a bounding box is defined by its opposite vertices: (x_1, y_1) and (x_2, y_2).
- show_series of gdxraydb: This function is used to display several images of a series in only one figure (see example in Fig. 2.10 or Example 2.1).

- get_dir of gdxraydb: This function is used to ascertain the path of a series of GDXray+.

---

[6]GUI: Graphic User Interface.

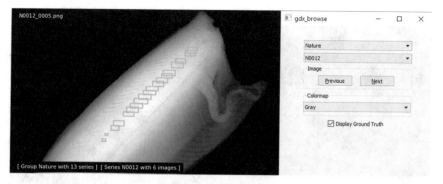

Fig. 2.19 Example of command gdx_browse that can be used to browse $\mathbb{GDX}$ray+. The user can click buttons [Previous] and [Next] to display the next groups, series or images. In addition, the ground truth option can be used to display manual annotations when they are available. In this example, the fish bones of a salmon filet are highlighted. For colored images, see Fig. 2.20

Fig. 2.20 Example of command gdx_browse using pseudo coloring of a wood X-ray image. For another example in grayscale, see Fig. 2.19

- gdx_stats of gdxraydb: This function is used to compute some statistics of $\mathbb{GDX}$ray+. The output is Table 2.1.

- load_image of gdxraydb: This function is used to load an image of $\mathbb{GDX}$ray+. For example, N0012_0004.png can be stored in matrix img using the following commands:

```
from pyxvis.io import gdxraydb
image_set = gdxraydb.Nature()
img = image_set.load_image(12, 4)
```

- load_data of gdxraydb: This function is used to load a file into a workspace. For instance, the ground truth data of series N0012 can be stored in matrix gt using the following commands:

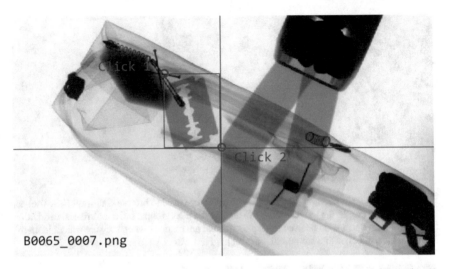

B0065_0007.png

Fig. 2.21 Annotation: there are some tools that can be used to manually annotate the ground truth of a series. In this example, the user is annotating the razor blades of series B0065

```
from pyxvis.io import gdxraydb
image_set = gdxraydb.Nature()
gt = load_set.load_image(12, 'ground_truth.txt')
```

For annotation, there are some open- source tools that can be used to manually annotate bounding boxes of a series of GDXray+. An example isn Fig. 2.21.[7]

2.9 Summary

In this chapter, we presented the details of a new public dataset called GDXray+. It consists of more than 21,100 X-ray images. The database includes five groups of X-ray images: castings, welds, baggage, natural objects, and settings. Each group has several series and X-ray images with many labels and annotations that can be used for training and testing purposes in computer vision algorithms. To the best knowledge of the author, up until now, there have not been any public databases of digital X-ray images for general purposes in X-ray testing.

In this chapter, we explained the structure of the GDXray+ database, we gave a description for each group (with some series examples), we presented some examples of applications that have been published using images of GDXray+, and some examples in Python with pyxvis Library, that can be used to manipulate the database.

---

[7]See, for example, LabelMe http://labelme.csail.mit.edu/Release3.0/ developed by the Computer Science and Artificial Intelligence Laboratory at MIT.

We believe that GDXray+ represents a relevant contribution to the X-ray testing community. On the one hand, students, researchers, and engineers can use these X-ray images to develop, test, and evaluate image analysis and computer vision algorithms without purchasing expensive X-ray equipment. On the other hand, these images can be used as a benchmark in order to test and compare the performance of different approaches on the same data. Moreover, the database can be used in the training programs of human inspectors.

References

1. Aydin, I., Karakose, M., Erhan, A.: A new approach for baggage inspection by using deep convolutional neural networks. In: 2018 International Conference on Artificial Intelligence and Data Processing (IDAP), pp. 1–6. IEEE (2018)
2. Carrasco, M., Mery, D.: Segmentation of welding defects using a robust algorithm. Mater. Eval. **62**(11), 1142–1147 (2004)
3. Carrasco, M., Mery, D.: Automatic multiple view inspection using geometrical tracking and feature analysis in aluminum wheels. Mach. Vis. Appl. **22**(1), 157–170 (2011)
4. Damashek, A., Doherty, J.: Detecting guns using parametric edge matching. Technical Report Project for Computer Vision Course: CS231A, Stanford University (2015)
5. Ferguson, M., Ak, R., Lee, Y.T.T., Law, K.H.: Automatic localization of casting defects with convolutional neural networks. In: 2017 IEEE International Conference on Big Data (Big Data), pp. 1726–1735. IEEE (2017)
6. Ferguson, M.K., Ronay, A., Lee, Y.T.T., Law, K.H.: Detection and segmentation of manufacturing defects with convolutional neural networks and transfer learning. Smart Sustain. Manuf. Syst. **2** (2018)
7. Ghoreyshi, A., Vidal, R., Mery, D.: Segmentation of circular casting defects using a robust algorithm. Insight-Non-Destr. Testing Cond. Monit. **47**(10), 615–617 (2005)
8. Hernández, S., Sáez, D., Mery, D., da Silva, R., Sequeira, M.: Automated defect detection in aluminium castings and welds using neuro-fuzzy classifiers. In: Proceedings of the 16th World Conference on Non-Destructive Testing (WCNDT–2004). Montreal (2004)
9. Hou, W., Wei, Y., Guo, J., Jin, Y., et al.: Automatic detection of welding defects using deep neural network. In: Journal of Physics: Conference Series, vol. 933, p. 012006. IOP Publishing (2018)
10. Hou, W., Wei, Y., Jin, Y., Zhu, C.: Deep features based on a dcnn model for classifying imbalanced weld flaw types. Measurement **131**, 482–489 (2019)
11. Huang, Q., Wu, Y., Baruch, J., Jiang, P., Peng, Y.: A template model for defect simulation for evaluating nondestructive testing in X-radiography. IEEE Trans. Syst. Man Cybern. Part A: Syst. Hum. **39**(2), 466–475 (2009)
12. Liu, L., Cao, D., Wu, Y., Wei, T.: Defective samples simulation through adversarial training for automatic surface inspection. Neurocomputing **360**, 230–245 (2019)
13. Mery, D.: A new algorithm for flaw simulation in castings by superimposing projections of 3D models onto X-ray images. In: Proceedings of the XXI International Conference of the Chilean Computer Science Society (SCCC-2001), pp. 193–202. IEEE Computer Society Press, Punta Arenas (2001)
14. Mery, D.: Crossing line profile: a new approach to detecting defects in aluminium castings. In: Proceedings of the Scandinavian Conference on Image Analysis (SCIA 2003), Lecture Notes in Computer Science, vol. 2749, pp. 725–732 (2003)
15. Mery, D.: Explicit geometric model of a radioscopic imaging system. NDT & E Intern. **36**(8), 587–599 (2003)

16. Mery, D.: Automated detection in complex objects using a tracking algorithm in multiple X-ray views. In: Proceedings of the 8th IEEE Workshop on Object Tracking and Classification Beyond the Visible Spectrum (OTCBVS 2011), in Conjunction with CVPR 2011, Colorado Springs, pp. 41–48 (2011)
17. Mery, D.: Automated detection of welding defects without segmentation. Mater. Eval. **69**(6), 657–663 (2011)
18. Mery, D.: Inspection of complex objects using multiple-X-ray views. IEEE/ASME Trans. Mechatron. **20**(1), 338–347 (2015)
19. Mery, D.: Aluminum casting inspection using deep learning: A method based on convolutional neural networks. J. Nondestr. Eval. **39**(1), 12 (2020)
20. Mery, D., Arteta, C.: Automatic defect recognition in X-ray testing using computer vision. In: 2017 IEEE Winter Conference on Applications of Computer Vision (WACV), pp. 1026–1035. IEEE (2017)
21. Mery, D., Berti, M.A.: Automatic detection of welding defects using texture features. Insight-Non-Destr. Testing Cond. Monit. **45**(10), 676–681 (2003)
22. Mery, D., Carrasco, M.: Automated multiple view inspection based on uncalibrated image sequences. Lecture Notes in Computer Science, vol. 3540, pp. 1238–1247 (2005)
23. Mery, D., Chacón, M., Munoz, L., González, L.: Automated inspection of aluminium castings using fusion strategies. Mater. Eval. (2005). In Press
24. Mery, D., Filbert, D.: The epipolar geometry in the radioscopy: Theory and application. at - Automatisierungstechnik **48**(12), 588–596 (2000). (in German)
25. Mery, D., Filbert, D.: A fast non-iterative algorithm for the removal of blur caused by uniform linear motion in X-ray images. In: Proceedings of the 15th World Conference on Non-Destructive Testing (WCNDT–2000). Rome (2000)
26. Mery, D., Filbert, D.: Automated flaw detection in aluminum castings based on the tracking of potential defects in a radioscopic image sequence. IEEE Trans. Robot. Autom. **18**(6), 890–901 (2002)
27. Mery, D., Filbert, D.: Automated inspection of moving aluminium castings. In: 8th European Conference on Non-Destructive Testing (ECNDT 2002). Barcelona (2002)
28. Mery, D., Hahn, D., Hitschfeld, N.: Simulation of defects in aluminum castings using cad models of flaws and real X-ray images. Insight **47**(10), 618–624 (2005)
29. Mery, D., Katsaggelos, A.K.: A logarithmic X-ray imaging model for baggage inspection: simulation and object detection. In: Proceedings of the IEEE Conference on Computer Vision and Pattern Recognition Workshops, pp. 57–65 (2017)
30. Mery, D., Lillo, I., Riffo, V., Soto, A., Cipriano, A., Aguilera, J.: Automated fish bone detection using X-ray testing. J. Food Eng. **2011**(105), 485–492 (2011)
31. Mery, D., Pedreschi, F., Soto, A.: Automated design of a computer vision system for visual food quality evaluation. Food Bioprocess Technol. **6**(8), 2093–2108 (2013)
32. Mery, D., Riffo, V., Mondragon, G., Zuccar, I.: Detection of regular objects in baggages using multiple X-ray views. Insight **55**(1), 16–21 (2013)
33. Mery, D., Riffo, V., Zscherpel, U., Mondragón, G., Lillo, I., Zuccar, I., Lobel, H., Carrasco, M.: GDXray: The database of X-ray images for nondestructive testing. J. Nondestr. Eval. **34**(4), 1–12 (2015)
34. Mery, D., Riffo, V., Zuccar, I., Pieringer, C.: Automated X-ray object recognition using an efficient search algorithm in multiple views. In: Proceedings of the 9th IEEE CVPR workshop on Perception Beyond the Visible Spectrum, Portland (2013)
35. Mery, D., Svec, E., Arias, M., Riffo, V., Saavedra, J.M., Banerjee, S.: Modern computer vision techniques for X-ray testing in baggage inspection. IEEE Trans. Syst. Man Cybern.: Syst. **47**(4), 682–692 (2016)
36. Miao, C., Xie, L., Wan, F., Su, C., Liu, H., Jiao, J., Ye, Q.: Sixray: A large-scale security inspection X-ray benchmark for prohibited item discovery in overlapping images. In: Proceedings of the IEEE Conference on Computer Vision and Pattern Recognition, pp. 2119–2128 (2019)
37. Mondragón, G., Leiva, G., Aguilera, J., Mery, D.: Automated detection of softening and hard columella in kiwifruits during postharvest using X-ray testing. In: Proceedings of International Congress on Engineering and Food (2011)

38. Perner, P., Zscherpel, U., Jacobsen, C.: A comparison between neural networks and decision trees based on data from industrial radiographic testing. Pattern Recognit. Lett. **22**(1), 47–54 (2001)
39. Pieringer, C., Mery, D.: Flaw detection in aluminium die castings using simultaneous combination of multiple views. Insight **52**(10), 548–552 (2010)
40. Pizarro, L., Mery, D., Delpiano, R., Carrasco, M.: Robust automated multiple view inspection. Pattern Anal. Appl. **11**(1), 21–32 (2008)
41. Ramírez, F., Allende, H.: Detection of flaws in aluminium castings: a comparative study between generative and discriminant approaches. Insight-Non-Destr. Testing Cond. Monit. **55**(7), 366–371 (2013)
42. Riffo, V., Flores, S., Mery, D.: Threat objects detection in X-ray images using an active vision approach. J. Nondestr. Eval. **36**(3), 44 (2017)
43. Riffo, V., Godoy, I., Mery, D.: Handgun detection in single-spectrum multiple X-ray views based on 3d object recognition. J. Nondestruct. Eval. **38**(3), 66 (2019)
44. Riffo, V., Mery, D.: Active X-ray testing of complex objects. Insight **54**(1), 28–35 (2012)
45. Saavedra, D., Banerjee, S., Mery, D.: Detection of threat objects in baggage inspection with X-ray images using deep learning. Neural Comput. Appl. pp. 1–17. Springer (2020)
46. da Silva, R.R., Siqueira, M.H., de Souza, M.P.V., Rebello, J.M., Calôba, L.P.: Estimated accuracy of classification of defects detected in welded joints by radiographic tests. NDT & E Intern. **38**(5), 335–343 (2005)
47. Sizyakin, R., Voronin, V., Gapon, N., Zelensky, A., Pižurica, A.: Automatic detection of welding defects using the convolutional neural network. In: Automated Visual Inspection and Machine Vision III, vol. 11061, p. 110610E. International Society for Optics and Photonics (2019)
48. Wang, Y., Shi, F., Tong, X.: A welding defect identification approach in X-ray images based on deep convolutional neural networks. In: International Conference on Intelligent Computing, pp. 53–64. Springer (2019)
49. Yahaghi, E., Movafeghi, A., Mirzapour, M., Rokrok, B.: Defect detections in industrial radiography images by a multi-scale lmmse estimation scheme. Radiat. Phys. Chem. **168**, 108560 (2020)

Chapter 3
Geometry in X-ray Testing

Abstract Geometry is of basic importance for understanding in X-ray testing. In this chapter we present a mathematical background of the monocular and multiple view geometry which is normally used in X-ray computer vision systems. The chapter describes an explicit model which relates the 3D coordinates of an object to the 2D coordinates of the digital X-ray image pixel, the calibration problem, the geometric and algebraic constraints between two, three, and more X-ray images taken at different projections of the object, and the problem of 3D reconstruction from n views.

Cover image: *Average of X-ray images of a wheel in motion (series* C0008 *colored with 'parula' colormap).*

© Springer Nature Switzerland AG 2021
D. Mery and C. Pieringer, *Computer Vision for X-Ray Testing*,
https://doi.org/10.1007/978-3-030-56769-9_3

3.1   Introduction

In certain nondestructive testing and evaluation applications, it is necessary to deal with some geometric problems. For example, the geometric distortion of an image amplifier must be reduced; 3D information of the object under test must be inferred; or multiple view X-ray images of the same object from different points of view must be analyzed. Multiple view information is required for example, for inspecting the internal and external geometry of an object under test; for locating its features using stereoscopic techniques and for finding regions of interest—such as defects—using correspondences of multiple views.

In this chapter we present a background of geometry which is normally used in X-ray computer vision systems. We start by presenting in Sect. 3.2 projective transformations that are very common in X-ray imaging. In Sect. 3.3, a model which relates the 3D coordinates of an object to the 2D coordinates of the digital X-ray image pixel. In Sect. 3.4, different approaches that can be used to estimate the parameters of the geometric model are outlined. In Sect. 3.5, we establish the geometric and algebraic constraints between two, three, and more X-ray images obtained as different projections of the object. The problem of the 3D reconstruction is explained in Sect. 3.6.

3.2   Geometric Transformations

Before we begin a detailed description of the geometric model of our X-ray computer vision system, it is worthwhile to outline certain geometric transformations that are used by the model.

3.2.1   Homogeneous Coordinates

We are familiar with Cartesian coordinates in 2D (x, y) and in 3D (X, Y, Z). As we will see in this section, in an X-ray computer vision system the geometric transformations between different coordinate systems can be handled in an easy way if *homogeneous coordinates* are used [2]. In this approach, the commonly used Cartesian coordinates are called *non-homogenous* coordinates.

In general, a point $\mathbf{a} \in \mathbb{R}^N$ given in non-homogeneous coordinates can be expressed as a point $\mathbf{b} \in \mathbb{R}^{N+1}$ in homogeneous coordinates as follows:

$$(a_1, a_2, \ldots, a_N) \rightarrow (b_1, b_2, \ldots, b_N, b_{N+1}) \tag{3.1}$$

where $a_i = b_i / b_{N+1}$ for $i = 1, \ldots N$.

Table 3.1 Transformation non-homogeneous ↔ homogeneous coordinates

Non-homogeneous coordinates	↔		Homogeneous coordinates
2D:			
(x, y)	→		$\lambda(x, y, 1)$
$(x = b_1/b_3, y = b_2/b_3)$	←		(b_1, b_2, b_3)
3D:			
(X, Y, Z)	→		$\lambda(X, Y, Z, 1)$
$(X = b_1/b_4, Y = b_2/b_4, Z = b_3/b_4)$	←		(b_1, b_2, b_3, b_4)

Using (3.1), a 2D point (x, y) is expressed as a homogeneous vector with three elements (b_1, b_2, b_3), where $x = b_1/b_3$ and $y = b_2/b_3$. Thus, we can convert a non-homogeneous point (x, y) into a homogeneous point as $(x, y, 1)$, or as $\lambda(x, y, 1)$ where λ is a scalar $\lambda \neq 0$. It is worth noting that the homogeneous coordinates $(4, 8, 2)$ and $(6, 12, 3)$ represent the same 2D non-homogeneous point because they can be expressed as $2 \cdot (2, 4, 1)$ and $3 \cdot (2, 4, 1)$ respectively. That means, $x = 2$ and $y = 4$ in non-homogeneous coordinates.

Similar examples could be given for a 3D point (X, Y, Z). The transformations between homogeneous and non-homogeneous coordinates are summarized in Table 3.1 for 2D and 3D.

In this book we use the notation of Faugeras [2], where we differentiate between the projective geometric objects themselves and their representations, e.g., a point in the 2D space will be denoted by m whereas its vector in homogeneous coordinates will be denoted by $\mathbf{m}$.

We can use homogeneous coordinates to represent points and lines as well. For instance, a point m and a line ℓ in 2D space can be represented as $\mathbf{m} = [x\ y\ 1]^T$ and $\ell = [a\ b\ 1]^T$ respectively. Thus, if m lies on ℓ then $\mathbf{m} \cdot \ell = \mathbf{m}^T \ell = 0$. It is worth noting that $\lambda\mathbf{m}$ for $\lambda \neq 0$ represents the same 2D point and $k\ell$ for $k \neq 0$ represents the same line, and they fulfill $\mathbf{m}^T \ell = 0$.

Two 2D points m_1 and m_2 that lie on line ℓ fulfill $\mathbf{m}_1^T \ell = 0$ and $\mathbf{m}_2^T \ell = 0$, where $\mathbf{m}_1$ and $\mathbf{m}_2$ are homogeneous representations of points m_1 and m_2 respectively (see Fig. 3.1). Using cross-product, we find a new vector $\mathbf{w} = \mathbf{m}_1 \times \mathbf{m}_2$ with following properties: *(i)* $\mathbf{w}$ is a 3D vector, *(ii)* $\mathbf{m}_1 \perp \mathbf{w}$, *(iii)* $\mathbf{m}_2 \perp \mathbf{w}$. According to properties *(ii)* and *(iii)*, $\mathbf{m}_1^T \mathbf{w} = 0$ and $\mathbf{m}_2^T \mathbf{w} = 0$, interestingly that means that $\mathbf{w} = \ell$. Thus, given $\mathbf{m}_1$ and $\mathbf{m}_2$ the homogeneous representation of the line that contains both points can be easily calculated by:

$$\ell = \mathbf{m}_1 \times \mathbf{m}_2. \tag{3.2}$$

The reader can demonstrate that given ℓ_1 and ℓ_2 (the homogeneous representations of two lines in 2D space), the homogenous representation of the intersection of both

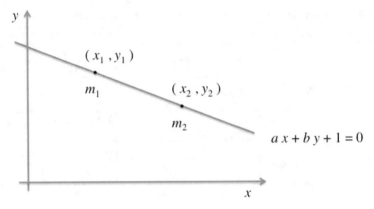

Fig. 3.1 Two points on a line in 2D space

lines can be computed by

$$\mathbf{m} = \ell_1 \times \ell_2. \tag{3.3}$$

3.2.2 2D → 2D Transformation

Sometimes, a 2D point that is given in a coordinate system (x', y'), must be expressed in another coordinate system (x, y) as illustrated in Fig. 3.2. In this example, there is a rotation θ and a translation (t_x, t_y). It is the same 2D point m, however, it is defined in two different coordinate systems. It is easy to demonstrate that the transformation between both coordinate systems is given in non-homogeneous coordinates by

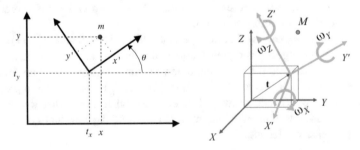

Fig. 3.2 Euclidean transformation: (Left) 2D [→ Example 3.1 🐍], (Right) 3D [→ Example 3.2 🐍]

$$\begin{bmatrix} x \\ y \end{bmatrix} = \underbrace{\begin{bmatrix} +\cos(\theta) & -\sin(\theta) \\ +\sin(\theta) & +\cos(\theta) \end{bmatrix}}_{R} \begin{bmatrix} x' \\ y' \end{bmatrix} + \underbrace{\begin{bmatrix} t_x \\ t_y \end{bmatrix}}_{t} = \mathbf{R} \begin{bmatrix} x' \\ y' \end{bmatrix} + \mathbf{t}. \tag{3.4}$$

Matrix $\mathbf{R}$ is known as the *rotation matrix* in 2D. It is an orthonormal matrix, i.e., $\mathbf{R}^\mathsf{T}\mathbf{R} = \mathbf{I}_{2\times2}$, where $\mathbf{I}_{2\times2}$ is the 2-by-2 identity matrix. The same transformation (3.4) can be expressed in homogenous coordinates as

$$\begin{bmatrix} x \\ y \\ 1 \end{bmatrix} = \begin{bmatrix} +\cos(\theta) & -\sin(\theta) & t_x \\ +\sin(\theta) & +\cos(\theta) & t_y \\ 0 & 0 & 1 \end{bmatrix} \begin{bmatrix} x' \\ y' \\ 1 \end{bmatrix} \tag{3.5}$$

or using a matrix notation as

$$\mathbf{m} = \mathbf{H}\mathbf{m}', \tag{3.6}$$

where $\mathbf{m} = [x\ y\ 1]^\mathsf{T}$, $\mathbf{m}' = [x'\ y'\ 1]^\mathsf{T}$, and $\mathbf{H}$ is a 3×3 matrix defined as

$$\mathbf{H} = \begin{bmatrix} \mathbf{R} & \mathbf{t} \\ \mathbf{0} & 1 \end{bmatrix}, \tag{3.7}$$

where $\mathbf{0} = [0\ 0]$.

Equation (3.6) defines the transformation $\mathbf{m}' \to \mathbf{m}$. The inverse transformation $\mathbf{m} \to \mathbf{m}'$ can be established by computing the inverse of matrix $\mathbf{H}$, i.e., $\mathbf{m}' = \mathbf{H}^{-1}\mathbf{m}$. Since $\mathbf{R}$ is orthonormal, $\mathbf{R}^{-1} = \mathbf{R}^\mathsf{T}$. Thus, the inverse of $\mathbf{H}$ is

$$\mathbf{H}^{-1} = \begin{bmatrix} \mathbf{R}^\mathsf{T} & -\mathbf{R}^\mathsf{T}\mathbf{t} \\ \mathbf{0} & 1 \end{bmatrix}. \tag{3.8}$$

Python Example 3.1: In Fig. 3.2, $t_x = 4.1$ cm, $t_y = 3.2$ cm and $\theta = 35°$. The coordinates of m are given as $x = 4.9$ cm and $y = 5.5$ cm. If we want to find the coordinates of this point in (x', y') coordinate system, we can use the following Python code, where m and m' in homogeneous coordinates are defined in Python variables m and mp respectively:

Listing 3.1 : Euclidean transformation 2D → 2D.

```
import numpy as np

from pyxvis.geometry.projective import rotation_matrix_2d

th = 35.0 / 180.0 * np.pi #Rotation in radians
t = np.array([4.1, 3.2]).reshape(2, -1)  # Translation tx,ty in cm

# The same can also be done using newaxis
# t = np.array([4.1, 3.2])
# t = t[:, np.newaxis]

R = rotation_matrix_2d(th)  # Generate the rotation matrix R
```

```
# Euclidean transformation matrix H
H = np.hstack([R, t])
H = np.vstack([H, np.array([0, 0, 1])])

x = 4.9  # x coordinate
y = 5.5  # y coordinate

# A 2D point in homogeneous coordinates
m = np.array([x, y, 1])
m = m[:, np.newaxis]

mp = np.dot(np.linalg.inv(H), m)   # Transformation m to mp
mp = mp / mp[-1]  # Homogeneous coordinates requires to be normalized by the matrix
      element (3,3)

xp = mp.item(0)  # x' coordinate
yp = mp.item(1)  # y' coordinate

print('xp = {:1.4f} cm — yp = {:1.4f} cm'.format(xp, yp))
```

The output of this code is

xp = 1.9745 cm - yp = 1.4252 cm.

In this code we use function rotation_matrix_2d of pyxvis Library. This function computes matrix $\mathbf{R}$ as defined in (3.4). It is worth noting that the division by the third element of m' (element mp[-1]) in this case is not necessary because it is 1, since the last row of $\mathbf{H}$ is [0 0 1].   □

This projective transformation is known as *Euclidean* or *isometric* transformation because the Euclidean distance between two points in both coordinate systems, (x, y) and (x', y'), is invariant. That means, the distance d' between two points in the first coordinate systems $\mathbf{m}'_i = [x'_i \ y'_i \ 1]^\mathsf{T}$, for $i = 1, 2$, is equal to the distance d between the two transformed points in the second coordinate systems $\mathbf{m}_i = [x_i \ y_i \ 1]^\mathsf{T}$ using (3.6): $\mathbf{m}_i = \mathbf{H}\mathbf{m}_i$. The distances are the same ($d' = d$), and they can be calculated by

$$d = \sqrt{(x_1 - x_2)^2 + (y_1 - y_2)^2} \text{ and } d' = \sqrt{(x'_1 - x'_2)^2 + (y'_1 - y'_2)^2}. \quad (3.9)$$

Other projective transformations are

- *Similarity* transformation, in which matrix $\mathbf{R}$ is replaced by $s\mathbf{R}$ in (3.7). Factor s is a scalar that is used to change the scale of the original coordinate system. Thus, using (3.9) if the distance between two points in (x', y') coordinate system is d', in the transformed coordinate system (x, y) the distance between the transformed points will be $d = sd'$. In this transformation, matrix $s\mathbf{R}$ is no longer orthonormal but still orthogonal: $[s\mathbf{R}]^\mathsf{T}[s\mathbf{R}] = s^2\mathbf{I}_{2\times2}$.
- *Affine* transformation, in which matrix $\mathbf{R}$ is replaced by any rank 2 matrix $\mathbf{A}$ in (3.7). In affine transformation, parallel lines in (x', y') coordinate system remain parallel in the transformed coordinate system (x, y).
- *General* transformation, in which matrix $\mathbf{H}$ in (3.7) is a general non-singular matrix. In general transformation, a straight line in (x', y') coordinate system remain a straight line in the transformed coordinate system (x, y). This transformation is known as *homography* [13].

3.2.3 3D → 3D Transformation

A 3D point M can be defined as (X, Y, Z) and (X', Y', Z') in two different coordinate systems as shown in Fig. 3.2, where the axes are translates and rotated. Using homogeneous coordinates (similar to 2D → 2D transformation), a 3D Euclidean transformation can be expressed by

$$\mathbf{M} = \mathbf{H}\mathbf{M}' \tag{3.10}$$

where $\mathbf{M} = [X\ Y\ Z\ 1]^\mathsf{T}$, $\mathbf{M}' = [X'\ Y'\ Z'\ 1]^\mathsf{T}$, and $\mathbf{H}$ is a 4×4 matrix defined as

$$\mathbf{H} = \begin{bmatrix} \mathbf{R} & \mathbf{t} \\ \mathbf{0} & 1 \end{bmatrix}, \tag{3.11}$$

where $\mathbf{0} = [0\ 0\ 0]$. Again we differentiate between M and $\mathbf{M}$. The first notation is a 3D point in space, where the second notation is its homogenous representation in a specific coordinate system. Note that $\mathbf{M}$ and $\mathbf{M}'$ represent the same 3D point M but in different coordinate systems.

The transformation is considered as rigid displacement of the coordinate system (X', Y', Z') represented by a 3×1 translation vector $\mathbf{t} = [t_X\ t_Y\ t_Z]^\mathsf{T}$ and a 3×3 rotation matrix $\mathbf{R}$. Matrix $\mathbf{R}$ considers three rotations as shown in Fig. 3.3. Each rotation can be modeled using a rotation matrix of two coordinates as shown in Table 3.2. Thus, matrix $\mathbf{R}$ can be modeled as a rotation about axis Z, then a rotation in Y axis and finally a rotation in X axis:

$$\mathbf{R}(\omega_X, \omega_Y, \omega_Z) = \mathbf{R}_X(\omega_X)\mathbf{R}_Y(\omega_Y)\mathbf{R}_Z(\omega_Z) = \begin{bmatrix} R_{11} & R_{12} & R_{13} \\ R_{21} & R_{22} & R_{23} \\ R_{31} & R_{32} & R_{33} \end{bmatrix}, \tag{3.12}$$

where the elements R_{ij} can be expressed as a function of cosine and sine of the Euler angles ω_X, ω_Y, and ω_Z that describe the rotation of the X, Y, and Z axes respectively [26]:

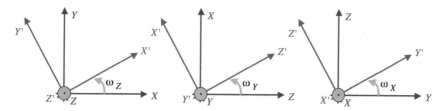

Fig. 3.3 Rotation of axes Z, Y, and X

Table 3.2 Rotation matrices of axes Z, Y, and X

Rotation	Rotation matrix
Axis Z	$\mathbf{R_Z} = \begin{bmatrix} \cos(\omega_Z) & -\sin(\omega_Z) & 0 \\ \sin(\omega_Z) & \cos(\omega_Z) & 0 \\ 0 & 0 & 1 \end{bmatrix}$
Axis Y	$\mathbf{R_Y} = \begin{bmatrix} \cos(\omega_Y) & 0 & \sin(\omega_Y) \\ 0 & 1 & 0 \\ -\sin(\omega_Y) & 0 & \cos(\omega_Y) \end{bmatrix}$
Axis X	$\mathbf{R_X} = \begin{bmatrix} 1 & 0 & 0 \\ 0 & \cos(\omega_X) & -\sin(\omega_X) \\ 0 & \sin(\omega_X) & \cos(\omega_X) \end{bmatrix}$

$$
\begin{aligned}
R_{11} &= \cos(\omega_Y)\cos(\omega_Z) \\
R_{12} &= -\cos(\omega_Y)\sin(\omega_Z) \\
R_{13} &= \sin(\omega_Y) \\
R_{21} &= \sin(\omega_X)\sin(\omega_Y)\cos(\omega_Z) + \cos(\omega_X)\sin(\omega_Z) \\
R_{22} &= -\sin(\omega_X)\sin(\omega_Y)\sin(\omega_Z) + \cos(\omega_X)\cos(\omega_Z). \\
R_{23} &= -\sin(\omega_X)\cos(\omega_Y) \\
R_{31} &= -\cos(\omega_X)\sin(\omega_Y)\cos(\omega_Z) + \sin(\omega_X)\sin(\omega_Z) \\
R_{32} &= \cos(\omega_X)\sin(\omega_Y)\sin(\omega_Z) + \sin(\omega_X)\cos(\omega_Z) \\
R_{33} &= \cos(\omega_X)\cos(\omega_Y)
\end{aligned}
\tag{3.13}
$$

Python Example 3.2: In Fig. 3.2, $t_X = 1$ mm, $t_Y = 3$ mm, $t_Z = 2$ mm, $\omega_X = 35°$, $\omega_Y = 0°$, and $\omega_Z = 0°$. The coordinates of the blue point are given as $X' = 0$ mm, $Y' = 1$ mm and $Z' = 1$ mm. If we want to find the coordinates of this point in (X, Y, Z) coordinate system, we can use the following Python code:

Listing 3.2 : Euclidean transformation 3D → 3D.

```python
import numpy as np

from pyxvis.geometry.projective import rotation_matrix_3d

w = (35.0 / 180.0) * np.pi #Rotation in radians

# Translation tx,ty in cm
t = np.array([1.0, 3.0, 2.0])
t = t[:, np.newaxis]

R = rotation_matrix_3d(w, 0, 0)  # Generate the rotation matrix R

# Euclidean transformation matrix H = [R t; 0 0 1]
H = np.hstack([R, t])
H = np.vstack([H, np.array([0, 0, 0, 1])])

Xp = 0  # x coordinate
Yp = 1  # y coordinate
```

```
Zp = 1

Mp = np.array([Xp, Yp, Zp, 1])  # A 2D point in homogeneous coordinates
Mp = Mp[:, np.newaxis]

M = np.dot(H, Mp)  # Transformation m to mp

M = M / M[-1]  # Normalize by matrix element (1, 4) into homogeneous coordinates

X = M.item(0)  # X coordinate
Y = M.item(1)  # Y coordinate
Z = M.item(2)  # Z coordinate

print('X = {:1.4f} mm — Y = {:1.4f} mm — Z = {:1.4f} mm'.format(X, Y, Z))
```

The output of this code is

X = 1.0000 mm - Y = 3.2456 mm - Z = 3.3927 mm.

In this code we use function rotation_matrix_3d of pyxvis Library. This function computes matrix **R** as defined in (3.13).

It is worth noting that the division by the fourth element of M (element M[-1]) in this case is not necessary because it is 1, since the last row of **H** is [0 0 0 1].

3.2.4   3D → 2D Transformation

In an X-ray computer vision system, a 3D point is projected using a *perspective* transformation as illustrated in Fig. 3.4. Besides applying different physical princi-

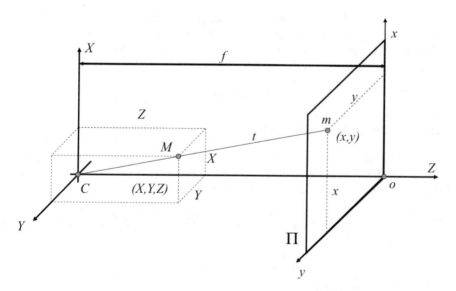

Fig. 3.4 Perspective transformation in an X-ray computer vision system. [→ Example 3.3 📥]

ples and technologies, it is common in X-ray testing to use terminology as introduced for optical imaging, such as optical axis, focal length, and so forth. In this model, a 3D point M is defined in (X, Y, Z) coordinate system, which is projected into projection plane $Z = f$ (called the retinal plane Π), where f is the focal length. All X-rays come from optical center C defined in $(X = 0, Y = 0, Z = 0)$. We define t as the straight line on which C and M lie (see Fig. 3.4). This line will be denoted as $\langle C, M \rangle$. Thus, the projection point m defined in (x, y) coordinate system is given by the intersection of t with the projection plane Π. This operation is called *central projection* [13]. The origin o of (x, y) coordinate system is pierced by the optical axis (Z-axis). After intercept theorem, it should be clear that

$$\frac{Z}{f} = \frac{Y}{y} = \frac{X}{x}, \tag{3.14}$$

that can be expressed as

$$\begin{cases} Zx = fX \\ Zy = fY \end{cases} \tag{3.15}$$

or using a matrix notation:

$$Z \begin{bmatrix} x \\ y \\ 1 \end{bmatrix} = \underbrace{\begin{bmatrix} f & 0 & 0 & 0 \\ 0 & f & 0 & 0 \\ 0 & 0 & 1 & 0 \end{bmatrix}}_{\mathbf{P}} \begin{bmatrix} X \\ Y \\ Z \\ 1 \end{bmatrix}. \tag{3.16}$$

Matrix $\mathbf{P}$ is a 3×4 matrix known as the perspective projection matrix. Thus, the projected point given in homogeneous coordinates $\mathbf{m} = [x\ y\ 1]^\mathsf{T}$ is proportional to $\mathbf{PM}$, where $\mathbf{M}$ is the 3D point given in homogeneous coordinates $\mathbf{M} = [X\ Y\ Z\ 1]^\mathsf{T}$. Usually, (3.16) is written in the following form:

$$\lambda \mathbf{m} = \mathbf{PM}, \tag{3.17}$$

where λ is a scale factor with $\lambda \neq 0$.

Python Example 3.3: In Fig. 3.4, $f = 100\,\text{cm}$, $X = 20\,\text{cm}$, $Y = 30\,\text{cm}$ and $Z = 50\,\text{cm}$. If we want to find the coordinates of projected point in (x, y) coordinate system, we can use the following Python code:

Listing 3.3 : Euclidean transformation 3D → 2D.

```python
import numpy as np

from pyxvis.geometry.projective import get_matrix_p

f = 100  # Focal distance in cm
X = 20   # X coordiante in cm
Y = 30   # Y coordinate in cm
```

```
Z = 50    # Z coordinate in cm

M = np.array([X, Y, Z, 1])  # A 3D point in homogeneous coordinates
M = M[:, np.newaxis]

P = get_matrix_p(f)  # Create the projection matrix P

m = np.dot(P, M)  # Transformation M to m
m = m / m[-1]  # Homogeneous coordinates requires to be normalized by the matrix element
      (3, 1)

x = m.item(0)  # x coordinate
y = m.item(1)  # y coordinate

print('x = {:1.1f} cm — y = {:1.1f} cm'.format(x, y))
```

The output of this code is

x = 40.0 cm - y = 60.0 cm

In this code we use function get_matrix_p of pyxvis Library. This function computes
matrix **P** as defined in (3.16).   □

As we can see, if non-homogeneous coordinates are used, the perspective projection will be non-linear (see (3.14)), however, it is linear in homogeneous coordinates (see (3.17)). In addition, all explained transformations are linear in homogeneous coordinates. This is the reason why we use homogeneous coordinates in X-ray
computer vision systems. Thus, we can handle all projective transformation easily.
For instance, if M is given in another coordinate system (X', Y', Z') ($\mathbf{M} = \mathbf{HM'}$ as
shown in (3.11)), it is very simple to replace this transformation in (3.17) yielding
$\lambda \mathbf{m} = \mathbf{PHM'}$. The reader can note that the same is valid for point m that can be
given in another coordinate system.

3.3   Geometric Model of an X-ray Computer Vision System

The geometric model of the X-ray computer vision system establishes the relationship between 3D coordinates of the object under test and their corresponding 2D
digital X-ray image coordinates. The model is required by both reconstructing 3D
information from image coordinates and reprojecting 2D image coordinates from
3D information. As explained in Sect. 1.4, the principal aspects of an X-ray computer vision system are shown in Fig. 1.8, where an X-ray image of a casting is
taken. Typically, it comprises the following five steps: *(i)* a manipulator for handling the test piece, *(ii)* an X-ray source, which irradiates the test piece with a con-
ical beam to generate an X-ray image of the test piece, *(iii)* an image intensifier
which transforms the invisible X-ray image into a visible one, *(iv)* a CCD-camera
which records the visible X-ray image, and *(v)* a computer to process the digital
image of the X-ray image and then classify the test piece by accepting or reject-
ing it. Steps *(iii)* and *(iv)* can be replaced by a flat panel. Flat amorphous silicon
detectors can be used as image sensors in some industrial inspection systems. In
such detectors, using a semi-conductor, energy from the X-ray is converted directly
into an electrical signal (without image intensifier). Nevertheless, NDT using flat

Fig. 3.5 Geometric and electromagnetic distortion obtained in an X-ray image of a regular object using an image intensifier

detectors is less feasible due to their higher cost in comparison to image intensifiers. In this section, we will present a geometric model for computer vision systems for both flat detectors and image intensifiers. Image intensifiers suffer from two significant distortions: geometric and electromagnetic field distortions (see an example in Fig. 3.5). On the other hand, computer vision systems based on flat detectors do not suffer from these distortions, and they can be easily modeled with a simple pinhole camera model [2].

In this section, we will give a geometric viewpoint about how an X-ray computer vision system can be explicitly modeled. When using explicit models, the physical parameters of the computer vision system, like image center, focal length, etc., are considered independently [27]. The model presented in this section maps the 3D object into a digital X-ray image using two transformations as shown in Fig. 1.8: *(i)* linear central projection in the X-ray projection; *(ii)* digital image formation, i.e., perspective transformation in the image intensifier and 2D projective transformation in the CCD-camera, or a single 2D projective transformation when using a flat panel. When modeling the image intensifier a high accuracy explicit model is presented, which takes into account the non-linear distortion caused by the curved input screen of the image intensifier (see Fig. 1.8), and the non-linear projection in the image intensifier caused by electromagnetic fields.

3.3.1 A General Model

In this section we present a general model which relates the 3D coordinates of the test object to the 2D coordinates of the digitalized X-ray image pixel. The model consists of two parts as shown in Fig. 1.8: X-ray projection and digital image formation. The coordinate systems used in our approach are illustrated in Fig. 3.6.

First we will describe how a 3D point M is projected onto a projection plane Π, called the retinal plane of the X-ray projection, in which the X-ray image is formed through central projection. In case of image intensifiers, the retinal plane is fictitious and is located tangentially to the input screen of the image intensifier, as shown in Fig. 3.7. The optical center C of the central projection corresponds to the X-ray

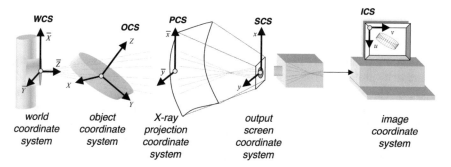

Fig. 3.6 Diagram of the coordinate systems (see Fig. 1.8)

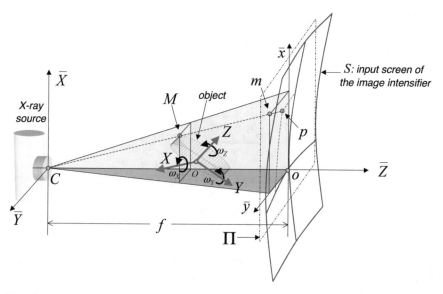

Fig. 3.7 X-ray projection using an image intensifier (see S surface) or a flat panel (see Π plane)

source, modeled as a point.[1] The optical center is located at a distance f, the focal length of the retinal plane. The central projection of M onto projection plane Π is the point m. It is defined as the intersection of the line that contains the points C and M with the retinal plane Π. The optical axis is defined as the line going through the optical center C and perpendicular to the retinal plane Π.

We define a 3D *world coordinate system* (WCS) in the optical center C of the central projection. The coordinates of this coordinate system are $\bar{X}$, $\bar{Y}$, and $\bar{Z}$, where

---

[1]Although industrial X-ray generators use standard tubes with larger focal size that blur the X-ray images slightly, the assumption that the X-ray source can be modeled as a point is valid for geometrical measurements. This is because the position of a point in the X-ray image can still be estimated as the center of the blurred point [28].

the $\bar{Z}$-axis coincides with the optical axis, as represented in Fig. 3.7. In WCS, the retinal plane Π is defined by $\bar{Z} = f$. The coordinates of the 3D point M are denoted by $(\bar{X}, \bar{Y}, \bar{Z})$ in this coordinate system.

Now, we define a 3D *object coordinate system* (OCS) that is attached to the object to be projected. The coordinates of the 3D point M are denoted by (X, Y, Z) in OCS. The center of the object is assumed to be at the origin O of this coordinate system, as shown in Fig. 3.7. The OCS is then considered as a rigid displacement of the WCS represented by a translation 3-component vector $\mathbf{t} = [t_X \ t_Y \ t_Z]^{\mathsf{T}}$ and a 3×3 rotation matrix $\mathbf{R}$ as explained in Sect. 3.2.3. Vector $\mathbf{t}$ represents the origin of OCS given in coordinates of WCS, and matrix $\mathbf{R}$ depends on the Euler angles ω_X, ω_Y and ω_Z as explained in (3.13).

The perspective projection of M onto the projection plane is the 2D point m that is represented as $(\bar{x}, \bar{y})$ in a new 2D coordinate system called the *X-ray projection coordinate system* (PCS). The $\bar{x}, \bar{y}$-axes are parallel to the $\bar{X}, \bar{Y}$-axes respectively. Applying intercept theorem, the coordinates of m in this 2D system are $\bar{x} = f\bar{X}/\bar{Z}$ and $\bar{y} = f\bar{Y}/\bar{Z}$. Using homogenous coordinates as in Sects. 3.2.3 and 3.2.4 we obtain

$$\lambda \begin{bmatrix} \bar{x} \\ \bar{y} \\ 1 \end{bmatrix} = \underbrace{\begin{bmatrix} f & 0 & 0 & 0 \\ 0 & f & 0 & 0 \\ 0 & 0 & 1 & 0 \end{bmatrix}}_{\mathbf{P}} \underbrace{\begin{bmatrix} \mathbf{R} & \mathbf{t} \\ \mathbf{0} & 1 \end{bmatrix}}_{\mathbf{H}} \begin{bmatrix} X \\ Y \\ Z \\ 1 \end{bmatrix}, \tag{3.18}$$

where $\mathbf{0} = [0 \ 0 \ 0]$, and λ is a scale factor $\lambda \neq 0$. Equation (3.18) can be rewritten in matrix form as

$$\lambda\mathbf{m} = \mathbf{AM}, \tag{3.19}$$

where $\mathbf{A} = \mathbf{PH}$, and the 3-component vector $\mathbf{m}$ and the 4-component vector $\mathbf{M}$ are homogeneous representations of $(\bar{x}, \bar{y})$ and (X, Y, Z) respectively (e.g., $\mathbf{m} = [\bar{x} \ \bar{y} \ 1]^{\mathsf{T}}$ and $\mathbf{M} = [X \ Y \ Z \ 1]^{\mathsf{T}}$). Equation (3.19) is a linear equation that maps object coordinates to projection plane coordinates. This equation depends on seven parameters:

$$\theta_{ext} = [f \ \omega_X \ \omega_Y \ \omega_Z \ t_X \ t_Y \ t_Z]^{\mathsf{T}}. \tag{3.20}$$

They are called the *extrinsic* parameters of the X-ray computer vision system. Thus, $\mathbf{A} := \mathbf{A}(\theta_{ext})$.

Finally , we introduce the 2D *image coordinate system* (ICS) to represent the pixel coordinates (u, v) of the digital image. The point (u, v) in ICS can be calculated from the point $(\bar{x}, \bar{y})$ in PCS using a function γ:

$$\mathbf{w} = \gamma(\mathbf{m}, \theta_{int}), \tag{3.21}$$

where the 3-component vectors $\mathbf{w}$ and $\mathbf{m}$ are homogeneous representations of (u, v) and $(\bar{x}, \bar{y})$ respectively, and θ_{int} is a vector with the parameters of the transformation called the *intrinsic* parameters. Several linear and non-linear models of γ, that were developed for X-ray computer vision systems and CCD-cameras, will be discussed in Sect. 3.3.2. On the one hand, the geometric model of image formation using a flat panel is linear and can be modeled using a 2D → 2D geometric transformation as explained in Sect. 3.2.2. On the other hand, image intensifiers must be modeled using non-linear transformations due to geometric and electromagnetic distortions.

To summarize, using (3.19) for the perspective projection and (3.21) for the digital image formation, an object point M, whose homogeneous coordinates are $\mathbf{M} = [X\ Y\ Z\ 1]^\mathsf{T}$ (in OCS), can be mapped into a 2D point of the digital X-ray image as w, the homogeneous coordinates of which are $\mathbf{w} = [u\ v\ 1]^\mathsf{T}$ (in ICS) using the following expression:

$$\mathbf{w} = \gamma(\mathbf{A}(\theta_{ext})\mathbf{M}, \theta_{int}) := \mathbf{F}(\theta, \mathbf{M}), \qquad (3.22)$$

where $\theta^\mathsf{T} = [\theta_{ext}\ ;\ \theta_{int}]$ is the vector of parameters involved in the projection model.

As we will explain in Sect. 3.4, in a process termed *calibration*, we estimate the parameters θ of the model based on n points whose object coordinates $\mathbf{M}_i = [X_i\ Y_i\ Z_i\ 1]^\mathsf{T}$ are known and whose image coordinates $\tilde{\mathbf{w}}_i = [\tilde{u}_i\ \tilde{v}_i\ 1]^\mathsf{T}$ are measured, for $i = 1, \ldots, n$. Using (3.22) we obtain the *reprojected* points $\mathbf{w}_i = [u_i\ v_i\ 1]^\mathsf{T}$, i.e., the inferred projections in the digital image computed from the calibration points $\mathbf{M}_i$ and the parameter vector θ. The parameter vector is then estimated by minimizing the distance between measured points ($\tilde{\mathbf{w}}_i$) and inferred points ($\mathbf{w}_i = \mathbf{F}(\theta, \mathbf{M}_i)$). Thus, the calibration is performed by minimizing the objective function $\mu(\theta)$ defined as the mean-square discrepancy between these points:

$$\mu(\theta) = \frac{1}{n} \sum_{i=1}^{n} \|\ \tilde{\mathbf{w}}_i - \mathbf{w}_i\ \| \to \min. \qquad (3.23)$$

The calibration problem is a non-linear optimization problem. Generally, the minimization of $\mu(\theta)$ has no closed-form solution. For this reason, the objective function must be iteratively minimized starting with an initial guess θ^0 that can be obtained from nominal values or preliminary reference measurements.

3.3.2   Geometric Models of the Computer Vision System

In this section, we present seven existing models that can be used to calibrate an X-ray computer vision system. Five models were conceived to calibrate cameras with and without distortion. The others were developed to calibrate computer vision systems with image intensifiers. In all these models, the perspective projection OCS → PCS is done using (3.19). For this reason, in this section only the transformation PCS → ICS will be described. We use the definition given in (3.21), where a point $\mathbf{m} = [\bar{x}\ \bar{y}\ 1]^\mathsf{T}$ in PCS is transformed by a function γ into a point $\mathbf{w} = [u\ v\ 1]^\mathsf{T}$ in

ICS. Recall that the parameters of γ are the intrinsic parameters of the computer vision system.

Camera Models

Faugeras and Toscani present in [5] a linear model without considering distortion:

$$
\begin{bmatrix} u \\ v \\ 1 \end{bmatrix} = \begin{bmatrix} k_u & s & u_0 \\ 0 & k_v & v_0 \\ 0 & 0 & 1 \end{bmatrix} \begin{bmatrix} \bar{x} \\ \bar{y} \\ 1 \end{bmatrix}. \tag{3.24}
$$

The five (intrinsic) parameters of the model consider scale factors (k_u, k_v) in each ordinate, a skew factor (s) that models non-orthogonal u, v-axes, and a translation of the origin (u_0, v_0) that represents the projection of $(\bar{x}, \bar{y}) = (0, 0)$ in ICS. This model can be used to model an X-ray computer vision system with a flat panel. In this linear model, the focal length is normalized to $f = 1$. A linear approach based on a least squares technique is proposed in [5] to estimate the intrinsic and extrinsic parameters in a closed-form. However, Faugeras in [2] proposes minimizing the distances between the observations and the model in ICS using the objective function μ of (3.23). Faugeras reported that this non-linear method clearly appears to be more robust than the linear method of Faugeras and Toscani when the measured data is perturbed by noise.

In order to model the distortion, a positional error (δ_u, δ_v) can be introduced:

$$
\underbrace{\begin{bmatrix} u \\ v \\ 1 \end{bmatrix}}_{\mathbf{w}} = \underbrace{\begin{bmatrix} k_u & 0 & u_0 \\ 0 & k_v & v_0 \\ 0 & 0 & 1 \end{bmatrix} \begin{bmatrix} \bar{x} \\ \bar{y} \\ 1 \end{bmatrix}}_{\mathbf{w}'} + \begin{bmatrix} \delta_u(\bar{x}, \bar{y}) \\ \delta_v(\bar{x}, \bar{y}) \\ 0 \end{bmatrix}. \tag{3.25}
$$

In this model, the ideal non-observable position w' is displaced to the real position w as illustrated in Fig. 3.8. The amount of the displacements, δ_u and δ_v, usually depends on the point position $(\bar{x}, \bar{y})$. Several models for the positional error were reported in the literature to calibrate a camera [14, 25, 26, 28]. In these models, the skew s is zero, because in modern digital cameras the u, v-axes can be considered as orthogonal.

The distortion is decomposed into two components: *radial* and *tangential* distortions as shown in Fig. 3.8.

Radial and tangential distortion depend on r and ϕ respectively, where (r, ϕ) are the polar coordinates of the ideal position $(\bar{x}, \bar{y})$ represented in PCS. Tsai in [26] uses a simple radial distortion model with only one additional parameter, because his experience with cameras shows that only radial distortion, which is principally caused by flawed radial curvature of the lens elements, needs to be considered.

Weng et al. propose in [28] an implicit model that includes radial, decentering and prism distortion. Decentering distortion arises when the optical centers of the lens elements are not exactly collinear, whereas the prism distortion occurs from

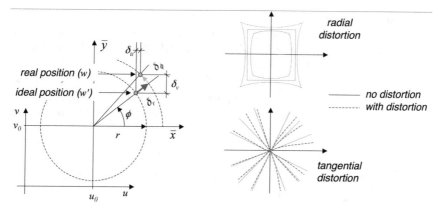

Fig. 3.8 Radial and tangential distortions [28]

imperfection in lens design, manufacturing, and camera assembly. The last two distortions, modeled with five parameters, have both radial and tangential components.

Heikkilä introduces in [14] an implicit model for radial and decentering distortion that takes into account an inverse distortion model to express the distorted image coordinates in terms of their undistorted coordinates. The number of parameters of this model is four.

Swaminathan and Nayar present in [25] a model for wide-angle lenses and poly-cameras. The model considers a shift of the optical center, radial distortion, and decentering distortion. A shift of the optical center means a shift of the image detector in the image plane. The suggested total distortion includes four parameters.

Image Intensifier Models

Two models were reported in the literature to calibrate an X-ray computer vision system composed by image intensifier and CCD-camera. The first model was proposed independently by Jaeger in [17] and Brack et al. in [1]. They propose an implicit model between **PCS** and **ICS**. The transfer function γ (3.21) is a third degree polynomial with twenty intrinsic parameters $(a_i, b_i, i = 0, \ldots 9)$ given by:

$$\begin{bmatrix} u \\ v \end{bmatrix} = \begin{bmatrix} a_0 \, a_1 \, \ldots \, a_9 \\ b_0 \, a_2 \, \ldots \, b_9 \end{bmatrix} \begin{bmatrix} 1 \, \bar{x} \, \bar{y} \, \bar{x}\bar{y} \, \bar{x}^2 \, \bar{y}^2 \, \bar{y}\bar{x}^2 \, \bar{x}\bar{y}^2 \, \bar{x}^3 \, \bar{y}^3 \end{bmatrix}^\mathsf{T}. \tag{3.26}$$

This cubic function can model not only the distortion caused by the (curved) input screen, but also the distortion introduced by electromagnetic fields present around the image intensifier. An example of this model is shown in Example 3.4.

The second model was developed by Mery and Filbert in [22, 23], in which a hyperbolic surface is used to model the input screen of the image intensifier [6] that is defined by

$$\bar{Z} = S(\bar{X}, \bar{Y}) = f\sqrt{1 + (\bar{X}/a)^2 + (\bar{Y}/b)^2} \tag{3.27}$$

with f (the focal length of the X-ray projection) being the real half axis of the hyperboloid; and a and b the imaginary half axes. The projection of point M onto the input screen of the image intensifier is denoted by p. It is calculated as the intersection of the line that contains points C, M, and m with the 3D surface S (see Fig. 3.7). Its coordinates are given by: $x' = \bar{x}/k(\bar{x}, \bar{y})$ and $y' = \bar{y}/k(\bar{x}, \bar{y})$, with $k(\bar{x}, \bar{y}) = \sqrt{1 - (\bar{x}/a)^2 - (\bar{y}/b)^2}$. The point p is imaged at the CCD-camera as w, whose coordinates can be estimated approximately using an affine transformation [2]:

$$
\begin{bmatrix} u \\ v \\ 1 \end{bmatrix} = \begin{bmatrix} k_u & 0 & u_0 \\ 0 & k_v & v_0 \\ 0 & 0 & 1 \end{bmatrix} \begin{bmatrix} +\cos(\alpha) & +\sin(\alpha) & 0 \\ -\sin(\alpha) & +\cos(\alpha) & 0 \\ 0 & 0 & 1 \end{bmatrix} \begin{bmatrix} \bar{x}/k(\bar{x}, \bar{y}) \\ \bar{y}/k(\bar{x}, \bar{y}) \\ 1 \end{bmatrix}, \qquad (3.28)
$$

where α represents rotation between $\bar{x}$, $\bar{y}$-and u, v-axes. This model has only three additional parameters a, b, and α.

Python Example 3.4: In Fig. 3.5, the holes of the calibration plate are uniformly distributed in a grid. The horizontal and vertical distance between two consecutive holes is 1 cm. The parameters of the cubic model (3.26) can be obtained using a regression approach[2] as illustrated in the following Python code:

Listing 3.4 : Cubic model of an image intensifier

```python
import numpy as np
import numpy.matlib

import matplotlib.pylab as plt
from pyxvis.io import gdxraydb

image_set = gdxraydb.Settings()

img = image_set.load_image(2, 1)  # Input image
data = image_set.load_data(2, 'points')  # Load data for the this image set

# Calibration coordinates in the image domain
um = data['ii'].flatten()  # This can also be done as um = data['ii'][::]
vm = data['jj'].flatten()

# Coordinates of holes in cm.
xb = np.tile(np.r_[-6.5:7.5], [11, 1]).flatten()
xb = xb[:, np.newaxis]

yb = np.r_[-5.0:6.0]
yb = yb[:, np.newaxis]
yb = np.tile(yb, [1, 14]).flatten()
yb = yb[:, np.newaxis]

n = xb.shape[0]

# Build the design matrix
XX = np.hstack(
    [ np.ones((n, 1)), xb, yb, xb * yb, xb**2, yb**2, yb * (xb**2), xb * (yb**2), xb**3,
        yb**3]
```

---

[2]For details of numpy function `np.linalg.lstsq` see https://numpy.org/doc/stable/reference/generated/numpy.linalg.lstsq.html.

```
)

a = np.linalg.lstsq(XX, um, rcond=None)[0]  # rcond=None silence warning for future
    deprecation
b = np.linalg.lstsq(XX, vm, rcond=None)[0]  # We refer the reader to the Numpy
    documentation.

# Also, you can compute Least squere regression as follow:
#a = np.dot(np.dot(np.linalg.inv(np.dot(XX.T, XX)), XX.T), um)

us = np.dot(a, XX.T)
vs = np.dot(b, XX.T)

d = np.array([um-us, vm-vs])
err = np.mean(np.sqrt(np.sum(d ** 2, axis=0)))

# Display the input image and points
fig, ax = plt.subplots(1, 1, figsize=(18, 14))
ax.imshow(img, cmap='gray')
ax.scatter(vm, um, facecolor='none', edgecolor='g', s=120, label='Detected points')
ax.scatter(vs.flatten(), us.flatten(), facecolor='r', marker='+', s=100, label='
    Reprojected points')

# Plot vertical and horizontal lines
lines = np.vstack([us, vs])  # Stack reprojected points
for i in range(11):
    ax.plot(lines[1, (14 * i):(14 * (i + 1))], lines[0, (14 * i):(14 * (i + 1))], 'r:')
        # Reprojected points

# Reshape and stack reprojected point
lines = np.vstack([us.reshape(-1, 14).T.flatten(), vs.reshape(-1, 14).T.flatten()])
for i in range(14):
    ax.plot(lines[1, (11 * i):(11 * (i + 1))], lines[0, (11 * i):(11 * (i + 1))], 'r:')
        # Reprojected points

ax.axis('off')
ax.set_title('Cubic model reprojection error: {:0.4f} pixels'.format(err))
plt.legend(loc=1)
plt.show()
```

The output of this code is shown in Fig. 3.9. The detected (or measured) points correspond to the centers of mass of the holes. They were found using an image processing algorithm. Their coordinates are stored in variable `data`. The mean error between measured and modeled points is 0.7699 pixels.   □

Python Example 3.5: In Fig. 3.10 we show how a 3D point M is projected onto 5 different X-ray images of an aluminum wheel that has been rotated. The coordinates of M given in the object coordinate system are the same for each projection. In this example we model the image intensifier using the hyperbolic model explained in this section:

Listing 3.5 : Transformation 3D → 2D using hyperbolic model

```
import numpy as np
import numpy.matlib

import matplotlib.pylab as plt
from pyxvis.io import gdxraydb
from pyxvis.geometry.projective import rotation_matrix_3d
from pyxvis.geometry.projective import hyperproj
```

Cubic model reprojection error: 0.7699 pixels

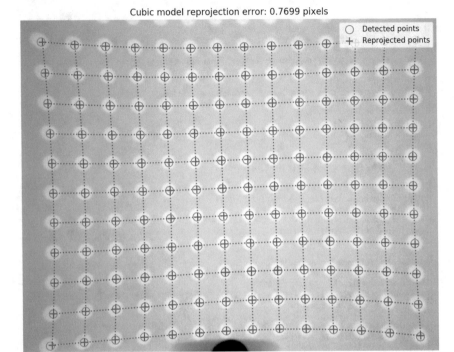

Fig. 3.9 Cubic model of the X-ray projection of regular grid (see Fig. 3.5). [→ Example 3.4]

```python
image_set = gdxraydb.Castings()  # Load the image set

# Load data for the the image set
hyp_model = image_set.load_data(1, 'HyperbolicModel.txt')
man_pos = image_set.load_data(1, 'ManipulatorPosition.txt')

M = np.array([55, 55, -40, 1])
M = M[:, np.newaxis]

# Display the input image and points
fig, ax = plt.subplots(1, 5, figsize=(18, 14))

for i, p in enumerate(range(38, 47, 2)):

    t = np.hstack([man_pos[p, j] for j in range(3)])
    Rp = rotation_matrix_3d(man_pos[p, 3], man_pos[p, 4], man_pos[p, 5])
    Hp = np.vstack([np.hstack([Rp, t[:, np.newaxis]]), np.array([0, 0, 0, 1])])

    w = hyperproj(M, hyp_model, Hp)

    img = image_set.load_image(1, p)

    ax[i].imshow(img, cmap='gray')
    ax[i].scatter(w[1], w[0], facecolor='r', edgecolor='r', s=50)
    ax[i].axis('off')
    ax[i].set_title('Image {0}'.format(p))

plt.show()
```

Image 38   Image 40   Image 42   Image 44   Image 46

Fig. 3.10 Projection of a 3D point onto 5 different X-ray images of the same object in 5 different positions. [→ Example 3.5 🐍]

The output of this code is Fig. 3.10. In this code we use function hyperproj of pyxvis Library. This function computes the transformation 3D → 2D defined in (3.28). □

3.3.3 Explicit Geometric Model Using an Image Intensifier

In this section, we present an explicit model [21] based on the hyperbolic model of Mery and Filbert [22, 23] to perform the transformation **PCS** → **ICS** that takes place in the image intensifier and CCD-camera. The original hyperbolic model, presented in the previous section, does not take into account the non-linear projection between input screen and output screen of the image intensifier, because it is considered as an affine transformation. Additionally, there is no decentering point, since in this model the optical axis of the X-ray projection coincides with the optical axis of the image intensifier. Furthermore, the skew factor of the CCD-camera is not included. Finally, the distortion that arises when electromagnetic fields are present around the image intensifier is not considered. In this section, we propose a complete model that incorporates the mentioned distortion effects.

Image Intensifier
The image intensifier converts the X-ray image into a bright visual image (see Sect. 1.4.3), that can be captured by a CCD-camera [9]. Due to the curvature of the input screen of the image intensifier, the X-ray image received at the output screen is deformed, especially at the corners of the image. An additional distortion can be caused by electromagnetic fields that perform a non-linear projection. An example of these distortion effects is shown in Fig. 3.5, where an X-ray image of a plate containing holes that have been placed in a regular grid manner is illustrated.

First, we will consider a model without electromagnetic field distortion. The geometry of the model used to compute the distortion perspective projection is shown in Fig. 3.11. It consists of a (curved) input screen S and an output screen Φ, on which the image is projected. The output screen Φ coincides with the retinal plane of this projection.[3] We have shown in Sect. 3.3.1, how the 3D object point M is projected onto plane Π as point m. Thus, the perspective X-ray projection OCS

---

[3]The reader should note that at this moment there are two retinal planes: Π for the central projection and Φ for the image intensifier.

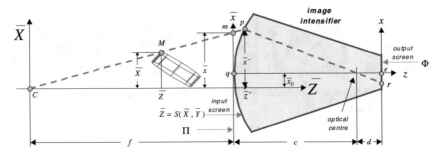

Fig. 3.11 Geometric model of the image intensifier (axes parallel to $\bar{Y}$ are not shown)

$\rightarrow$ **PCS** is given by (3.19). In this section we will calculate, how point m is projected onto input screen S as point p and then onto the retinal plane Φ as point r.

The X-ray image present on the input screen is projected onto the output screen through an optical center of the image intensifier. We may assume without loss of generality that the optical axis of the image intensifier (z-axis) is parallel to the optical axis of the X-ray projection ($\bar{Z}$-axis), because, in a central X-ray projection, there is always a ray that is parallel to the optical axis of the image intensifier. However, the displacement of these axes must be determined. For this reason, we modify the hyperbolic surface of (3.27) by introducing a shift of the center of the hyperboloid as shown in Fig. 3.11. Therefore, the hyperbolic surface S is defined in **WCS** by

$$\bar{Z} = S(\bar{X}, \bar{Y}) = f\sqrt{1 + \frac{(\bar{X} - \bar{x}_0)^2}{a^2} + \frac{(\bar{Y} - \bar{y}_0)^2}{b^2}} \qquad (3.29)$$

with f being the real half axis of the hyperboloid; a and b the imaginary half axes; and $(\bar{x}_0, \bar{y}_0)$ the coordinates of the center of the hyperboloid. The focal length of the X-ray projection (f), defined in Sect. 3.3.1, is the minimal value that takes the surface S. This occurs in $(\bar{x}_0, \bar{y}_0)$, that is represented as q in Fig. 3.11. The displacement between $\bar{Z}$- and z-axis is given by $(\bar{x}_0, \bar{y}_0)$.

The projection of point M onto the input screen of the image intensifier is calculated as the intersection of the line that contains points C, M and m with the 3D surface S. This intersection is denoted by p in Fig. 3.11, whose coordinates in **WCS** are given by $(\bar{x}', \bar{y}', \bar{z}')$:

$$\bar{x}' = \bar{z}'\bar{x}/f, \quad \bar{y}' = \bar{z}'\bar{y}/f \quad \text{and} \quad \bar{z}' = \frac{-B + \sqrt{B^2 - 4AC}}{2A} \qquad (3.30)$$

with

$$A = \frac{1}{f^2}\left(1 - \frac{\bar{x}^2}{a^2} - \frac{\bar{y}^2}{b^2}\right), \quad B = \frac{2}{f}\left(\frac{\bar{x}\bar{x}_0}{a^2} + \frac{\bar{y}\bar{y}_0}{b^2}\right), \quad C = -\left(1 + \frac{\bar{x}_0^2}{a^2} + \frac{\bar{y}_0^2}{b^2}\right).$$

The coordinates of point p depend on the coordinates $(\bar{x}, \bar{y})$ of point m in PCS. Using homogeneous coordinates, p can be expressed as follows:

$$\mathbf{p} = \mathbf{g(m)}, \tag{3.31}$$

where $\mathbf{p} = [\bar{x}'\ \bar{y}'\ \bar{z}'\ 1]^{\mathsf{T}}$, $\mathbf{m}$ is a homogeneous representation of $(\bar{x}, \bar{y})$, and $\mathbf{g}$ is the non-linear function defined from (3.30).

As illustrated in Fig. 3.11, point p is projected through the optical center of the image intensifier onto the output screen Φ as point r. The projected point r has coordinates (x, y) in a new 2D coordinate system, called the *output screen coordinate system* (SCS). This coordinate system is centered in e, and its x, y-axes are parallel to the $\bar{x}$, $\bar{y}$-axes of PCS. We can conclude from consideration of similar triangles that

$$\lambda \begin{bmatrix} x \\ y \\ 1 \end{bmatrix} = \begin{bmatrix} d & 0 & 0 & -d\bar{x}_0 \\ 0 & d & 0 & -d\bar{y}_0 \\ 0 & 0 & 1 & -(f+c) \end{bmatrix} \begin{bmatrix} \bar{x}' \\ \bar{y}' \\ \bar{z}' \\ 1 \end{bmatrix}, \tag{3.32}$$

where λ is a scale factor, and c and d are the distances of the input and output screen to the optical center of the image intensifier (see Fig. 3.11). This equation can be expressed in matrix form as

$$\lambda \mathbf{r} = \mathbf{Dp}, \tag{3.33}$$

where the 3-component vector $\mathbf{r}$ is a homogeneous representation of (x, y), and $\mathbf{D}$ is the 3×4 projective matrix of the image intensifier expressed in (3.32). From (3.31) and (3.33) we obtain the non-linear equation, which depends on six parameters: a, b, c, d, $\bar{x}_0$ and $\bar{y}_0$, that maps a projected point on the retinal plane Π of the X-ray projection onto a point on the retinal plane Φ of the image intensifier:

$$\lambda \mathbf{r} = \mathbf{Dg(m)}. \tag{3.34}$$

To model the effect of the electromagnetic distortion we propose an empirical model, in which a point r on plane Φ will be transformed into a new point r'. We observed that the projection of a regular grid seems to have an additional harmonic signal (see Fig. 3.5). For this reason, we can empirically model this distortion with sinusoidal functions.

The electromagnetic distortion is modeled in two steps. The first step introduces a distortion in the x-direction and the second one in the y-direction. Thus, x is firstly transformed into x' from (x, y) and secondly, y is transformed into y' from (x', y) as follows:

$$\begin{aligned} x' &= x + A_1 \sin(B_1 y + C_1) \\ y' &= y + A_2 \sin(B_2 x' + C_2), \end{aligned} \tag{3.35}$$

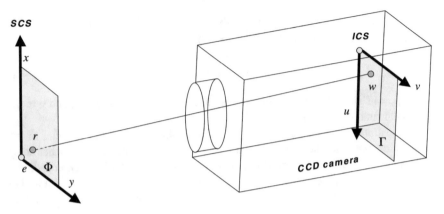

Fig. 3.12   Imaging process in the CCD-camera

where A_i, B_i and C_i, $i = 1, 2$, are the parameters of the electromagnetic distortion model. Formally, r' can be expressed using homogeneous coordinates as follows:

$$\mathbf{r}' = \mathbf{f}(\mathbf{r}) = [x'\ y'\ 1]^{\mathsf{T}}, \tag{3.36}$$

where $\mathbf{f}$ is the non-linear function defined from (3.35).

Other sinusoidal functions can be used to model the distortion introduced by electromagnetic fields. The reason why we use a two-step based model is because Eq. (3.35) can be back-projected in a closed-form as shown previously.

CCD-Camera
The 2D image coordinate system (ICS) is used to represent the pixel coordinates of the X-ray image captured by the CCD-camera. The point r (or r' if we consider the electromagnetic field distortion) at the output screen of the image intensifier (see Fig. 3.11) is projected onto the retinal plane Γ of the CCD-array[4] as point w as shown in Fig. 3.12.

The camera could be modeled as a general pinhole camera [2], in which a projective mapping from a 3D point of the space to a 2D projective space takes place. However, in our model the 3D points to be mapped belong to a plane, namely the retinal plane Φ. For this reason, in this work we use a homography, i.e., a 2D → 2D general projective transformation as explained in Sect. 3.2.2, which relates the coordinates of retinal plane Φ to retinal plane Γ of the camera. This transformation is defined by

$$\lambda\mathbf{w} = \mathbf{Hr}, \tag{3.37}$$

---

[4]The reader should note that Γ, the retinal plane of the CCD-camera, is the third retinal plane of our model (see footnote 3).

where the 3-component vectors $\mathbf{r}$ and $\mathbf{w}$ are homogeneous representations of (x, y) and (u, v) (coordinates of r in SCS and w in ICS) respectively. Matrix $\mathbf{H}$ is a homogeneous 3×3 matrix that causes a general perspective transformation where rotation, translation, scaling, skew and perspective distortion are considered. Matrix $\mathbf{H}$ has nine elements where only their ratio is significant, so the transformation is defined by only eight parameters, e.g., , $h_{11}, h_{12}, \ldots, h_{32}$. Parameter h_{33} can be defined as $h_{33} = 1$, or $\mathbf{H}$ can be constrained by $\| \mathbf{H} \| = 1$ [13].

Summary
In this section we described a model which relates the transformation 3D $\rightarrow$ 2D, from a 3D point M of the test object to a 2D point w of the digitalized X-ray image pixel using homogeneous coordinates. Therefore, the transformation is expressed by $\mathbf{M} \rightarrow \mathbf{w}$, where $\mathbf{M} = [X \ Y \ Z \ 1]^\mathsf{T}$ and $\mathbf{w} = [u \ v \ 1]^\mathsf{T}$. There are two possibilities for performing the transformation, namely without and with considering the electromagnetic distortion. In the first case, the transformation is given by $\mathbf{M} \rightarrow \mathbf{m} \rightarrow \mathbf{r} \rightarrow \mathbf{w}$ using Eqs. (3.19), (3.34) and (3.37) respectively:

$$\lambda \mathbf{w} = \mathbf{HDg(PM)}. \tag{3.38}$$

This model has seven extrinsic parameters (defined in (3.20)) and fourteen intrinsic parameters: $a, b, c, d, \bar{x}_0, \bar{y}_0, h_{11}, h_{12}, \ldots, h_{31}$ and h_{32}.

In the second case, where the electromagnetic distortion is modeled, the transformation is expressed by $\mathbf{M} \rightarrow \mathbf{m} \rightarrow \mathbf{r} \rightarrow \mathbf{r}' \rightarrow \mathbf{w}$ using Eqs. (3.19), (3.34), (3.36) and (3.37) respectively:

$$\lambda \mathbf{w} = \mathbf{HDf(g(PM))}. \tag{3.39}$$

In comparison with the first case model, the consideration of the electromagnetic distortion requires six more intrinsic parameters: A_i, B_i and C_i, for $i = 1, 2$.

3.3.4   Multiple View Model

In many applications, a single view of a test object is not enough because there are for example occluded parts or intricate projections that cannot be observed with a single view. For this reason the test object must be analyzed from n points of views (with $n \geq 2$). In this section, we present a geometric model that can be used when dealing with *multiple views*, i.e., a geometric model that relates the transformation of a 3D point of the test object into the 2D coordinates of each X-ray projection. For multiple view analysis, i.e., 3D reconstruction or analysis of a part from different points of view, it is required that the n geometric models must share the same 3D object coordinate system (OCS). That means the 3D coordinates (X, Y, Z) of each projection, for $p = 1 \ldots n$, are the same, and we are interested to find the location of the projection of this unique 3D point in each 2D view. Using (3.22) for each view

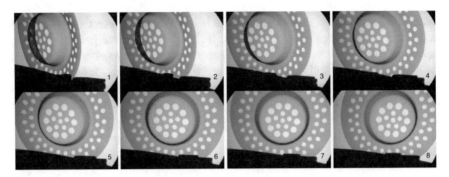

Fig. 3.13 Multiple views of an object acquired using a manipulator that rotates the object around its vertical axis (series S0007 of GDXray+)

we obtain

$$
\begin{cases}
\mathbf{w}_1 = \mathbf{F}(\mathbf{M}, \theta_1) \\
\mathbf{w}_2 = \mathbf{F}(\mathbf{M}, \theta_2) \\
\quad \vdots \\
\mathbf{w}_n = \mathbf{F}(\mathbf{M}, \theta_n)
\end{cases},
\tag{3.40}
$$

where $\mathbf{M} = [X \ Y \ Z \ 1]^\mathsf{T}$ are the homogenous coordinates of the 3D points in OCS, θ_p are the parameters of the geometric model for pth projection, and $\mathbf{w}_p = [u_p \ v_p \ 1]^\mathsf{T}$ are the homogenous coordinates in pixels in pth X-ray image.

Two views can be simultaneously achieved using two different X-ray detectors. There are some X-ray computer vision systems with three or four detectors as well (see, for example, [7]). In medicine, for instance, it is a common practice to take two X-ray images simultaneously (from two different points of views) of certain organs that change their shape and size because they are in motion (e.g., , X-ray stereo angiography [8]). In this case, we have independent geometric models (one for each view) that can be obtained using the theory outlined in the previous section. That means, in (3.40), the parameters of the models are independent from each other. Usually in X-ray testing, however, there is a manipulator that is able to locate the test object in different positions, and different views are obtained in different times using a single detector. Given that we are capturing X-ray images of a rigid test object it is not necessary to acquire the images simultaneously (see Fig. 3.13). In this case, the parameters of the models are not independent from each other because they share the same intrinsic parameters as there is only one X-ray detector. For example, if a manipulator rotates the test object as shown in Fig. 3.6, and for each position a new X-ray image is acquired, it is clear therefore that the transformation from OCS to word coordinate system (WCS) is different for each projection, however, the projection from WCS into image coordinate system (ICS) is exactly the same.

In an X-ray computer vision system with a manipulator and a flat panel (that can be modeled by a simple model with no distortion), the following equation can be used for pth view ($\mathbf{w}_p = \mathbf{F}(\mathbf{M}, \theta_p)$) according to (3.18) and (3.24):

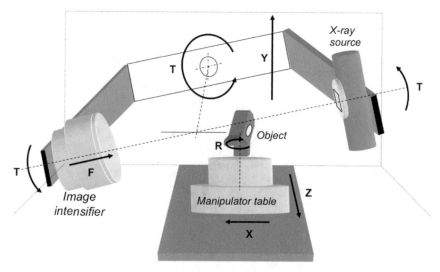

Fig. 3.14 X-ray computer vision system with a manipulator with several degrees of freedom

$$
\lambda_p \begin{bmatrix} u_p \\ v_p \\ 1 \end{bmatrix} = \begin{bmatrix} k_u & s & u_0 \\ 0 & k_v & v_0 \\ 0 & 0 & 1 \end{bmatrix} \begin{bmatrix} f & 0 & 0 & 0 \\ 0 & f & 0 & 0 \\ 0 & 0 & 1 & 0 \end{bmatrix} \underbrace{\begin{bmatrix} \mathbf{R}_p & \mathbf{t}_p \\ \mathbf{0} & 1 \end{bmatrix}}_{\mathbf{H}_p} \begin{bmatrix} X \\ Y \\ Z \\ 1 \end{bmatrix}, \tag{3.41}
$$

where 4×4 matrix $\mathbf{H}_p$, that includes rotation matrix $\mathbf{R}_p$ and translation vector $\mathbf{t}_p$, define the Euclidean transformation of pth position from OCS to WCS. In this simple model, the vector parameter θ_p for pth projection includes intrinsic parameters (k_u, k_v, s, u_0, v_0) and extrinsic parameters focal length f for the central projection, $(\omega_{Xp}, \omega_{Yp}, \omega_{Zp})$ for the pth rotation and (t_{Xp}, t_{Yp}, t_{Zp}) for the pth translation. It is clear, that the intrinsic parameters are the same for each projection whereas the extrinsic parameters are different. In many cases, however, the test object is rotated around one axis only, that means that (t_{Xp}, t_{Yp}, t_{Zp}) is constant and only one ω angle changes for each position. This is the case of example of Fig. 3.13 where $\omega_{Xp} = \omega_X$, $\omega_{Yp} = \omega_Y$ and $\omega_{Zp} = \omega_Z + p5^0$, where the rotation around the vertical axis between two consecutive frames is 5^0.

The transformation defined by $\mathbf{H}_p$ (OCS $\rightarrow$ WCS) can be modeled using a manipulator coordinate system (MCS): OCS $\rightarrow$ MCS $\rightarrow$ WCS. Thus, for pth view the transformation OCS $\rightarrow$ MCS is constant (with some constant rotation and translation), whereas MCS $\rightarrow$ WCS has a different rotation and translation for each position.

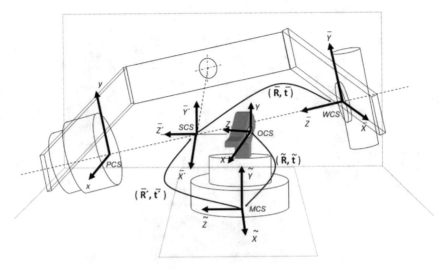

Fig. 3.15 Coordinate systems used in the geometric model of Fig. 3.14

This methodology can be extended to more complex manipulators with several degrees of freedom. For example in Fig. 3.14 such a manipulator system is presented. This system can be used in the inspection of aluminum castings. The object can be rotated around its vertical axis using rotation R. In addition, it can be translated in X and Z direction using the manipulator table. Moreover, the whole computer vision system can be translated in Y and rotated using rotation T. In order to model the transformation $\mathsf{OCS} \rightarrow \mathsf{WCS}$ of this manipulator system, we can include additional coordinate systems as illustrated in Fig. 3.15: $\mathsf{OCS} \rightarrow \mathsf{MCS} \rightarrow \mathsf{SCS} \rightarrow \mathsf{WCS}$. Thus, there are three 3D Euclidean transformations. Each one is modeled using a 4×4 transformation matrix that includes a 3×3 rotation matrix and a 3×1 translation vector as explained in Sect. 3.2.3. That means, the whole transformation $\mathsf{OCS} \rightarrow \mathsf{WCS}$ is a 4×4 matrix computed as the multiplication of these three matrices. This matrix corresponds to $\mathbf{H}_p$ in (3.41). The reader can find more details of this model in [20].

3.4 Calibration

The calibration of an X-ray computer vision system—in the context of 3D machine vision—is the process of estimating the parameters of the model, which is used to determine the projection of the 3D object under test into its 2D digital X-ray image (Fig. 3.16). This relationship 3D $\rightarrow$ 2D can be modeled with the transfer function $\mathbf{F} : \mathbb{R}^3 \rightarrow \mathbb{R}^2$ expressed in (3.22).

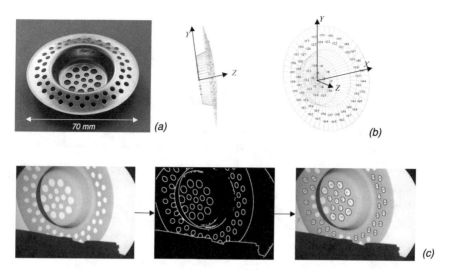

Fig. 3.16 Calibration object: **a** photography; **b** CAD-model; and **c** X-ray image of the calibration object and measured calibration points

There are several techniques developed to calibrate a computer vision system. They can be roughly classified into two categories: *photogrammetric calibration* and *self-calibration* [29]. The first one is a 3D reference object-based calibration, where the calibration is performed by observing a calibration object whose geometry in 3D space is known with high accuracy [2]. The second technique uses the identification of matching points in several views of a scene taken by the same camera. Self-calibration does not use a calibration object with known 3D geometry because it aims to identify the intrinsic parameters of the computer vision system and to reconstruct 3D structure up to a scale similarity [18]. Due to the high precision feature measurement of 3D geometry required in the NDT applications, it would be necessary to do a *true* reconstruction of the 3D space without a scale factor. For this reason, usually the calibration technique in X-ray testing belongs to the photogrammetric category.[5]

In calibration, we estimate the parameters of the model based on n points of a *calibration object* whose object coordinates $\mathbf{M}_i = [X_i \ Y_i \ Z_i \ 1]^\mathsf{T}$ are known and whose image coordinates $\tilde{\mathbf{w}}_i = [\tilde{u}_i \ \tilde{v}_i \ 1]^\mathsf{T}$ are measured, for $i = 1, \ldots, n$. In Fig. 3.16 an example of a calibration object is illustrated.

Using the geometric model explained in Sect. 3.3 (see (3.22)) we obtain the *reprojected* points $\mathbf{w}_i = [u_i \ v_i \ 1]^\mathsf{T}$, i.e., the inferred projections in the digital image computed from the calibration points $\mathbf{M}_i$ and the parameter vector θ:

$$\mathbf{w}_i = \gamma(\mathbf{A}\mathbf{M}_i) := \mathbf{F}(\theta, \mathbf{M}_i) \quad \text{for } i = 1 \ldots n, \tag{3.42}$$

---

[5]Nevertheless, in Sect. 9.4.3 the reader can find an interesting X-ray testing application where the 3D model is estimated using a self-calibration method based on *bundle adjustment*.

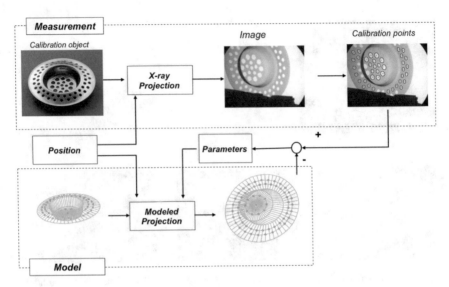

Fig. 3.17 Calibration process

where $\theta^{\mathsf{T}} = [\theta_{ext} \; ; \; \theta_{int}]$ is the vector of parameters involved in the projection model including both extrinsic and intrinsic parameters. The parameter vector is then estimated by minimizing the distance between measured points $\tilde{\mathbf{w}}_i$ (see Fig. 3.16c) and inferred points $\mathbf{w}_i = \mathbf{F}(\theta, \mathbf{M}_i)$ (see Fig. 3.16b). Thus, the calibration is performed by minimizing the objective function $\mu(\theta)$ defined as the mean-square discrepancy between these points:

$$\mu(\theta) = \frac{1}{n} \sum_{i=1}^{n} \| \tilde{\mathbf{w}}_i - \mathbf{F}(\theta, \mathbf{M}_i) \| \rightarrow \min. \qquad (3.43)$$

The whole calibration process is summarized in Fig. 3.17. The calibration problem is a non-linear optimization problem. Generally, the minimization of $\mu(\theta)$ has no closed-form solution. For this reason, the objective function must be iteratively minimized starting with an initial guess θ^0 that can be obtained from nominal values or preliminary reference measurements. In this section, we present two different methodologies that can be used to calibrate an X-ray computer vision system. The first one was proposed originally in [29] and it is implemented in OpenCV for Python.[6] This technique is very effective and it can be used in the calibration of computer vision with flat panels or with image intensifiers with low distortion. The second technique was proposed in [21] and can be used in computer vision systems with image intensifiers with high distortion.

---

[6]See https://opencv.org.

3.4.1   *Calibration Using Python*

OpenCV Library can be used to easily calibrate an X-ray computer vision system. In our experiments, we use `calibrateCamera` based on Tsai's calibration method [26]. It requires a checkerboard as calibration object, and at least two X-ray images taken by the computer vision system to be calibrated. For best results, however, it is recommended to acquire between 10 and 20 images. An example of 18 X-ray images of a calibration pattern is shown in Fig. 2.17.

Python Example 3.6: For the calibration of an X-ray computer vision system the X-ray images of 2.17 and the following Python code can be used. In this example a 3D Gaussian bell is superimposed onto an X-ray image in order to show how we can use the obtained geometric model to reproject 3D points onto the original image.

Listing 3.6 :   Calibration of an X-ray computer vision system.

```python
import cv2 as cv
import numpy as np

from pyxvis.io import gdxraydb
from pyxvis.io.visualization import project_edges_on_chessboard, gaussian_superimposition

image_set = gdxraydb.Settings()

# Termination criteria
criteria = (cv.TERM_CRITERIA_EPS + cv.TERM_CRITERIA_MAX_ITER, 30, 0.001)

nx = 10 #  number of inside corners per row
ny = 10 #  number of inside corners per column

# Prepare object points, like (0,0,0), (1,0,0), (2,0,0) ....,(6,5,0)
objp = np.zeros((nx*ny, 3), np.float32)
objp[:, :2] = np.mgrid[0:nx, 0:ny].T.reshape(-1,2)
objp = 250 * objp

# Arrays to store object points and image points from all the images.
obj_points = [] # 3d point in real world space
img_points = [] # 2d points in image plane.

img_boards = []

for i in range(1, 19):

    print('Find chessboard corners in image {}: '.format(i), end='')

    img = image_set.load_image(8, i)
    img_h = img.copy()

    # Keep a copy of the image but using three color channels. Just for visualization.
    img = cv.cvtColor(img, cv.COLOR_GRAY2BGR)

    ret, corners = cv.findChessboardCorners(img_h, (nx, ny), flags=cv.
        CALIB_CB_ADAPTIVE_THRESH)

    print('{}'.format(ret))

    if ret:
        obj_points.append(objp)
```

```
        corners = cv.cornerSubPix(img_h, corners, (11, 11), (-1, -1), criteria)
        img = cv.drawChessboardCorners(img, (nx, ny), corners, ret)
        img_points.append(corners)
        img_boards.append({'img': img, 'idx': i})

# Calibrate the camera
ret, mtx, dist, rvecs, tvecs = cv.calibrateCamera(obj_points, img_points, img_h.shape
    [::-1], None, None)

# Show parameters
print('ret: ', ret)
print('Mtx: ', mtx)
print('Dist: ', dist)
print('Rvecs: ', rvecs)
print('Tvecs: ', tvecs)

i = 6
img = image_set.load_image(8, i)

# Projection matrix of image i. Remember that for this example
# indexation of rotation and translation matrices starts at 0.
R, _ = cv.Rodrigues(rvecs[i-1])  # Rotation matrix 3x3
t = tvecs[i - 1]  # Translation vector 3x1

H = np.hstack([R, t])
H = np.vstack([H, np.array([0, 0, 0, 1])])

# Projection matrix
P = np.hstack([mtx, np.zeros((3, 1))])
P = np.dot(P, H)

project_edges_on_chessboard(img, P, square_size=250)

gaussian_superimposition(img, P, square_size=250, n_points=30)
```

The output of this code is shown in Fig. 3.18. The reader that is interested in the computer graphics details of the superimposition can study the program developed in pyxvis Library under the name `gaussian_superimposition`.   □

Fig. 3.18 Details of image 6: (left) measured and reprojected points, (right) simulation of a 3D Gaussian bell superimposed onto the checkerboard. [→ Example 3.6 🐍]

Table 3.3 Characterization of the implemented models for calibration

Model	Name	Reference	Intrinsic parameters	Back-projection	Calibration	Distortion	Model
1. Faugeras	Linear	[2]	5	Direct	Iterative	None	Explicit
2. Tsai	Radial	[26]	5	Indirect	Iterative	Radial	Implicit
3. Weng et al.	Rad-Tan-1	[28]	9	Indirect	Iterative	Radial, tangential	Implicit
4. Heilikkä	Rad-Tan-2	[14]	8	Direct	Iterative	Radial, tangential	Implicit
5. Swami-nathan & Nayar	Rad-Tan-3	[25]	8	Indirect	Iterative	Radial, tangential	Implicit
6. Jaeger / Brack et al.	Cubic	[17] [1]	20	Indirect	Iterative	Cubic	Implicit
7. Mery & Filbert	Hyp-Simple	[22, 23]	7	Direct	Iterative	Hyperbolic simple	Explicit
8. Mery (without EFD)	Hyp-Full	[21]	14	Direct	Iterative	Hyperbolic	Explicit
9. Mery 2 (with EFD)	Hyp-EFD	[21]	20	Direct	Iterative	Hyperbolic, sinusoidal	Explicit

EFD electromagnetic field distortion, *Hyp* Hyperbolic, *Rad* Radial, *Tan* Tangential

3.4.2 Experiments of Calibration

In this section, we present the experiments which we did in order to evaluate the performance of the different models used to calibrate X-ray computer vision systems. The tested models and their principal features are summarized in Table 3.3. They are the seven models outlined in Sect. 3.3.2 and the two proposed models of Sect. 3.3.3 (without and with considering electromagnetic distortion). In the presentation of the results, each model will be identified by the name given in the second column of Table 3.3.

As we explained in the introduction of Sect. 3.4, the calibration process estimates the parameters of a model based on points whose object coordinates are known, and whose image coordinates are measured. The calibration object used in our experiments is shown in Fig. 3.16a. It is an aluminum object with an external diameter of 70 mm. A CAD-model was developed by measurement of the calibration object (see Fig. 3.16b). It has seventy small holes ($\phi = 3 - 5$ mm) distributed on four rings and the center. As we can see in Fig. 3.16, the centers of gravity of the holes are arranged in three heights.

The search for the calibration points within the X-ray image is carried out with a simple procedure that detects regions with high contrast and defined size for the area. The centers of gravity of the detected regions, computed with sub-pixel accuracy, are defined to be the calibration points. Only complete enclosed regions will be segmented. Figure 3.16c shows an example of the search for the calibration points within an X-ray image. The reader can use series S0007 of

GDXray+ with detected points stored in file `ground_truth.txt` and 3D points in file `points_object_3D.txt`. The correspondence between the 3D object points and their images was established manually. The image intensifier used in the experiments was the XRS 232[7] with a 22 cm input screen. The size of the images was 576 × 768 pixels.

In our experiments, the calibration object was placed in different positions using a manipulator. The positions of the calibration object were achieved by rotating one of the axes of the manipulator. Some of the images obtained are illustrated in Fig. 3.19. In order to incorporate the pth position of the manipulator (for each X-ray image) into the geometric model, we modify Eq. (3.18) by

$$
\lambda \begin{bmatrix} \bar{x} \\ \bar{y} \\ 1 \end{bmatrix} = \underbrace{\begin{bmatrix} f & 0 & 0 & 0 \\ 0 & f & 0 & 0 \\ 0 & 0 & 1 & 0 \end{bmatrix}}_{\mathbf{P}} \underbrace{\begin{bmatrix} \bar{\mathbf{R}}_p & \bar{\mathbf{t}}_p \\ \mathbf{0}^\mathsf{T} & 1 \end{bmatrix}}_{\bar{\mathbf{H}}_p} \underbrace{\begin{bmatrix} \mathbf{R} & \mathbf{t} \\ \mathbf{0}^\mathsf{T} & 1 \end{bmatrix}}_{\mathbf{H}} \begin{bmatrix} X \\ Y \\ Z \\ 1 \end{bmatrix}. \tag{3.44}
$$

In this equation, we have two 4×4 matrices ($\mathbf{H}$ and $\bar{\mathbf{H}}_p$) that define respectively two 3D Euclidean transformations: *(i)* between object and manipulator coordinate systems, and *(ii)* between manipulator and world coordinate systems. The translation vectors ($\mathbf{t}$ and $\bar{\mathbf{t}}_p$) and the rotation matrices ($\mathbf{R}$ and $\bar{\mathbf{R}}_p$) are related to the corresponding translation and rotation of the mentioned transformations. Since the calibration object is fixed to manipulator, matrix $\mathbf{H}$ is constant for each position: $\mathbf{t} = [t_X \ t_Y \ t_Z]^\mathsf{T}$ and a matrix $\mathbf{R}$ is calculated from the Euler angles ω_X, ω_Y and ω_Z (see (3.12)). However, matrix $\bar{\mathbf{H}}_p$ depends on the position of the manipulator with respect to the world coordinate system. Matrix $\bar{\mathbf{H}}_p$ is defined by a translation vector $\bar{\mathbf{t}}_p = [\bar{t}_X \ \bar{t}_Y \ \bar{t}_Z]_p^\mathsf{T}$ and a rotation matrix $\bar{\mathbf{R}}_p$ computed from the Euler angles $\bar{\omega}_X$, $\bar{\omega}_Y$ and $\bar{\omega}_Z$. In our experiments, $\bar{t}_X$, $\bar{t}_Y$, $\bar{t}_Z$, $\bar{\omega}_X$ and $\bar{\omega}_Y$ were constant. Nevertheless, $\bar{\omega}_Z$ was incremented by the manipulator in constant steps. Thus, the rotation of this axis can be linearly modeled by $\bar{\omega}_Z(p) = \bar{\omega}_{Z0} + p\Delta\bar{\omega}_Z$, where p denotes the number of the position. This new model introduces seven additional extrinsic parameters ($\bar{t}_X$, $\bar{t}_Y$, $\bar{t}_Z$, $\bar{\omega}_X$, $\bar{\omega}_Y$, $\bar{\omega}_{Z0}$ and $\Delta\bar{\omega}_Z$) that must be estimated in the calibration process as well.

The calibration is performed by minimizing the mean reprojection error (μ) computed as the average of the distance between measured points ($\tilde{\mathbf{w}}_{ip}$) and inferred points ($\mathbf{w}_{ip}$)—in the image coordinate system (ICS)—obtained from the pth projection of the ith object point $\mathbf{M}_i$ according to the model of the computer vision system. As we can see, the calibration problem is a non-linear optimization problem, where the minimization of the objective function has no closed-form solution. For this reason, the objective function must be iteratively minimized starting with an initial estimated value for the parameters involved in the model. In our experiments, the estimation is achieved using the well-known algorithm for minimization

---

[7]Image intensifier developed by YXLON International Inc.

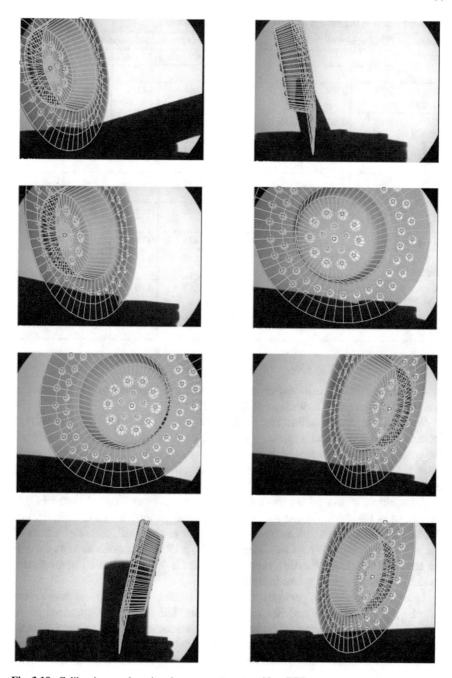

Fig. 3.19 Calibration results using the proposed method Hyp-EFD

Table 3.4 Error of the models in control (C) and test (T) points

Model ↓	2D reprojection (pixels)				3D reconstruction (mm)			
	μ		σ		μ		σ	
	C	T	C	T	C	T	C	T
1. Linear	2.52	2.45	1.57	1.53	0.47	0.41	0.229	0.157
2. Radial	1.70	1.64	0.96	0.83	0.18	0.17	0.087	0.075
3. Rad-Tan-1	1.40	1.39	0.85	0.76	0.15	0.15	0.076	0.071
4. Rad-Tan-2	1.62	1.66	0.87	0.87	0.25	0.21	0.123	0.092
5. Rad-Tan-3	1.64	1.62	0.92	0.81	0.20	0.18	0.102	0.079
6. Cubic	1.16	1.16	0.67	0.59	0.15	0.14	0.079	0.063
7. Hyp-Simple	1.36	1.40	0.77	0.73	0.18	0.17	0.090	0.074
8. Hyp-Full	1.25	1.29	0.72	0.69	0.17	0.16	0.096	0.072
9. Hyp-EFD	1.18	1.17	0.71	0.64	0.16	0.14	0.084	0.069

problems: the *BFGS Quasi-Newton* method,[8] which is implemented by MathWorks Inc. in the optimization toolbox of MATLAB [19].

We subdivided the calibration points into two groups: the points measured from seven images ($p = 1, 3, 5, \ldots 13$) were used as *control points* to calibrate the computer vision system, whereas the points extracted from seven other images ($p = 2, 4, 6, \ldots 14$) were used as *test points* in order to evaluate the accuracy of calibration.

An example of the calibration using our proposed model, called Hyp-EFD, is illustrated in Fig. 3.19. We can see that the modeled projection of a CAD-model of the calibration object coincides with the X-ray image very well. Although points of the top right and the bottom left images of Fig. 3.19 could not be used as control (or test) points because they are very intricate, the inferred projection of the CAD-model in these positions seems to be fine.

In order to assess the performance of each model, we carried out two experiments: 2D reprojection and 3D reconstruction. The results are summarized in Table 3.4. The first experiment estimates the parameters of each model by minimizing the average error of the reprojection error—in ICS given in pixels—of the control points. The accuracy is assessed with the reprojection error in the test points.

---

[8]This is a gradient method that uses the Broyden–Fletcher–Goldfarb–Shanno formula for updating the approximation of the Hessian matrix iteratively, which reduces the computational cost of the minimization.

Once the calibration is completed, the second experiment is performed using the parameters estimated in the first. The 3D reconstruction of the measured points was performed using a least square technique [2]. As a performance measurement of the second experiment, the Euclidean distance between measured and reconstructed points in OCS was calculated in millimeters. The mean μ and the standard deviation σ of the computed distances errors in control and test points were tabulated for each experiment. For emphasis, we remind the reader that the calibration is performed by minimizing the average of the reprojection error of the control points (first column in Table 3.4), i.e., the control points in the experiments of 3D reconstruction were not used to calibrate, but also as test points too.

As the results obtained on control and test points are very similar, our analysis will consider test point measurements only. The two values x/y given below correspond to the mean error values obtained by computing the 2D reprojection and 3D reconstruction given in pixels and millimeters respectively. We observed that the best results were obtained by Cubic and Hyp-EFD models in both experiments. In these cases the mean errors were in the order of 1.16~1.17/0.14, i.e., 1.16~1.17 pixels for the 2D reprojection, and 0.14 mm for the 3D reconstruction. Although the Cubic model obtains a fractionally better accuracy than model Hyp-EFD (see standard deviations), we must take into account that Cubic model uses an implicit model with 20 parameters for the projection, and 20 other parameters for the back-projection. On the other hand, model Hyp-EFD uses the same 20 parameters for both projection and back-projection.

In our experiments, the X-ray computer vision system could not be adequately modeled without consideration of the lens distortion or with only radial and tangential distortion. The models that were originally developed for cameras (Linear, Radial, Rad-Tan-1, Rad-Tan-2 and Rad-Tan-3, where the mean errors were 2.45/0.41, 1.64/0.17, 1.39/0.15, 1.66/0.21 and 1.62/0.18 respectively), did not work appropriately for our X-ray computer vision system. In many cases, the maximum reprojection error was greater than 4 pixels. The reason for this is that the distortion introduced by the image intensifiers is different from the distortion introduced by a camera lens and the camera models do not consider the electromagnetic distortion in the image intensifier.

On the other hand, hyperbolic models are used by Hyp-Simple, Hyp-Full and Hyp-EFD. The results obtained with model Hyp-Simple are comparable with the best results obtained from camera models (Rad-Tan-1), where the mean errors were 1.40/0.17 and 1.39/0.15 respectively. In relation to model Hyp-Simple, model Hyp-Full introduces a decentering point and a non-linear transformation in the image intensifier. This additional complexity in the model has a significant decrease in the reprojection error (1.40/0.17 vs. 1.29/0.16). In addition, another important reduction of both errors is achieved by considering the electromagnetic field distortion in model Hyp-EFD (1.29/0.16 vs. 1.17/0.14).

We conclude that the proposed explicit model considers the physical parameters of the computer vision system, like image center, focal length, etc., independently. The model is able to map the 3D coordinates of a test object to the 2D coordinates of the corresponding pixel on the digital X-ray image. The model consists of three parts: X-ray projection, image intensifier, and CCD-camera. The distortion introduced by the image intensifier was modeled using a hyperbolic surface for the input screen and sinusoidal functions for electromagnetic fields. Using our explicit model, the back-projection function—required for 3D reconstruction—can be calculated directly using a closed-form solution.

The suggested model was experimentally compared with seven other models, which are normally used to calibrate a computer vision system with and without lens distortion. Fourteen X-ray images were taken of a calibration object in different positions. Seven of them were used to calibrate the computer vision system and the other seven were employed to test the accuracy of calibration. The results show that the consideration of only radial and tangential distortions is not good enough if we are working with image intensifiers. In this case, other models must be used for high accuracy requirements. For this reason, Cubic or Hyp-EFD models are recommended. Their mean errors are very similar as shown in Table 3.4. However, for the back-projection, it is convenient to use the proposed model Hyp-EFD because the same parameters are used for both the projection and the back-projection model.

3.5   Geometric Correspondence in Multiple Views

As explained in Sect. 3.3.4, in certain X-ray testing applications it is necessary to analyze multiple views of a test object. In general, in this kind of computer vision applications only the images (2D projections) are available and no 3D information of the test object is known.

In multiple view analysis, it is very important to find *corresponding points* because they can be used for 3D reconstruction, or for analysis of the test object from different points of views. Corresponding points are those 2D points (in different views) that are projections of the same 3D point (see Fig. 3.20). An example of two views is shown in Fig. 3.22, in which we can see two perspective projections of a 3D point M. This stereo rig consists of projection p and projection q. It is built using two monocular perspective projection models (Fig. 3.4). In this example, m_p and m_q are corresponding points because they are projections of the same 3D point M.

In this section, we consider geometric and algebraic constraints to solve the *correspondence problem* between X-ray images obtained as different projections of the test object. We will consider the correspondence in two (Sect. 3.5.1), three (Sect. 3.5.2), and more views (Sect. 3.5.3). In order to model the perspective projection in each view, we will use linear model (3.19) with no distortion:

$$\begin{cases} \lambda_p \mathbf{m}_p = \mathbf{P}_p \mathbf{M} \\ \lambda_q \mathbf{m}_q = \mathbf{P}_q \mathbf{M} \\ \lambda_r \mathbf{m}_r = \mathbf{P}_r \mathbf{M} \\ \quad \vdots \end{cases} \tag{3.45}$$

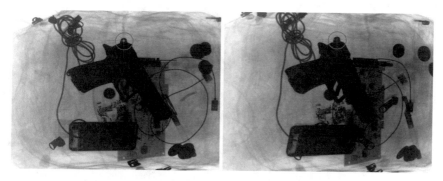

Fig. 3.20 Corresponding points in two different views

for different views $p, q, r \ldots$ In general, we assume that there are n views, and indices $p, q, r \cdots \in \{1 \ldots n\}$. It is worth noting that the coordinates of $\mathbf{M}$ are given in the same 3D coordinate system for each projection. That means, $\mathbf{M} = [X \; Y \; Z \; 1]^\mathsf{T}$ in each equation of (3.45).

The correspondence problem with non-linear projection models will be considered for two views only. The reader, however, will be able to establish correspondences with non-linear models in more views using the methodology of two views.

3.5.1   Correspondence Between Two Views

Now, the correspondence between two points m_p and m_q (in the X-ray projection coordinate system) is considered. The first point is obtained by projecting the object point M at position p, and the second one at position q:

$$\begin{cases} \lambda_p \mathbf{m}_p = \mathbf{P}_p \mathbf{M} \\ \lambda_q \mathbf{m}_q = \mathbf{P}_q \mathbf{M} \end{cases} \tag{3.46}$$

for $\mathbf{M} = [X \; Y \; Z \; 1]^\mathsf{T}$ given in the same 3D coordinate system for each equation. The correspondence problem in two views p and q can be stated as follows: given $\mathbf{m}_p, \mathbf{P}_p$, and $\mathbf{P}_q$, is it possible to find $\mathbf{m}_q$? Note that in this problem $\mathbf{M}$ is unknown. Moreover, if we know $\mathbf{m}_p$ and $\mathbf{P}_p$, it is impossible to find an exact location for M. In this section, we will explain a geometric and algebraic approach that can be used to solve the correspondence problem. In addition, the section gives practical considerations and the use of non-linear projection models.

Epipolar Geometry
We do not know where M is exactly, however, it is known that M lies on the line $\langle C_p, m_p \rangle$ as illustrated in Fig. 3.21. Since m_q is the projection of M onto view q, we

Fig. 3.21 Projection in two views

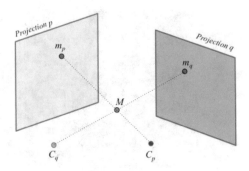

can affirm that m_q lies on line ℓ defined as the projection of $\langle C_p, m_p \rangle$ onto view q. Line ℓ is known as the *epipolar line* of m_p in view q.

Thus, to solve the correspondence problem in two views we use *epipolar geometry* [2, 13, 22]. The epipolar constraint is well known in stereo vision: for each projection point m_p at the position p, its corresponding projection point m_q at the position q lies on the epipolar line ℓ of m_p, as shown in Fig. 3.22, where C_p and C_q are the centers of projections p and q respectively. In this representation, a rotation and translation relative to the object coordinate system is assumed. The epipolar line ℓ can be defined as the projection of line $\langle C_p, m_p \rangle$ by the center of projection C_q onto projection plane q.

Epipolar line can be calculated in three simple steps (see Fig. 3.22):

(1) From m_p and $\mathbf{P}_p$, we find two 3D points M_{p1} and M_{p2} that lies on $\langle C_p, m_p \rangle$.

Obviously, one point that lies on $\langle C_p, m_p \rangle$ is C_p, i.e., $M_{p1} = C_p$. Since the location of C_p is unknown, we can find it by considering the following reasoning [13]: it is not possible to project C_p onto view p because C_p is the optical center of this projection. For this reason, if we use the first equation of (3.46) to project $\mathbf{C}_p$ (the homogeneous representation of C_p) we will obtain $\mathbf{P}_p \mathbf{C}_p$. Since this projected point is not defined it can be shown that its homogeneous representation is $[0\ 0\ 0]^{\mathsf{T}}$. It is not possible to estimate the non-homogeneous representation of this point because there is a division by zero. For this reason, $\mathbf{P}_p \mathbf{C}_p = [0\ 0\ 0]^{\mathsf{T}}$. Thus, $\mathbf{C}_p$ can be easily calculated as the null space of $\mathbf{P}_p$: For $\mathbf{A} = \mathbf{P}_p$ and $\mathbf{C}_p = [C_X\ C_Y\ C_Z\ 1]^{\mathsf{T}}$, $\mathbf{A}\mathbf{C}_p = \mathbf{0}$ can be reformulated as

$$\underbrace{\begin{bmatrix} a_{11} & a_{12} & a_{13} \\ a_{21} & a_{22} & a_{23} \\ a_{31} & a_{32} & a_{33} \end{bmatrix}}_{\mathbf{A}_1} \begin{bmatrix} C_X \\ C_Y \\ C_Z \end{bmatrix} = - \underbrace{\begin{bmatrix} a_{14} \\ a_{24} \\ a_{34} \end{bmatrix}}_{\mathbf{a}_4}. \tag{3.47}$$

Then the coordinates of C_p in 3D coordinate system are $[C_X\ C_Y\ C_Z]^{\mathsf{T}} = -\mathbf{A}_1^{-1}\mathbf{a}_4$.

Fig. 3.22 Estimation of
epipolar line ℓ in three steps

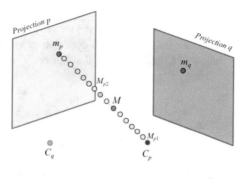

[STEP 1]

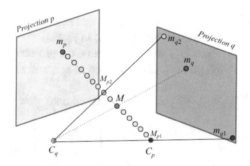

[STEP 2]

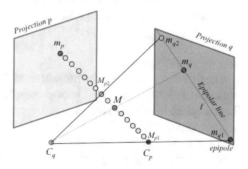

[STEP 3]

The second point should be m_p, however, we do not know the coordinates of m_p in 3D coordinate space, we only know $\mathbf{m}_p = [x_p \ y_p \ 1]^\mathsf{T}$, where (x_p, y_p) are coordinates of m_p in 2D coordinates system of view p. Nevertheless, it can be shown [13], that a point that lies on $\langle C_p, m_p \rangle$ is $\mathbf{M}^+$ calculated as

$$\mathbf{M}^+ = \mathbf{P}_p^+ \mathbf{m}_p, \tag{3.48}$$

where $\mathbf{P}_p^+$ is

$$\mathbf{P}_p^+ = \mathbf{P}_p^\mathsf{T}[\mathbf{P}_p \mathbf{P}_p^\mathsf{T}]^{-1}. \tag{3.49}$$

The 4×3 matrix $\mathbf{P}_p^+$ is known as the *pseudoinverse* of $\mathbf{P}_p$ because $\mathbf{P}_p \mathbf{P}_p^+ = \mathbf{I}_{3 \times 3}$. The reader can demonstrate that the projection of $\mathbf{M}^+$ onto view p is $\mathbf{m}_p$ by substituting (3.48) in the first equation of (3.46).

Thus, the two 3D points that lies on $\langle C_p, m_p \rangle$ are $\mathbf{M}_{p1} = \mathbf{C}_p$ (defined in (3.47)) and $\mathbf{M}_{p2} = \mathbf{M}^+$ (defined in (3.48)).

(2) From M_{p1} and M_{p2}, we find the projection of them onto view q using $\mathbf{P}_q$. These 2D points will be denoted as m_{q1} and m_{q2} respectively.

Both 3D points are projected onto view q using the second equation of (3.46):

$$\begin{cases} \lambda_{q1} \mathbf{m}_{q1} = \mathbf{P}_q \mathbf{M}_{p1} = \mathbf{P}_q \mathbf{C}_p \\ \lambda_{q2} \mathbf{m}_{q2} = \mathbf{P}_q \mathbf{M}_{p2} = \mathbf{P}_q \mathbf{P}_p^+ \mathbf{m}_p \end{cases}. \tag{3.50}$$

(3) From m_{q1} and m_{q2} we find line ℓ which contain them.

Since the epipolar line contains m_{q1} and m_{q2}, line ℓ can be computed in homogeneous coordinates using (3.2):

$$\ell = \mathbf{m}_{q1} \times \mathbf{m}_{q2}. \tag{3.51}$$

The first point $\mathbf{m}_{q1}$, i.e., the projection of C_p onto plane q, as defined in the first equation of (3.50), is the well known *epipole*.[9] The epipolar line is defined as the line that contains the epipole $\mathbf{m}_{q1}$ and the point $\mathbf{m}_{q2}$. We observe that the epipole belongs to any epipolar line obtained from any arbitrary point m_p. In other words, all epipolar lines share a common point: the epipole. Moreover, the epipole does not depend on m_p or m_q. It depends only on the two views geometry.

The projective representation of the epipolar line is obtained by taking the cross-product of these two points, i.e., $\ell = \mathbf{m}_{q1} \times \mathbf{m}_{q2}$. Line ℓ can be written using $[\mathbf{m}_{q1}]_\times$, the anti-symmetric matrix of $\mathbf{m}_{q1}$, where $\ell = [\mathbf{m}_{q1}]_\times \mathbf{m}_{q2}$. Matrix $[\mathbf{m}_{q1}]_\times$ is defined as the 3×3 matrix such that $[\mathbf{m}_{q1}]_\times \mathbf{s} = \mathbf{m}_{q1} \times \mathbf{s}$ for all vectors $\mathbf{s}$, i.e.,

---

[9]The word *epipole* comes from the Greek ἐπι (*epi*): over and πόλος (*polos*): attractor.

$$[\mathbf{m}_{q1}]_\times = \begin{bmatrix} 0 & +m_{q1}(3) & -m_{q1}(2) \\ -m_{q1}(3) & 0 & +m_{q1}(1) \\ +m_{q1}(2) & -m_{q1}(1) & 0, \end{bmatrix}.$$

where $\mathbf{m}_{q1} = [m_{q1}(1) \ m_{q1}(2) \ m_{q1}(3)]^\mathsf{T}$. Thus, using the anti-symmetric matrix, from (3.51) and (3.50), line ℓ can be computed by

$$\ell = \mathbf{F}_{pq}\mathbf{m}_p, \tag{3.52}$$

where $\mathbf{F}_{pq}$ is the *fundamental matrix* known from multiple view computer vision [3, 13] given by

$$\mathbf{F}_{pq} = [\mathbf{P}_q\mathbf{C}_p]_\times \mathbf{P}_q\mathbf{P}_p^+\mathbf{m}_p. \tag{3.53}$$

Since the point m_q belongs to the epipolar line ℓ, it follows that

$$\mathbf{m}_q^\mathsf{T}\ell = \mathbf{m}_q^\mathsf{T}\mathbf{F}_{pq}\mathbf{m}_p = 0. \tag{3.54}$$

Equation (3.54) is known as the *epipolar constraint*: If m_p and m_q are corresponding points, then m_q must lie on the epipolar line ℓ of m_p, i.e., $\mathbf{m}_q^\mathsf{T}\mathbf{F}_{pq}\mathbf{m}_p$ must be zero.

Python Example 3.7: This example shows epipolar lines in two views (p and q). We assume that the projection matrices $\mathbf{P}_p$ and $\mathbf{P}_q$ are known from a calibration process. The code computes the fundamental matrix. We select manually eight m_p points in view p. The code plots the epipolar lines of these points in view q.

Listing 3.7 : Epipolar lines for two views.

```python
import numpy as np
import matplotlib.pylab as plt

from pyxvis.geometry.epipolar import estimate_fundamental_matrix
from pyxvis.geometry.epipolar import plot_epipolar_line
from pyxvis.io import gdxraydb

image_set = gdxraydb.Baggages()

data = image_set.load_data(44, 'Pmatrices')  # Load projection matrices

p, q = (1, 82)  # indices p and q

Ip = image_set.load_image(44, p)
Iq = image_set.load_image(44, q)

Pp = data['P'][:, :, p]  # Projection matrix of view p
Pq = data['P'][:, :, q]  # Projection matrix of view q

F = estimate_fundamental_matrix(Pp, Pq, method='pseudo')

colors = 'bgrcmykw'  # Colors for each point-line pair

fig1, ax1 = plt.subplots(1, 1)
fig1.suptitle('Figure p')
ax1.imshow(Ip, cmap='gray')
ax1.axis('off')
```

```
fig2, ax2 = plt.subplots(1, 1, subplot_kw=dict(xlim=(0, Ip.shape[1]), ylim=(Iq.shape[0],
    1)))
fig2.suptitle('Figure q')
ax2.imshow(Iq, cmap='gray')
ax2.axis('off')
fig2.show()

for i in range(8):
    plt.figure(fig1.number)        # Focus on fig1 and get the mouse locations
    m = np.hstack([np.array(plt.ginput(1)), np.ones((1, 1))]).T  # Click
    ax1.plot(m[0, 0], m[1, 0], f'{colors[i]}*')  # Plot lines and plot on figures
    fig1.canvas.draw()
    ax2 = plot_epipolar_line(F, m, line_color=colors[i], ax=ax2)
    fig2.canvas.draw()

plt.show()
```

The output of this code is shown in Fig. 3.23. The code uses two functions of pyxvis Library: the first one is estimate_fundamental_matrix to compute the fundamental matrix and the second one is plot_epipolar_line to plot the epipolar lines onto view q. The example uses images $p = 1$ and $q = 82$ of series B0044 of GDXray+. In this set of images there are 178 different views (taken by rotating the test object around a quasi vertical axis in 2^0 between consecutive views). The reader that is interested in other views can change the code in order to define other values for p and q.   □

Bifocal Tensors

Another way to estimate the epipolar constraint is using *bifocal tensors* [11, 15], as explained next. This can be considered as an algebraic approach. From (3.46) the two projections can be expressed by

$$\begin{cases} \lambda_p \mathbf{m}_p = \mathbf{P}_p \mathbf{M} := \mathbf{A}\mathbf{M} \\ \lambda_q \mathbf{m}_q = \mathbf{P}_q \mathbf{M} := \mathbf{B}\mathbf{M} \end{cases} . \tag{3.55}$$

These two equations can also be written as

$$\underbrace{\begin{bmatrix} \mathbf{a}_1 & x_p & 0 \\ \mathbf{a}_2 & y_p & 0 \\ \mathbf{a}_3 & 1 & 0 \\ \mathbf{b}_1 & 0 & x_q \\ \mathbf{b}_2 & 0 & y_q \\ \mathbf{b}_3 & 0 & 1 \end{bmatrix}}_{\mathbf{G}} \underbrace{\begin{bmatrix} \mathbf{M} \\ -\lambda_p \\ -\lambda_q \end{bmatrix}}_{\mathbf{v}} = \begin{bmatrix} 0 \\ 0 \\ 0 \\ 0 \\ 0 \\ 0 \end{bmatrix}, \tag{3.56}$$

where $\mathbf{a}_i$ and $\mathbf{b}_i$ denote the ith row of matrices $\mathbf{A}$ and $\mathbf{B}$ respectively. If m_p and m_q are corresponding points, then the 3D point M exists. It follows that there must be a nontrivial solution for $\mathbf{v}$ in (3.56), i.e., the determinant of the 6×6 matrix $\mathbf{G}$ must be zero. Expanding the determinant of $\mathbf{G}$ we obtain

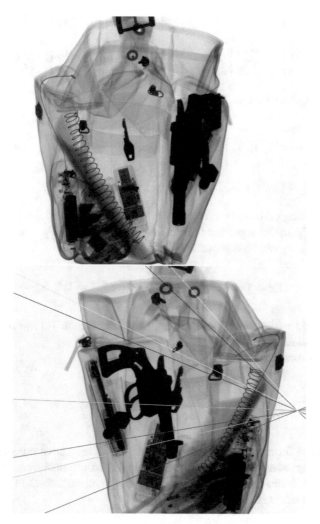

Fig. 3.23 Example of epipolar geometry: (Top) view p with eight points. (Bottom) view q with corresponding epipolar lines. It is clear that the corresponding points in view q lie on the corresponding epipolar lines. The intersection of the epipole lines defines the epipole. [$\rightarrow$ Example 3.7 🐍]

$$|\mathbf{G}| = x_p x_q \begin{vmatrix} \mathbf{a}_2 \\ \mathbf{a}_3 \\ \mathbf{b}_2 \\ \mathbf{b}_3 \end{vmatrix} - y_p x_q \begin{vmatrix} \mathbf{a}_1 \\ \mathbf{a}_3 \\ \mathbf{b}_2 \\ \mathbf{b}_3 \end{vmatrix} + x_q \begin{vmatrix} \mathbf{a}_1 \\ \mathbf{a}_2 \\ \mathbf{b}_2 \\ \mathbf{b}_3 \end{vmatrix} +$$

$$-x_p y_q \begin{vmatrix} \mathbf{a}_2 \\ \mathbf{a}_3 \\ \mathbf{b}_1 \\ \mathbf{b}_3 \end{vmatrix} + y_p y_q \begin{vmatrix} \mathbf{a}_1 \\ \mathbf{a}_3 \\ \mathbf{b}_1 \\ \mathbf{b}_3 \end{vmatrix} - y_q \begin{vmatrix} \mathbf{a}_1 \\ \mathbf{a}_2 \\ \mathbf{b}_1 \\ \mathbf{b}_3 \end{vmatrix} + x_p \begin{vmatrix} \mathbf{a}_2 \\ \mathbf{a}_3 \\ \mathbf{b}_1 \\ \mathbf{b}_2 \end{vmatrix} - y_p \begin{vmatrix} \mathbf{a}_1 \\ \mathbf{a}_3 \\ \mathbf{b}_1 \\ \mathbf{b}_2 \end{vmatrix} + \begin{vmatrix} \mathbf{a}_1 \\ \mathbf{a}_2 \\ \mathbf{b}_1 \\ \mathbf{b}_2 \end{vmatrix} = 0$$

that can be expressed by

$$|\mathbf{G}| = [x_q \ y_q \ 1] \underbrace{\begin{bmatrix} F_{11} & F_{12} & F_{13} \\ F_{21} & F_{22} & F_{23} \\ F_{31} & F_{32} & F_{33} \end{bmatrix}}_{\mathbf{F}_{pq}} \begin{bmatrix} x_p \\ y_p \\ 1 \end{bmatrix} = \mathbf{m}_q^{\mathsf{T}} \mathbf{F}_{pq} \mathbf{m}_p = 0, \qquad (3.57)$$

where $\mathbf{F}_{pq}$ corresponds to the mentioned fundamental matrix of Eq. (3.53) for $\mathbf{A} = \mathbf{P}_p$ and $\mathbf{B} = \mathbf{P}_q$. In this algebraic formulation, the elements of $\mathbf{F}_{pq}$ are called *bifocal tensors* [13]. They can be computed as

$$F_{ij} = (-1)^{i+j} \begin{vmatrix} \sim \mathbf{a}_j \\ \sim \mathbf{b}_i \end{vmatrix} \qquad \text{for } i, j = 1, 2, 3, \qquad (3.58)$$

where $\sim\mathbf{a}_j$ and $\sim\mathbf{b}_i$ mean respectively matrix $\mathbf{A}$ without the jth row and matrix $\mathbf{B}$ without the ith row.

Usually, we can express matrix $\mathbf{A}$ in a canonical form:

$$\mathbf{A} = \begin{bmatrix} 1 & 0 & 0 & 0 \\ 0 & 1 & 0 & 0 \\ 0 & 0 & 1 & 0 \end{bmatrix} = [\mathbf{I} \mid \mathbf{0}]. \qquad (3.59)$$

The canonical form can be achieved using a general projective transformation of the object coordinate system: $\mathbf{M}' = \mathbf{HM}$, where $\mathbf{M}'$ is the transformation of $\mathbf{M}$, and $\mathbf{H}$ is a 4×4 non-singular matrix obtained by adding one extra row to $\mathbf{P}_p$ [10]. Thus, Eq. (3.55) can be expressed as

$$\begin{cases} \lambda_p \mathbf{m}_p = [\mathbf{I} \mid \mathbf{0}]\mathbf{M}' = \mathbf{AM}' \\ \lambda_q \mathbf{m}_q = \qquad\qquad \mathbf{BM}' \end{cases} \qquad (3.60)$$

with

$$\begin{aligned} \mathbf{M}' &= \mathbf{HM} \\ \mathbf{A} &= \mathbf{P}_p \mathbf{H}^{-1}. \\ \mathbf{B} &= \mathbf{P}_q \mathbf{H}^{-1} \end{aligned}$$

For the canonical form $\mathbf{A} = [\mathbf{I} \mid \mathbf{0}]$, the bifocal tensors may be expressed by

$$F_{ij} = b_{i\oplus1,j} b_{i\oplus2,4} - b_{i\oplus2,j} b_{i\oplus1,4}, \qquad (3.61)$$

where

$$i \oplus k = \begin{cases} i + k & \text{if } i + k \leq 3 \\ i + k - 3 & \text{otherwise} \end{cases}.$$

Practical Considerations

In practice, the projection points m_p and m_q can be corresponding points, if the perpendicular Euclidean distance from the epipolar line ℓ of the point m_p to the point m_q is smaller than a small number ε [12]:

$$d_2 = \frac{|\mathbf{m}_q^\mathsf{T} \mathbf{F}_{pq} \mathbf{m}_p|}{\sqrt{\ell_1^2 + \ell_2^2}} < \varepsilon, \tag{3.62}$$

where $\ell = \mathbf{F}_{pq} \mathbf{m}_p = [\ell_1 \, \ell_2 \, \ell_3]^\mathsf{T}$.

An additional criterion to establish the correspondence between two views is that the 3D point reconstructed from the projection points m_p and m_q must belong to the space occupied by the test object [23]. From m_p and m_q the corresponding 3D point $\hat{M}$ can be estimated using 3D reconstruction techniques (see Sect. 3.6). It is necessary to examine if $\hat{M}$ resides in the volume of the test object, the dimensions of which are usually known a priori (e.g., , a wheel is assumed to be a cylinder). This criterion implies that the epipolar is delimited as illustrated in Fig. 3.24. It is possible to use a CAD-model of the test object to evaluate this criterion in a more precise way.

Non-linear Projections

An example of corresponding points using a non-linear model that considers geometric distortions is illustrated in Fig. 3.25. The idea is simple. The projection model has a linear part (that corresponds to the perspective projection) a non-linear part (that corresponds to the image intensifier) as illustrated in Fig. 3.6 as PCS and ICS. The epipolar geometry explained in the previous section is defined for the linear part only (for PCS and not for ICS). That means, the epipolar lines are straight lines in PCS, however, they are curves in ICS. In order to use the theory of the epipolar geometry we need the inverse transformation from both coordinate systems: ICS $\rightarrow$ PCS. Thus a point in ICS defined as $\mathbf{w}$ is transformed into a point in PCS as $\mathbf{m}$ by $\mathbf{m} = \mathbf{f}^{-1}(\mathbf{w})$, where $\mathbf{w} = \mathbf{f}(\mathbf{m})$ is the direct transformation: PCS $\rightarrow$ ICS in homogeneous coordinates. Using the inverse transformation of this non-linear model, we can use the epipolar constraint (3.54). Thus the epipolar curves are given by

$$[\mathbf{f}^{-1}(\mathbf{w}_q)]^\mathsf{T} \mathbf{F}_{pq} [\mathbf{f}^{-1}(\mathbf{w}_p)] = 0. \tag{3.63}$$

In the example of Fig. 3.25, the epipolar lines were computed using a hyperbolic model (see Sect. 3.3.3). Other non-linear model can be used as well.

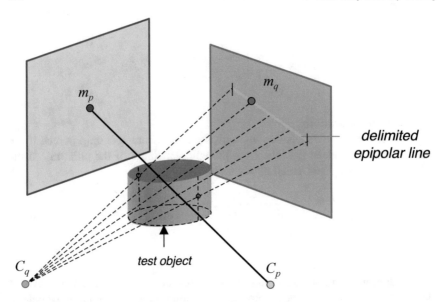

Fig. 3.24 Epipolar geometry in two views using 3D information of the test object

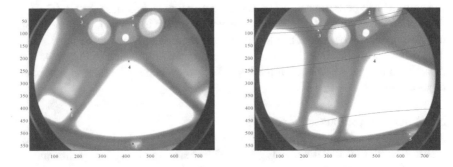

Fig. 3.25 Epipolar lines using a hyperbolic model [22]

It is clear that if the transformation between PCS → ICS is linear (for example using a flat panel), we have $\mathbf{w} = \mathbf{Hm}$. That means, we can substitute $\mathbf{m} = \mathbf{H}^{-1}\mathbf{w}$ in (3.57):

$$\mathbf{w}_q^\mathsf{T} \underbrace{[\mathbf{H}^{-\mathsf{T}}\mathbf{F}_{pq}\mathbf{H}^{-1}]}_{\mathbf{F}'_{pq}} \mathbf{w}_p = 0, \qquad (3.64)$$

where $\mathbf{F}'_{pq}$ is the fundamental matrix given in ICS.

Nevertheless, using a general non-linear model, as explained in Sect. 3.3.3, a point w, whose coordinates in ICS are (u, v), can be back-projected into a point m, whose coordinates in PCS are $(\bar{x}, \bar{y})$. The back-projection is carried out in two steps: transformation ICS → SCS and transformation SCS → PCS.

Transformation ICS → SCS: Without considering the electromagnetic distortion, the transformation ICS → SCS can be directly obtained from (3.37):

$$\lambda \mathbf{r} = \mathbf{H}^{-1}\mathbf{w}. \tag{3.65}$$

However, if the electromagnetic distortion is considered, the inverse function of (3.36) $\mathbf{r} = \mathbf{f}^{-1}(\mathbf{r}')$ must be obtained from (3.35):

$$y = y' - A_2 \sin(B_2 x' + C_2) \\ x = x' - A_1 \sin(B_1 y + C_1) \tag{3.66}$$

Therefore, it yields

$$\mathbf{r} = \mathbf{f}^{-1}(\mathbf{H}^{-1}\mathbf{w}). \tag{3.67}$$

Transformation SCS → PCS: The second transformation is non-linear because it takes into account the geometric distortion of the image intensifier. Given the coordinates (x, y) of point r in SCS, a point p on the surface S (see Fig. 3.11) that is the back-projection of r can be computed by finding the coordinates $(\bar{x}', \bar{y}', \bar{z}')$ that satisfy Eqs. (3.32) and (3.29) simultaneously. The solution is

$$\bar{x}' = \bar{x}_0 - \tfrac{x}{d}(f + c - \bar{z}')$$

$$\bar{y}' = \bar{y}_0 - \tfrac{y}{d}(f + c - \bar{z}'), \tag{3.68}$$

$$\bar{z}' = \frac{-B' + \sqrt{B'^2 - 4A'C'}}{2A'}$$

where

$$A' = \tfrac{g}{f^2} - 1, \quad B' = 2(f + c), \quad C' = -g - (f + c)^2, \quad g = \frac{d^2}{\frac{x^2}{a^2} + \frac{y^2}{b^2}},$$

or using matrix notation:

$$\mathbf{p} = \mathbf{h}(\mathbf{r}), \tag{3.69}$$

where $\mathbf{p} = [\bar{x}' \; \bar{y}' \; \bar{z}' \; 1]^\mathsf{T}$, $\mathbf{r}$ is a homogeneous representation of (x, y), and $\mathbf{h}$ is the non-linear function defined from (3.68).

Now, the coordinates of the back-projected point m on the projection plane can be calculated from (3.68) and the first two equations of (3.30):

$$\bar{x} = f\bar{x}'/\bar{z}' \quad \text{and} \quad \bar{y} = f\bar{y}'/\bar{z}'. \tag{3.70}$$

Equations (3.65), (3.69) and (3.70) can be joined in:

$$\lambda \mathbf{m} = \mathbf{E}\mathbf{h}(\mathbf{H}^{-1}\mathbf{w}), \tag{3.71}$$

where $\mathbf{E}$ is the 3×4 perspective projection matrix expressed in (3.18). However, if the electromagnetic distortion is taken into account, the homogeneous representation of m is from (3.67):

$$\lambda\mathbf{m} = \mathbf{Eh}(\mathbf{f}^{-1}(\mathbf{H}^{-1}\mathbf{w})). \tag{3.72}$$

3.5.2 Correspondence Between Three Views

In three views, we have the projection points m_p, m_q, and m_r at pth, qth, and rth positions respectively. The correspondence in three views can be established by calculating the epipolar lines of m_p and m_q in third view as shown in Fig. 3.26. If the intersection coincides with m_r, then the three points are corresponding. However, the intersection of epipolar lines in trifocal geometry is not well-defined when the epipolar lines are equal. This situation occurs in two cases: *(i)* when the 3D point M that has generated the points m_p, m_q, and m_r, lie in the plane defined by the three optical centers, and *(ii)* when the three optical centers are aligned [4].

In order to avoid these singularities, the relationships in three views are generally described using *trifocal tensors* [13]. Analogous to the two views case explained in Sect. 3.5.1, the three projection equations are

$$\begin{cases} \lambda_p\mathbf{m}_p = \mathbf{P}_p\mathbf{M} := \mathbf{AM} \\ \lambda_q\mathbf{m}_q = \mathbf{P}_q\mathbf{M} := \mathbf{BM} \\ \lambda_r\mathbf{m}_r = \mathbf{P}_r\mathbf{M} := \mathbf{CM} \end{cases}. \tag{3.73}$$

They can be written according to (3.56) by

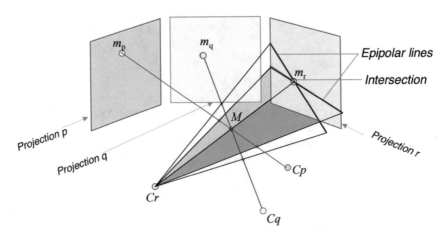

Fig. 3.26 Epipolar geometry in three views

$$\underbrace{\begin{bmatrix} \mathbf{a}_1 & x_p & 0 & 0 \\ \mathbf{a}_2 & y_p & 0 & 0 \\ \mathbf{a}_3 & 1 & 0 & 0 \\ \mathbf{b}_1 & 0 & x_q & 0 \\ \mathbf{b}_2 & 0 & y_q & 0 \\ \mathbf{b}_3 & 0 & 1 & 0 \\ \mathbf{c}_1 & 0 & 0 & x_r \\ \mathbf{c}_2 & 0 & 0 & y_r \\ \mathbf{c}_3 & 0 & 0 & 1 \end{bmatrix}}_{\mathbf{G}} \begin{bmatrix} \mathbf{M} \\ -\lambda_p \\ -\lambda_q \\ -\lambda_r \\ \mathbf{v} \end{bmatrix} = \begin{bmatrix} 0 \\ 0 \\ 0 \\ 0 \\ 0 \\ 0 \\ 0 \\ 0 \\ 0 \end{bmatrix}, \tag{3.74}$$

where $\mathbf{a}_i$, $\mathbf{b}_i$ and $\mathbf{c}_i$ denotes the ith row of matrix $\mathbf{A}$, $\mathbf{B}$ and $\mathbf{C}$ respectively. If m_p, m_q and m_r are corresponding points, then there must be a nontrivial solution for $\mathbf{v}$. It follows that the rank of the 9×7 matrix $\mathbf{G}$ must be at most 6. In other words, all 7×7 submatrices have vanishing determinants. The minors of $\mathbf{G}$ can be written using Laplace expansions as sums of products of determinants of four rows taken from the first four columns of $\mathbf{G}$ and products of image coordinates [16]. By expanding the determinants, we can find four linearly independent relationships:

$$\begin{cases} D_1 = (x_r \mathbf{T}^{13} - x_r x_q \mathbf{T}^{33} + x_q \mathbf{T}^{31} - \mathbf{T}^{11})m_p = 0 \\ D_2 = (y_r \mathbf{T}^{13} - y_r x_q \mathbf{T}^{33} + x_q \mathbf{T}^{32} - \mathbf{T}^{12})m_p = 0 \\ D_3 = (x_r \mathbf{T}^{23} - x_r y_q \mathbf{T}^{33} + y_q \mathbf{T}^{31} - \mathbf{T}^{21})m_p = 0 \\ D_4 = (y_r \mathbf{T}^{23} - y_r y_q \mathbf{T}^{33} + y_q \mathbf{T}^{32} - \mathbf{T}^{22})m_p = 0 \end{cases}, \tag{3.75}$$

where

$$\mathbf{T}^{jk} = [T_1^{jk} \; T_2^{jk} \; T_3^{jk}],$$

and

$$T_i^{jk} = (-1)^{i+1} \begin{vmatrix} \sim \mathbf{a}_i \\ \mathbf{b}_j \\ \mathbf{c}_k \end{vmatrix} = \qquad \text{for } i, j, k = 1, 2, 3, \tag{3.76}$$

where $\sim \mathbf{a}_i$ means the matrix $\mathbf{A}$ without row i. The elements T_i^{jk} are called the *trifocal tensors* for the images p, q and r [10, 11]. For the canonical form $\mathbf{A} = [\mathbf{I} \mid \mathbf{0}]$, the trifocal tensors may be easily obtained by

$$T_i^{jk} = b_{ji} c_{k4} - b_{j4} c_{ki} \qquad \text{for } i, j, k = 1, 2, 3. \tag{3.77}$$

The equations denoted by (3.75) above are known as the *trilinearities* [24]. They establish a linear relationship between the coordinates of points m_p, m_q y m_r to find the correspondence. If the three points satisfy the four trilinearities, then they are corresponding points. Equation (3.76) implies that the trifocal tensors do not depend on the points of the images, rather they are computed from the three projection matrices.

The *reprojection* of m_r, i.e., the coordinates $\hat{x}_r$ and $\hat{y}_r$ obtained from the points m_p and m_q, may be simply estimated from the trilinearities (3.75):

$$\lambda\hat{\mathbf{m}}_r = (\mathbf{T}^1 - x_q\mathbf{T}^3)\mathbf{m}_p = (\mathbf{T}^2 - y_q\mathbf{T}^3)\mathbf{m}_p, \tag{3.78}$$

where λ is a scale factor, $\hat{\mathbf{m}}_r = [\hat{x}_r \ \hat{y}_r \ 1]^\mathsf{T}$, and $\mathbf{T}^j$ is a 3×3 matrix with the (k, i)-element equal to T_i^{jk}.

In practice, given two corresponding points $\mathbf{m}_p$ and $\mathbf{m}_q$, the third one $\mathbf{m}_r$ can be considered as the corresponding point in third view, if the Euclidean distance between $\mathbf{m}_r$ and its reprojection $\hat{\mathbf{m}}_r$ is smaller than a small number ε:

$$d_3 = \|\hat{\mathbf{m}}_r - \mathbf{m}_r\| < \varepsilon. \tag{3.79}$$

Trifocal geometry is performed for the corresponding points in the X-ray projection coordinate system. In case that the X-ray computer vision system is modeled using a non-linear geometric model due to an image intensifier, as explained in Sect. 3.5.1, a point w_p, that is found in the pth image, is first transformed into the coordinates m_p of the X-ray projection coordinate system.

Python Example 3.8: This example shows how to estimate the coordinates of a corresponding point in view r if we know the trifocal tensors and corresponding points m_p and m_q in views p and q respectively. Epipolar lines in two views (p and q). This code computes the trifocal tensors from the projection matrices $\mathbf{P}_p$, $\mathbf{P}_q$ and $\mathbf{P}_r$. The reprojection is computed using (3.78).

Listing 3.8 : Reprojection of third point using trifocal tensors.

```python
import matplotlib.pylab as plt
import numpy as np

from pyxvis.geometry.epipolar import estimate_trifocal_tensor
from pyxvis.geometry.epipolar import reproject_trifocal
from pyxvis.io import gdxraydb

image_set = gdxraydb.Baggages()
data = image_set.load_data(44, 'Pmatrices')  # Load projection matrices

p, q, r = (1, 90, 170)  # Indices for p, q, and r

# Load projection matrices for views p, q, r
Pp = data['P'][:, :, p]
Pq = data['P'][:, :, q]
Pr = data['P'][:, :, r]

Ip = image_set.load_image(44, p)
Iq = image_set.load_image(44, q)
Ir = image_set.load_image(44, r)

T = estimate_trifocal_tensor(Pp, Pq, Pr)

# Plot lines and plot on figures
print('Click a point in Figure 1 ...')
fig1, ax1 = plt.subplots(1, 1, subplot_kw=dict(title='Figure p'))
ax1.imshow(Ip, cmap='gray')
```

```
ax1.axis('off')
mp = np.hstack([np.array(plt.ginput(1)), np.ones((1, 1))]).T  # Click
ax1.plot(mp[0], mp[1], 'r*')
fig1.canvas.draw()

print('Click a point in Figure 2 ...')
fig2, ax2 = plt.subplots(1, 1, subplot_kw=dict(title='Figure q'))
ax2.imshow(Iq, cmap='gray')
ax2.axis('off')
mq = np.hstack([np.array(plt.ginput(1)), np.ones((1, 1))]).T  # Click
ax2.plot(mq[0], mq[1], 'r*')
fig2.canvas.draw()

mr = reproject_trifocal(mp, mq, T)  # reprojection of mr from mp, mq and T

fig3, ax3 = plt.subplots(1, 1, subplot_kw=dict(title='Figure r'))
ax3.imshow(Ir, cmap='gray')
ax3.axis('off')
ax3.plot(mr[0, 0], mr[1, 0], 'r*')
fig3.canvas.draw()

plt.show()
```

The output of this code is shown in Fig. 3.27. The code uses two functions of pyxvis Library: estimate_trifocal_tensor to compute the trifocal tensors and reproject_trifocal to compute the reprojection of m_r. The example uses images $p = 1$, $q = 90$, and $r = 170$ of series B0044 of GDXray+. In this set of images there are 178 different views (taken by rotating the test object around a quasi vertical axis in 2^0 between consecutive views). The reader that is interested in other views can change the code in order to define other values for p, q, and r.   □

3.5.3   Correspondence Between Four Views or More

In the four views case we have the projection points m_p, m_q, m_r, and m_s at pth, qth, rth, and sth positions respectively. Similar to the previous sections we can write the four projection equations as a linear equation $\mathbf{Gv} = \mathbf{0}$. Once more, the existence of a nontrivial solution for $\mathbf{v}$ yields in this case to the condition that all 8×8 minors of $\mathbf{G}$ must be zero. Thus, we obtain the well known 81 *quadrifocal tensors* and the corresponding 16 *quadrilinearities* [11, 16].

In practice, the quadrilinearities are not used because they are redundant. Corresponding constraints in four views are obtained from the trilinearities. Thus, the points m_p, m_q, m_r, and m_s are corresponding if m_p, m_q, and m_r are corresponding, and m_q, m_r, and m_s are corresponding as well [20].

For more than four views, a similar approach can be used.

Fig. 3.27 Example of trifocal geometry: views p, q, and r with corresponding points. In this example the corresponding points in view p and q are known. Corresponding point m_r is estimated from m_p, m_q and the trifocal tensors **T** of these views. [$\rightarrow$ Example 3.8]

3.6 Three-Dimensional Reconstruction

In X-ray testing, three-dimensional reconstruction is usually related to computed tomography (CT). However, in Computer Vision the attempt is made to estimate only the location (and not the X-ray absorption coefficient) of 3D points in space. In this sense, the reconstruction is based on photogrammetric rather than tomographic methods.

The 3D reconstruction problem can be stated as follows. Given n corresponding points $\mathbf{m}_p$ for $p = 1 \ldots n$ with $n \geq 2$, and the projection matrices of each view $\mathbf{P}_p$, find the 'best' 3D point $\mathbf{M}$ which projected by $\mathbf{P}_p$ gives approximately $\mathbf{m}_p$. If we have the projection equation $\lambda_p \mathbf{m}_p = \mathbf{P}_p \mathbf{M}$, we can solve the following system of equation for $\mathbf{M}$:

$$\begin{cases} \lambda_1 \mathbf{m}_1 = \mathbf{P}_1 \mathbf{M} \\ \quad\vdots \\ \lambda_n \mathbf{m}_n = \mathbf{P}_n \mathbf{M} \end{cases} . \tag{3.80}$$

In this section, two approaches that perform the 3D reconstruction, in sense of locating in 3D space, will be described. The 3D reconstruction will be undertaken from corresponding points in the X-ray projection coordinate system. As explained in Sect. 3.5.1, a point w_p, that is found in the pth image, is first transformed into the coordinates m_p of the X-ray projection coordinate system.

3.6.1 Linear 3D Reconstruction from Two Views

Now, we estimate now the 3D point M from two corresponding points m_p and m_q using the linear approach introduced by Hartley [10]. Without loss of generality, the method employs the canonical form (see Eq. (3.60)) for the first projection:

$$\lambda_p \mathbf{m}_p = [\mathbf{I} \mid \mathbf{0}]\mathbf{M}'.$$

Thus, the transformed 3D point can be expressed by

$$\mathbf{M}' = \lambda_p [\mathbf{m}_p^\mathsf{T} \ 1/\lambda_p]^\mathsf{T}. \tag{3.81}$$

The second projection of this point yields

$$\lambda_q \mathbf{m}_q = \mathbf{B}\mathbf{M}' = \mathbf{B}\lambda_p [\mathbf{m}_p^\mathsf{T} \ 1/\lambda_p]^\mathsf{T}, \tag{3.82}$$

that is an equation system with three linear equations in the unknowns λ_p and λ_q. If m_p and m_q are corresponding points one may consider only two of these three equations. Taking for example the first two equations one may compute λ_p. Substituting the value of λ_p into (3.81) and after some simplifications we obtain

$$\mathbf{M} = \mathbf{H}^{-1}\mathbf{M}' = \alpha \mathbf{H}^{-1} \begin{bmatrix} (y_q b_{14} - x_q b_{24})\mathbf{m}_p \\ (x_q \mathbf{b}_2 - y_q \mathbf{b}_1)\mathbf{m}_p \end{bmatrix}, \tag{3.83}$$

where α is a scale factor.

3.6.2 3D Reconstruction from Two or More Views

We assume that we have $n \geq 2$ projections at n different positions. In these projections we have found the corresponding points m_p, $p = 1, \ldots, n$, with coordinates (x_p, y_p). To reconstruct the corresponding 3D point M that has produced these projection points, we use a least squares technique [2].

Each projection yields the equation $\lambda_p m_p = \mathbf{P}_p \mathbf{M}$ as shown in (3.80), with three linear equations in the unknowns (X, Y, Z) and λ_p:

$$
\begin{bmatrix}
\lambda_1 x_1 \\
\lambda_1 y_1 \\
\lambda_1 \\
\vdots \\
\lambda_n x_n \\
\lambda_n y_n \\
\lambda_n
\end{bmatrix}
=
\begin{bmatrix}
s_{11}^1 & s_{12}^1 & s_{13}^1 & s_{14}^1 \\
s_{21}^1 & s_{22}^1 & s_{23}^1 & s_{24}^1 \\
s_{31}^1 & s_{32}^1 & s_{33}^1 & s_{34}^1 \\
\vdots & \vdots & \vdots & \vdots \\
s_{11}^n & s_{12}^n & s_{13}^n & s_{14}^n \\
s_{21}^n & s_{22}^n & s_{23}^n & s_{24}^n \\
s_{31}^n & s_{32}^n & s_{33}^n & s_{34}^n
\end{bmatrix}
\begin{bmatrix}
X \\
Y \\
Z \\
1
\end{bmatrix},
\tag{3.84}
$$

where s_{ij}^p denotes the (i, j)-element of $\mathbf{P}_p$. With $\lambda_p = s_{31}^p X + s_{32}^p Y + s_{33}^p Z + s_{34}^p$ and after some slight rearranging we obtain

$$
\underbrace{\begin{bmatrix}
s_{31}^1 x_1 - s_{11}^1 & s_{32}^1 x_1 - s_{12}^1 & s_{33}^1 x_1 - s_{13}^1 \\
s_{31}^1 y_1 - s_{21}^1 & s_{32}^1 y_1 - s_{22}^1 & s_{33}^1 y_1 - s_{23}^1 \\
\vdots & \vdots & \vdots \\
s_{31}^n x_n - s_{11}^n & s_{32}^n x_n - s_{12}^n & s_{33}^n x_n - s_{13}^n \\
s_{31}^n y_n - s_{21}^n & s_{32}^n y_n - s_{22}^n & s_{33}^n y_n - s_{23}^n
\end{bmatrix}}_{\mathbf{Q}}
\begin{bmatrix}
X \\
Y \\
Z
\end{bmatrix}
=
\underbrace{\begin{bmatrix}
s_{14}^1 - s_{34}^1 x_1 \\
s_{24}^1 - s_{34}^1 y_1 \\
\vdots \\
s_{14}^n - s_{34}^n x_n \\
s_{24}^n - s_{34}^n y_n
\end{bmatrix}}_{\mathbf{r}}.
\tag{3.85}
$$

If $\mathrm{rank}(\mathbf{Q}) = 3$, the least squares solution for $\hat{\mathbf{M}} = [\hat{X}\ \hat{Y}\ \hat{Z}]^\mathsf{T}$ is then given by

$$
\hat{\mathbf{M}} = [\mathbf{Q}^\mathsf{T}\mathbf{Q}]^{-1}\mathbf{Q}^\mathsf{T}\mathbf{r}.
\tag{3.86}
$$

Python Example 3.9: In this example we estimate the length of an object in millimeters. There are three views p, q and r. Two points of the object are given by the user in each view (by mouse clicking). The code estimates the two 3D points and compute the 3D distance between them. Since the calibration of this X-ray computer vision system was implemented using a calibration object with dimensions measured in millimeters, it is clear that that the 3D reconstructed points are given in millimeters as well.

Listing 3.9 : 3D Reconstruction.

```
import numpy as np
import matplotlib.pylab as plt
```

```
from pyxvis.geometry.epipolar import recon_3dn
from pyxvis.io import gdxraydb

image_set = gdxraydb.Baggages()
data = image_set.load_data(44, 'Pmatrices')  # Load projection matrices

p, q, r = (1, 40, 90)  # indices for p, q and r

# Load projection matrices for views p, q, r
P1 = data['P'][:, :, p]  # Reprojection matrix of view p
P2 = data['P'][:, :, q]  # Reprojection matrix of view q
P3 = data['P'][:, :, r]  # Reprojection matrix of view r
P = np.vstack([P1, P2, P3])  # Join all projection matrices

Ip = image_set.load_image(44, p)
Iq = image_set.load_image(44, q)
Ir = image_set.load_image(44, r)

# Plot lines and plot on figures
fig1, ax1 = plt.subplots(1, 1, subplot_kw=dict(title='Figure p'))
ax1.imshow(Ip, cmap='gray')
ax1.axis('off')
print('Click first and second points in Figure 1 ...')
mp = np.hstack([np.array(plt.ginput(2)), np.ones((2, 1))]).T  # Click
ax1.plot(mp[0, :], mp[1, :], 'ro')
ax1.plot(mp[0, :], mp[1, :], 'g', linewidth=1.0)
fig1.canvas.draw()

fig2, ax2 = plt.subplots(1, 1, subplot_kw=dict(title='Figure q'))
ax2.imshow(Iq, cmap='gray')
ax2.axis('off')
print('Click first and second points in Figure 2 ...')
mq = np.hstack([np.array(plt.ginput(2)), np.ones((2, 1))]).T  # Click
ax2.plot(mq[0, :], mq[1, :], 'ro')
ax2.plot(mq[0, :], mq[1, :], 'g', linewidth=1.0)
fig2.canvas.draw()

fig3, ax3 = plt.subplots(1, 1, subplot_kw=dict(title='Figure r'))
ax3.imshow(Ir, cmap='gray')
ax3.axis('off'),
print('Click first and second points in Figure 3 ...')
mr = np.hstack([np.array(plt.ginput(2)), np.ones((2, 1))]).T  # Click
ax3.plot(mr[0, :], mr[1, :], 'ro')
ax3.plot(mr[0, :], mr[1, :], 'g', linewidth=1.0)
fig3.canvas.draw()

# 3D reprojection
mm_1 = np.vstack([mp[:, 0], mq[:, 0], mr[:, 0]]).T  # First 2D point in each view
mm_2 = np.vstack([mp[:, 1], mq[:, 1], mr[:, 1]]).T  # Second 2D point in each view
M1, d1, err1 = recon_3dn(mm_1, P)  # 3D reconstruction of first point
M2, d2, err2 = recon_3dn(mm_2, P)  # 3D reconstruction of second point

Md = M1.ravel()[:-1] - M2.ravel()[:-1]  # 3D vector from 1stt to 2nd point
dist = np.linalg.norm(Md)  # length of 3D vector in mm

print(f'Object size: {dist:0.3} mm')

plt.show()
```

The output of this code is shown in Fig. 3.28. The code uses the function of pyxvis Library: reco_3dn to compute the 3D reconstruction using (3.86). The estimated length in this example was dist = 46.2881mm. The reader that is interested in 3D reconstruction using (3.83) for two views can use command reco_3d2 from pyxvis Library.   □

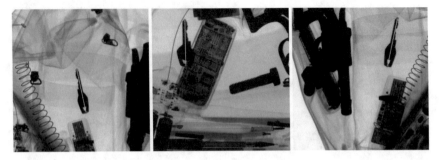

Fig. 3.28 Example of 3D reconstruction using three views. There are two corresponding points in each view (in this example the coordinates of the top and the bottom of the key were manually given). Two 3D points were reconstructed using (3.86) and the distance between these two 3D points was computed. In this example, the estimation of the length of the key was 46.29mm. [→ Example 3.9 🐍]

3.7   Summary

In this chapter we presented several methods that can be used when dealing with geometric problems in X-ray testing. We gave a theoretical background of geometry using homogenous coordinates. Thus, the projective transformations can be easily established. Linear and non-linear models for X-ray computer vision systems were outlined, in order to relate the 3D coordinates of an object to the 2D coordinates of the digital X-ray image pixel. In addition, calibration approaches that can be used to estimate the parameters of these models were studied. Finally, multiple view geometry was outlined. We presented geometric and algebraic constraints between two, three, and more X-ray images obtained as different projections of the object, and we explained the problem of the 3D reconstruction.

References

1. Brack, C., Götte, H., Gossé, F., Moctezuma, J., Roth, M., Schweikard, A.: Towards accurate X-ray-camera calibration in computer-assisted robotic surgery. In: Proceedings of the International Symposium on Computer Assisted Radiology (CAR), pp. 721–728. Paris (1996)
2. Faugeras, O.: Three-Dimensional Computer Vision: A Geometric Viewpoint. The MIT Press, Cambridge (1993)
3. Faugeras, O., Luong, Q.T., Papadopoulo, T.: The Geometry of Multiple Images: The Laws that Govern the Formation of Multiple Images of a Scene and Some of Their Applications. The MIT Press, Cambridge (2001)
4. Faugeras, O., Papadopulo, T.: A nonlinear method for estimating the projective geometry of 3 views. In: 6th International Conference on Computer Vision (ICCV-98), pp. 477–484. Bombay, India (1998)
5. Faugeras, O., Toscani, G.: The calibration problem for stereo. In: Proceedings IEEE Computer Vision and Pattern Recognition, pp. 15–20 (1986)

6. Felix, R., Ramm, B.: Das Röntgenbild, 3rd edn. Georg Thieme Verlag, Stuttgart, New York (1988)
7. Franzel, T., Schmidt, U., Roth, S.: Object detection in multi-view X-Ray images. Pattern Recogn. 144–154 (2012)
8. Grignon, B., Mainard, L., Delion, M., Hodez, C., Oldrini, G.: Recent advances in medical imaging: anatomical and clinical applications. Surgical Radiolog. Anatomy **34**(8), 675–686 (2012)
9. Halmshaw, R.: Non-Destructive-Testing, 2nd edn. Edward Arnold, London (1991)
10. Hartley, R.: A linear method for reconstruction from lines and points. In: 5th International Conference on Computer Vision (ICCV-95), pp. 882–887. Cambridge, MA (1995)
11. Hartley, R.: Multilinear relationships between coordinates of corresponding image points and lines. In: Proceedings of the International Workshop on Computer Vision and Applied Geometry. International Sophus Lie Center, Nordfjordeid, Norway (1995)
12. Hartley, R.: Lines and points in three views and the trifocal tensor. Int. J. Comput. Vis. **22**(2), 125–150 (1997)
13. Hartley, R.I., Zisserman, A.: Multiple view geometry in computer vision, 2nd edn. Cambridge University Press, Cambridge (2003)
14. Heikkilä, J.: Geometric camera calibration using circular control points. IEEE Trans. Patt. Anal. Mach. Intell **22**(10), 1066–1077 (2000)
15. Heyden, A.: A common framework for multiple view tensors. In: 5th European Conference on Computer Vision (ECCV-98), pp. 3–19 (1998)
16. Heyden, A.: Multiple view geometry using multifocal tensors. In: DSAGM. Köpenhamn (1999)
17. Jaeger, T.: Methods for rectification of geometric distortion in radioscopic images. Master theses, Institute for Measurement and Automation, Faculty of Electrical Engineering, Technical University of Berlin (1990). (in German)
18. Luong, Q.T., Faugeras, O.: Self calibration of a moving camera from point correspondences and fundamental matrices. Int. J. Comput. Vis. **22**(3), 261–289 (1997)
19. MathWorks: Optimization Toolbox for Use with MATLAB: User's Guide. The MathWorks Inc. (2014)
20. Mery, D.: Automated Flaw Detection in Castings from Digital Radioscopic Image Sequences. Verlag Dr. Köster, Berlin (2001). (Ph.D. Thesis in German)
21. Mery, D.: Explicit geometric model of a radioscopic imaging system. NDT & E Int. **36**(8), 587–599 (2003)
22. Mery, D., Filbert, D.: The epipolar geometry in the radioscopy: theory and application. at - Automatisierungstechnik **48**(12), 588–596 (2000). (in German)
23. Mery, D., Filbert, D.: Automated flaw detection in aluminum castings based on the tracking of potential defects in a radioscopic image sequence. IEEE Trans. Robot. Autom. **18**(6), 890–901 (2002)
24. Shashua, A., Werman, M.: Trilinearity of three perspective views and its associated tensor. In: 5th International Conference on Computer Vision (ICCV-95). Boston (1995)
25. Swaminathan, R., Nayar, S.: Nonmetric calibration of wide-angle lenses and polycameras. IEEE Trans. Pattern Anal. Mach. Intell. **22**(10), 1172–1178 (2000)
26. Tsai, R.: A versatile camera calibration technique for high-accuracy 3D machine vision metrology using off-the-shelf TV cameras and lenses. IEEE Trans. Robot. Autom. **RA-3**(4), 323–344 (1987)
27. Wei, G.Q., Ma, S.: Implicit and explicit camera calibration: theory and experiments. IEEE Trans. Pattern Anal. Mach. Intell. **16**(5), 469–480 (1994)
28. Weng, J., Cohen, P., Herniou, M.: Camera calibration with distorsion models and accuracy evaluation. IEEE Trans. Pattern Anal. Mach. Intell. **4**(10), 965–980 (1992)
29. Zhang, Z.: A flexible new technique for camera calibration. IEEE Trans. Pattern Anal. Mach. Intell. **22**(11), 1330–1334 (2000)

Chapter 4
X-Ray Image Processing

Abstract In this chapter, we cover the main techniques of image processing used in X-ray testing. They are (i) image processing to enhance details, (ii) image filtering to remove noise or detect high-frequency details, (iii) edge detection to identify the boundaries of the objects, (iv) image segmentation to isolate the regions of interest, and (v) to remove the blurriness of the X-ray image. The chapter provides an overview and presents several methodologies with examples using real and simulated X-ray images.

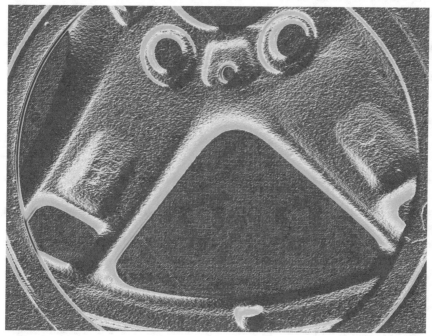

Gradient of an X-ray image of a wheel (from X-ray image `C0001_0001` *colored with 'jet' colormap).*

© Springer Nature Switzerland AG 2021
D. Mery and C. Pieringer, *Computer Vision for X-Ray Testing*,
https://doi.org/10.1007/978-3-030-56769-9_4

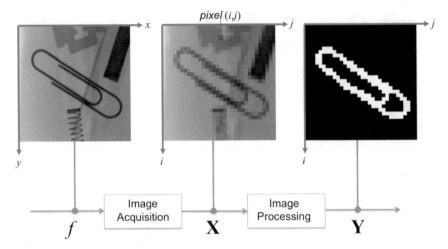

Fig. 4.1 Image processing: input is digital image **X**, output is digital image **Y**

4.1 Introduction

Image processing manipulates a digital image in order to obtain a *new* digital image, i.e., in this process the input is an image and the output is another image. A typical example is *segmentation* as shown in Fig. 4.1, where the input is a grayscale image that contains a clip and the output is a binary image where the pixels that belong to the clip are detected. In our book, we distinguish image processing from image analysis, in which the output is rather an interpretation, a recognition or a measurement of the input image. We will perform image analysis further on, when we learn pattern recognition techniques such as feature extraction (see Sect. 5) and classification (see Sect. 6).

In this chapter, we cover the following image processing techniques that are used in X-ray testing.

- Image preprocessing: The quality of the X-ray image is improved in order to enhance its details.
- Image Filtering: Mainly used to remove noise and detect high-frequency details of the X-ray image.
- Edge detection: The details of the images can be highlighted by detecting the boundaries of the objects of the X-ray image.
- Image segmentation: Regions of interest of the X-ray image are identified and isolated from their surroundings.
- Image restoration: This involves recovering details in blurred images.

In this chapter, we provide an overview of these five techniques. Methodologies and principles will also be outlined, and some application examples followed by limitations to the applicability of the used methodologies will be presented.

In image processing methodology, we have a continuous image f defined in a coordinate system (x, y). Image f is digitalized. The obtained image is a digital image which is stored in matrix $\mathbf{X}$ of size $M \times N$ pixels. The gray value of pixel (i, j) of image $\mathbf{X}$ is $X(i, j)$. Image $\mathbf{X}$ is processed digitally. The output image of this process is image $\mathbf{Y}$, usually a matrix of the same size of $\mathbf{X}$. In this example, the output is a binary image, that means $Y(i, j)$ is '1' (white) and '0' black. Image $\mathbf{Y}$ corresponds to the segmentation of a clip (that is the *object of interest* in this example).

4.2 Image Preprocessing

The X-ray image taken must be preprocessed to improve the quality of the image before it is analyzed. In this section, we will discuss preprocessing techniques that can remove noise, enhance contrast, correct the shading effect, and restore blur deformation in X-ray images.

4.2.1 Noise Removal

Noise in an X-ray image can prove a significant source of image degradation and must be taken into account during image processing and analysis. In an X-ray imaging system, *photon noise* occurs given the quantum nature of X-rays. If we have a system that receives μ photons per pixel in a time ΔT on average, the number of photons striking any particular pixel in any time ΔT will be random. At low levels, however, the noise follows a Poisson law, characterized by the probability:

$$p(x|\mu) = \frac{e^{-\mu}}{\mu^x x!} \tag{4.1}$$

to obtain a value x of photons given its average μ photons in a time ΔT. The standard deviation of this distribution is equal to the square root of the mean.[1] This means that the photon noise amplitude is signal-dependent.

Integration (or averaging) is used to remove X-ray image noise. This technique requires n stationary X-ray images. It computes the filtered image as follows:

$$Y(i, j) = \frac{1}{n} \sum_{k=1}^{n} X_k(i, j), \tag{4.2}$$

[1] At high levels, the Poisson distribution approaches the Gaussian with a standard deviation equal to the square root of the mean: $\sigma = \sqrt{\mu}$.

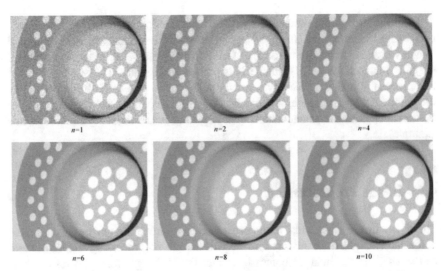

Fig. 4.2 Noise removal after an averaging of n frames. The noise is reduced by factor $\sqrt{n}$

where $X_k(i, j)$ is pixel (i, j) of k-th stationary image and $Y(i, j)$ is the corresponding pixel of the filtered image.

In this technique, the X-ray image noise is modeled using two components: the stationary component (that is constant throughout the n images) and the noise component (that varies from one image to the next). If the noise component has zero mean, by averaging the n images the stationary component is unchanged, while the noise pattern decreases by increasing n. Integrating n stationary X-ray images improves the signal-to-noise ratio by a factor of $\sqrt{n}$ [1, 3].

The effect of image integration is illustrated in Fig. 4.2 that uses n stationary images of an aluminum casting and shows the improvement in the quality of the X-ray image. The larger the number of stationary images n, the better the improvement. Normally, between 10 and 16 stationary images are taken ($10 \leq n \leq 16$).

Python Example 4.1: In this example, we have 20 noisy X-ray images obtained from a very thin wood piece. The following Python code uses averaging to effectively remove X-ray image noise (4.2):

Listing 4.1 : Noise removal by averaging.

```python
import numpy as np
import matplotlib.pylab as plt

from pyxvis.io import gdxraydb

image_set = gdxraydb.Nature()
s = np.double(image_set.load_image(4, 1))

n = 20
for i in range(2, n+1):  # For loops in Python runs until n—1
    xk = np.double(image_set.load_image(4, i))
```

```
        s += xk

    y = s / n

    fig1, ax = plt.subplots(1, 2, figsize=(16, 8))
    ax[0].imshow(s, cmap='gray'), ax[0].axis('off')
    ax[1].imshow(y, cmap='gray'), ax[1].axis('off')
    plt.show()
```

The output of this code is shown in Fig. 4.3. The reduction of noise is not perfect but very satisfactory. The reader can test this approach on series C0034 and C0041 of GDXray+, in which 37 noisy X-ray images of an aluminum wheel with no motion are taken.   □

4.2.2 Contrast Enhancement

The gray values in some X-ray images lie in a relatively narrow range of the grayscale. In this case, enhancing the contrast will amplify the differences in the gray values of the image.

We compute the gray value histogram to investigate how an X-ray image uses the grayscale. The function summarizes the gray value information of an X-ray image. The histogram is a function $h(x)$ that denotes the number of pixels in the X-ray image that have a gray value equal to x. Figure 4.4 shows how each histogram represents the distribution of gray values in the X-ray images.

A transformation can be applied to modify the distribution of gray value in an X-ray image. Simple contrast enhancement can be achieved if we use a linear transformation which sets the minimal and maximal gray values of the X-ray image to the minimal and maximal gray value of the grayscale respectively. Thus, the histogram is expanded to occupy the full range of the grayscale. Mathematically, for a scale between 0 and 255, this transformation is expressed as follows:

$$Y(i, j) = 255 \cdot \frac{X(i, j) - x_{min}}{x_{max} - x_{min}}, \qquad (4.3)$$

where x_{min} and x_{max} denote the minimal and maximal gray value of the input X-ray image. The output image is stored in matrix **Y**. This simple function is implemented in command linimg from pyxvis Library. Figure 4.4b shows the result of the transformation applied to the X-ray image in Fig. 4.4a. We observe in the histogram of the enhanced X-ray image, how the gray values expand from '0' to '255'. The mapping is linear and means that a gray value equal to $\frac{1}{2}(x_{max} - x_{min})$ will be mapped to 255/2. This linear transformation is illustrated in Fig. 4.5a, where the abscissa is the input gray value and the ordinate is the output gray value.

In a similar fashion, gray input image values can be mapped using a non-linear transformation $y = f(x)$, as illustrated in Fig. 4.5b and c, the results of which are shown in Fig. 4.4c and d respectively. Here, x and y are the gray values of the input and output images respectively. The non-linear transformation is usually performed

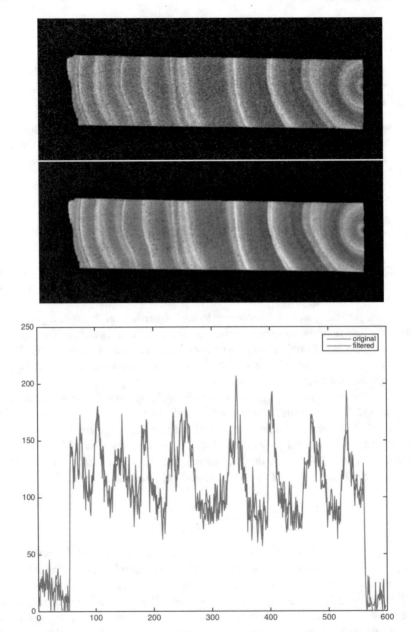

Fig. 4.3 Noise removal of an X-ray image of a wood piece after an averaging of 20 frames. (Top) one of the 20 images. (Middle) filtered image. (Bottom) row 100 of each image. [→ Example 4.1 🐍]

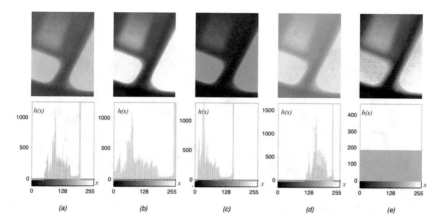

Fig. 4.4 Contrast enhancement: **a** original image, **b** linear transformation ($\gamma = 1$), **c** non-linear transformation ($\gamma = 2$), **d** non-linear transformation ($\gamma = 1/2$), **e** gray values uniformly distributed

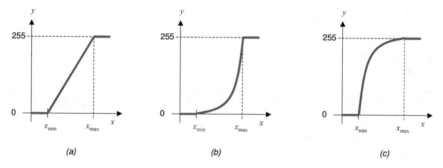

Fig. 4.5 Plots showing different transformations of the gray values: **a** linear transformation ($\gamma = 1$), **b** non-linear transformation with $\gamma > 1$, **c** non-linear transformation with $\gamma < 1$

with a γ *correction* [9]. In these examples, if $\gamma > 1$ the mapping is weighted toward darker output values, and if $\gamma < 1$ the mapping is weighted toward brighter output values. Gamma transformation can be expressed as follows:

$$
Y(i, j) = \begin{cases} 0 & \text{for } X(i, j) < x_{\min} \\ 255 \cdot \left[\frac{X(i,j) - x_{\min}}{x_{\max} - x_{\min}} \right]^{\gamma} & \text{for } x_{\min} \leq X(i, j) \leq x_{\max} \\ 255 & \text{for } X(i, j) > x_{\max} \end{cases}
\tag{4.4}
$$

Finally, we present a contrast enhancement equalizing the histogram. Here, we can alter the gray value distribution in order to obtain a desired histogram. A typical equalization corresponds to the uniform histogram as shown in Fig. 4.4d. We see that the number of pixels in the X-ray image for each gray value is constant.

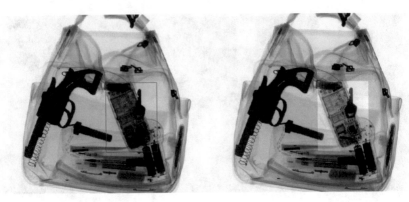

Fig. 4.6 Contrast enhancement by uniforming a histogram of the selected area. [→ Example 4.2 🐍]

🐍 Phython Example 4.2: In this example, we have an X-ray image of a baggage with very dark zones. The user defines a zone to be enhanced by clicking two opposite corners of a rectangle. The code forces the histogram of this zone to be uniform:

Listing 4.2 : Contrast enhancement of a selected area.

```
import numpy as np
import matplotlib.pylab as plt

from pyxvis.io import gdxraydb
from pyxvis.processing.images import hist_forceuni

image_set = gdxraydb.Baggages()
img = np.double(image_set.load_image(44, 130))

x_box = img[750:2000, 1250:2000]
x_box = hist_forceuni(x_box)
img2 = img.copy()
img2[750:2000, 1250:2000] = x_box

fig1, ax = plt.subplots(1, 2, figsize=(16, 8))
ax[0].imshow(img, cmap='gray'), ax[0].axis('off')
ax[1].imshow(img2, cmap='gray'), ax[1].axis('off')
plt.show()
```

The output of this code is shown in Fig. 4.6. For the equalization, the code uses function hist_forceuni from pyxvis Library.   □

4.2.3   Shading Correction

A decrease in the angular intensity in the projection of the X-rays causes low spatial frequency variations in X-ray images [1, 7]. An example is illustrated in Fig. 4.7a,

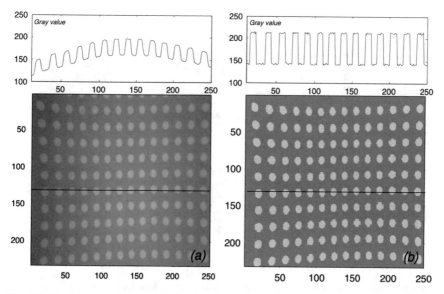

Fig. 4.7 Shading correction: **a** original image, **b** image after shading correction. The corresponding gray values profiles of row number 130 are shown above the images

which shows an X-ray image of an aluminum plate with holes in it. Since the plate is of a constant thickness, we would expect to see a constant gray value for the aluminum part and another constant gray value for the holes. In fact, the X-ray image is darker at the corners. This deficiency can be overcome by using a linear shading correction.

In this technique, we take two images as shown in Fig. 4.8. The first one, $\mathbf{r}_1$, of a thin plate, and the second one, $\mathbf{r}_2$, of a thick plate. We define i_1 and i_2 as the ideal gray values for the first and second images respectively. From $\mathbf{r}_1$, $\mathbf{r}_2$, i_1, and i_2, offset and gain, correction matrices $\mathbf{a}$ and $\mathbf{b}$ are calculated assuming a linear transformation between the original X-ray image $\mathbf{x}$ and corrected X-ray image $\mathbf{y}$:

$$Y(i, j) = a(i, j) \cdot X(i, j) + b(i, j), \tag{4.5}$$

where the offset and gain matrices are computed as follows:

$$a(i, j) = \frac{i_2 - i_1}{r_2(i, j) - r_1(i, j)} \qquad b(i, j) = i_1 - r_1(i, j) \cdot a(i, j). \tag{4.6}$$

An example of this technique is illustrated in Fig. 4.7b. In this case, we obtain only two gray values (with noise) one for the aluminum part and another for the holes of the plate.

Python Example 4.3: In this example, we simulate images $\mathbf{X}$ (a plate with a square cavity). In addition, we simulate X-ray images $\mathbf{r}_1$ (a thin plate) and $\mathbf{r}_2$ (a thick

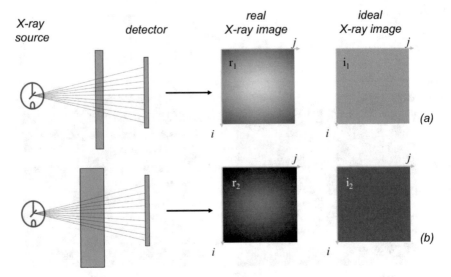

Fig. 4.8 Shading correction: **a** X-ray image for a thin plate, **b** X-ray image for a thick plate. Ideal X-ray images have a constant gray value

plate) as illustrated in Fig. 4.8. The following Python code shows how the shading effect of **X** can be corrected:

Listing 4.3 : Shading correction.

```python
import numpy as np
import matplotlib.pylab as plt

from pyxvis.processing.images import shading, fspecial

mat_r1 = fspecial('gaussian', 256, 80)
mat_r1 = mat_r1 / np.max(mat_r1.flatten()) * 0.8

mat_r2 = fspecial('gaussian', 256, 60)
mat_r2 = mat_r2 / np.max(mat_r2.flatten()) * 0.4

i1 = 0.8
i2 = 0.4

mat_x = fspecial('gaussian', 256, 70)
mat_x = mat_x / np.max(mat_x.flatten()) * 0.7
mat_x[30:80, 30:80] = mat_x[30:80, 30:80] * 1.5

mat_y = shading(mat_x, mat_r1, mat_r2, i1, i2)

fig, ax = plt.subplots(1, 2, figsize=(10, 10))
ax[0].imshow(mat_x, cmap='gray')
ax[0].axis('off');
ax[1].imshow(mat_y, cmap='gray')
ax[1].axis('off');

plt.show()
```

The output of this code is shown in Fig. 4.9. The correction is evident: the appearance of the background is homogenous, whereas the square is more distinguishable. In this code, we use function shading of pyxvis Library. This function computes shading correction as defined in (4.5).   □

4.3   Image Filtering

2D image filtering is performed in digital image processing using a small neighborhood of a pixel $X(i, j)$ in an input image to produce a new gray value $Y(i, j)$ in the output image, as shown in Fig. 4.10. A *filter mask* defines the input pixels to be processed by an *operator* f. The resulting value is the output pixel. The output for the entire image is obtained by shifting the mask over the input image. Mathematically, the image filtering is expressed as follows:

$$Y(i, j) = f[\underbrace{X(i - p, j - p), \ldots, X(i, j), \ldots, X(i + p, j + p)}_{\text{input pixels}}], \qquad (4.7)$$

for $i = p + 1 \ldots M - p$ and $j = p + 1 \ldots N - p$, where M and N are the number of rows and columns of the input and output images. The size of the filter mask is, in this case, $(2p + 1) \times (2p + 1)$. The operator f can be linear or non-linear. In this section, the most important linear and non-linear filters for X-ray testing are outlined.

Fig. 4.9 Simulation of shading correction: (Left) X-ray image for a plate with a square cavity (**X**), (Right) corrected image (**Y**). [→ Example 4.3 🖱]

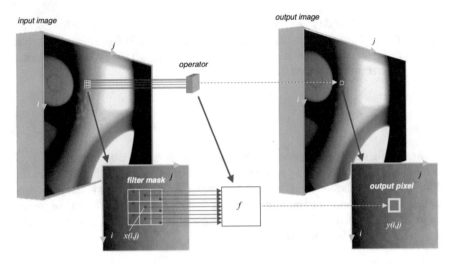

Fig. 4.10 Image filtering

4.3.1 Linear Filtering

The operator f is linear, if the resulting value $Y(i, j)$ is calculated as a linear combination of the input pixels:

$$Y(i, j) = \sum_{m=-p}^{p} \sum_{n=-p}^{p} h(m, n) \cdot X(i - m, j - n), \qquad (4.8)$$

where **h** is called the *convolution mask*. The elements of **h** weight the input pixels. The convolution of an image **X** with a mask **h** can be written as follows:

$$\mathbf{Y} = \mathbf{X} * \mathbf{h}. \qquad (4.9)$$

Averaging is a simple example of linear filtering. For a 3×3 neighborhood, the convolution mask is

$$\mathbf{h} = \frac{1}{9} \begin{bmatrix} 1 & 1 & 1 \\ 1 & 1 & 1 \\ 1 & 1 & 1 \end{bmatrix}$$

Gaussian mask can be used as well

$$h(m, n) = \frac{1}{2\pi\sigma^2} \cdot e^{-\frac{m^2+n^2}{2\sigma^2}} \qquad (4.10)$$

scale factor $1/(2\pi\sigma^2)$ ensures $\sum_{m,n} h(m,n) = 1$ over all elements of **h**. Average and Gaussian filtering are implemented respectively as functions im_average and im_gaussian in pyxvis Library.

A common application of filtering in X-ray testing is defect detection, e.g., in castings and welds. Filtering out defects detected in an X-ray image will provide a reference *defect-free* image. The defects are detected by finding deviations in the original image from the reference image. The problem is how one can generate a defect-free image from the original X-ray image. Assuming that the defects will be smaller than the regular structure of the test piece, one can use a low-pass filter that does not consider the high-frequency components of the image. However, if a linear filter is used for this task, the edges of the regular structure of the specimen are not necessarily preserved and many false alarms are raised at the edges of regular structures. Consequently, a non-linear filter is used.

4.3.2   Non-linear Filtering

In order to avoid the mentioned problems of linear filters, non-linear filters are used. Defect discrimination can be performed with a *median filter*. The median filter is a ranking operator (and thus non-linear), where the output value is the middle value of the input values ordered in a rising sequence [5]. For an even number of input numbers, the median value is the arithmetic mean of the two middle values.

The application of a median filter is useful for generating the reference image because it smoothes noise yet preserves sharp edges, whereas other linear low-pass filters blur such edges (see a comparison with linear filters in Fig. 4.11). Hence, it follows that small defects can be suppressed, while the regular structures are preserved. Figure 4.12 shows this phenomenon for a 1D example. The input signal x is filtered using a median filter with nine input elements, and the resulting signal is y. We can see that structures of length n greater than four cannot be eliminated. The third column shows the detection $x - y$. Large structures of $n \geq 5$ are not detected as presented in the last two cases.

If the background captured by the median filter is constant, foreground structures could be suppressed if the number of values belonging to the structure is less than one half of the input value to the filter. This characteristic is utilized to suppress the defect structures and to preserve the design features of the test piece in the image.

An example for the application of a median filter on 2D signals (images) is shown in Fig. 4.13 and includes different structures and mask sizes compared to the effects of two linear low-pass filters. One can appreciate that only the median filter manages to suppress the relatively small structures completely, whereas the large patterns retain their gray values and sharp edges.

The goal of the background image function, therefore, is to create a defect-free image from the test image. A real example is shown in Fig. 4.14. In this example, from an original X-ray image **X**, we generate a filtered image **Y** and a difference image $|\mathbf{X} - \mathbf{Y}|$. By setting a threshold, we obtain a binary image whose pixels are

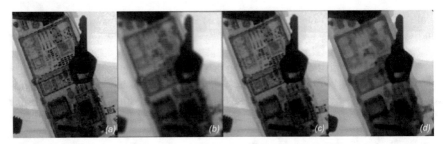

Fig. 4.11 Example of filtering of **a** an X-ray image of 600 × 700 pixels using **b** arithmetic, **c** Gaussian and **d** median filters with a mask of 19 × 19 pixels. The filtered images where obtained using commands im_average, im_gaussian, and im_median of **pyxvis** Library

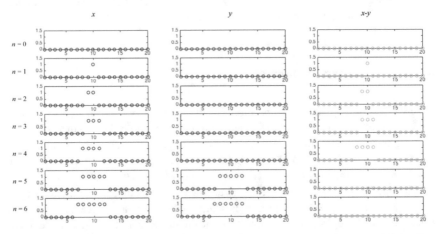

Fig. 4.12 Median filter application on a 1D signal x. The filtered signal is y (the size of the median mask is 9). Structures of length n less than 9/2 are eliminated in y. This filter can be used to detect small structures ($n \leq 4$)

Fig. 4.13 Median filter application on an $n \times n$ structure using an $m \times m$ quadratic mask compared to average and Gauss low-pass filter application

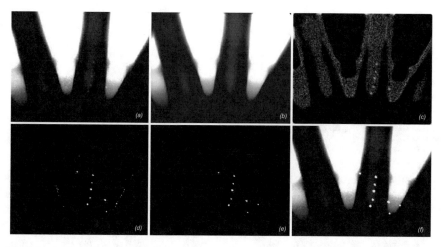

Fig. 4.14 Defect detection using median filtering: **a** original X-ray image of an aluminum wheel with small defects, **b** filtered X-ray image, **c** difference image, **d** binary image using a threshold, **e** elimination of small regions, **f** detection superimposed onto original image. [→ Example 4.4 🐍]

'1' (or white), where the gray values in the difference image are greater than the selected threshold. Finally, we eliminate very small regions. The remaining pixels correspond to the detected flaws.

🐍 Python Example 4.4: In this example, we detect small defects of an aluminum wheel. First, a reference defect-free image is estimated from the original image itself using median filtering. Second, the difference between original and reference images is computed. Finally, defects are detected when the difference in gray values is high enough and the size of the detected region is large enough:

Listing 4.4 : Defect detection using median filtering

```python
import numpy as np
import matplotlib.pylab as plt

from skimage.morphology import remove_small_objects, binary_dilation
from skimage.segmentation import clear_border

from pyxvis.io import gdxraydb
from pyxvis.io.visualization import binview
from pyxvis.processing.images import im_gaussian, im_median

image_set = gdxraydb.Castings()

X = image_set.load_image(21, 29)  # Original image
X = im_gaussian(X, k=5)  # Low pass filtering

fig1, ax1 = plt.subplots(1, 1, figsize=(6, 6))
ax1.set_title('Original Image with defects')
ax1.imshow(X, cmap='gray')
ax1.axis('off')
plt.show()
```

```
Y0 = im_median(X, k=23)
fig2, ax2 = plt.subplots(1, 1, figsize=(6, 6))
ax2.set_title('Median filter')
ax2.imshow(Y0, cmap='gray')
ax2.axis('off')
plt.show()

Y1 = np.abs(np.double(X) - np.double(Y0))
fig3, ax3 = plt.subplots(1, 1, figsize=(6, 6))
ax3.set_title('Difference Image')
ax3.imshow(np.log10(Y1 + 1), cmap='gray')
ax3.axis('off')
plt.show()

Y2 = Y1 > 18
fig4, ax4 = plt.subplots(1, 1, figsize=(6, 6))
ax4.set_title('Binary')
ax4.imshow(Y2, cmap='gray')
ax4.axis('off')
plt.show()

Y3 = remove_small_objects(Y2, 20)
fig5, ax5 = plt.subplots(1, 1, figsize=(6, 6))
ax5.set_title('Binary')
ax5.imshow(Y3, cmap='gray')
ax5.axis('off')
plt.show()

Y = clear_border(binary_dilation(Y3, np.ones((3, 3))))
fig6, ax6 = plt.subplots(1, 1, figsize=(6, 6))
ax6.set_title('Small region are eliminated')
ax6.imshow(Y, cmap='gray')
ax6.axis('off')
plt.show()

blend_mask = binview(X, Y, 'y', 1)
fig6, ax6 = plt.subplots(1, 1, figsize=(6, 6))
ax6.set_title('Small region are eliminated')
ax6.imshow(blend_mask, cmap='gray')
ax6.axis('off')
plt.show()
```

The output of this code—step by step—is shown in the last row of Fig. 4.14.   □

4.4   Edge Detection

In this section, we will study how the *edges* of an X-ray image can be detected. The edges correspond to pixels of the image in which the gray value changes significantly over a short distance [3]. Since edges are discontinuities in the intensity of the X-ray image, they are normally estimated by maximizing the gradient of the image. Edge detection image corresponds to a binary image (of the same size of the X-ray image), where a pixel is '1' if it belongs to an edge; otherwise, it is '0', as shown in Fig. 4.15. Before we begin a more detailed description of edge detection, it is worthwhile to highlight some aspects of its relevance in the analysis of X-ray images.

The edges of an X-ray image should show the boundary of objects, e.g., boundaries of defects in control quality of aluminum castings, boundaries of the weld

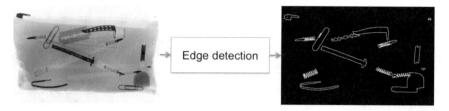

Fig. 4.15 Edge detection of an X-ray image of a pen case. The edges correspond the boundaries of the objects that are inside the pen case. [→ Example 4.11 🐍]

in welding inspection and boundaries of objects in baggage screening (Fig. 4.15). Thus, the input X-ray image is transformed into a binary image which shows structural properties of the X-ray image. The key idea is to detect objects of interest, such as defects in case of quality control or threatening objects in case of baggage screening, based on the information provided by edge detection.

In this section, we will review some   basic edge detection techniques that have been used in X-ray testing: gradient estimation (Sect. 4.4.1), LoG, Laplacian-of-Gaussian (Sect. 4.4.2), and Canny (Sect. 4.4.3). Segmentation techniques based on edge detection will be outlined in Sect. 4.5.

4.4.1 Gradient Estimation

The gradient for a 1D function $f(x)$ is defined by

$$f'(x) = \frac{\partial f}{\partial x} = \lim_{\Delta x \to 0} \frac{f(x + \Delta x) - f(x)}{\Delta x} \tag{4.11}$$

and for a 2D function $f(x, y)$ is defined by a vector of two elements, one in x direction, and the another one in y direction:

$$\nabla f(x, y) = \left[\frac{\partial f}{\partial x}, \frac{\partial f}{\partial y} \right]. \tag{4.12}$$

In digital images, after digitalization of $f(x, y)$, however, corresponding Δx or Δy values cannot be less than one pixel. A simple way to compute the gradient of image $\mathbf{X}$ in i and j direction can be respectively:

$$G_i(i, j) = X(i + 1, j) - X(i, j) \quad \text{and} \quad G_j(i, j) = X(i, j + 1) - X(i, j). \tag{4.13}$$

Thus, the magnitude of the gradient can be computed as follows:

$$G(i, j) = \sqrt{(G_i(i, j))^2 + (G_j(i, j))^2} \tag{4.14}$$

and the direction of the gradient as follows:

$$A(i, j) = \arctan \frac{G_j(i, j)}{G_i(i, j)}. \tag{4.15}$$

In this formulation, gradient images $\mathbf{G}_i$ and $\mathbf{G}_j$ can be easily calculated by convolution (4.9). Thus,

$$\mathbf{G}_i = \mathbf{X} * \mathbf{h}^\mathsf{T} \quad \text{and} \quad \mathbf{G}_j = \mathbf{X} * \mathbf{h}, \tag{4.16}$$

where $\mathbf{h}$ is the mask used to compute the gradient in horizontal direction. For instance, if we compute the gradient using the simple way (4.13), we can use $\mathbf{h} = [-1 \ +1]$ in (4.16). Nevertheless, for noisy images, larger masks are suggested for (4.16). Sobel and Prewitt masks are commonly used in image processing [5]. They are defined as follows:

$$\mathbf{h}_{\text{Sobel}} = \begin{bmatrix} -1 & 0 & +1 \\ -2 & 0 & +2 \\ -1 & 0 & +1 \end{bmatrix} \quad \text{and} \quad \mathbf{h}_{\text{Prewitt}} = \begin{bmatrix} -1 & 0 & +1 \\ -1 & 0 & +1 \\ -1 & 0 & +1 \end{bmatrix}. \tag{4.17}$$

For severe noise, it is recommended to use Gaussian filtering before applying gradient operators. Since Gaussian and gradient operations are linear, the Gaussian gradient operator can be defined by taking the derivative of the Gaussian (4.10):

$$h_{\text{Gauss}}(m, n) = m \cdot e^{-\frac{m^2 + n^2}{2\sigma^2}}. \tag{4.18}$$

It should be noted that edges are detected when the magnitude of the gradient is maximal. That means the location of edge pixels will not be modified if a mask $\mathbf{h}$ is replaced by $\lambda \mathbf{h}$ with $\lambda \neq 0$. Moreover, the direction of the gradient does not become modified either. For this reason, the elements of $\mathbf{h}$ are usually shown in its simplest way.

An example of the estimation of gradient using the explained masks is illustrated in Fig. 4.16. After the gradient image is calculated, the edges are detected by thresholding. Thus, if the magnitude of the gradient is greater than a certain threshold, then the pixel of the output image is set as an edge pixel. The output for the mentioned example is illustrated in Fig. 4.17. We can see how the boundaries are detected, especially for those objects that are very dark in comparison with their background.

Python Example 4.5: In this example, we show the edge detection of an X-ray image of a pen case using the gradient operators according to the method explained in this Sect. 4.5.1:

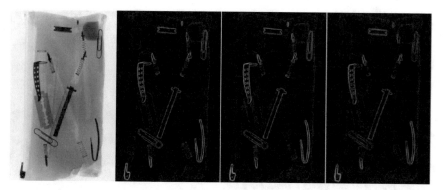

Fig. 4.16 Gradient of an X-ray of a pen case using different masks (Sobel, Prewitt, and Gaussian). See edge detection in Fig. 4.17 [→ Example 4.5 🐍]

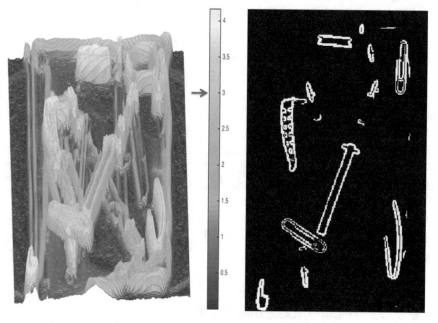

Fig. 4.17 Edge detection by thresholding a Gaussian gradient image of Fig. 4.16. The edges are detected for gradients greater than 3. In this representation, a logarithmical scale for the gray values was used. [→ Example 4.5 🐍]

Listing 4.5 : Gradient with different masks

```
import numpy as np
import matplotlib.pylab as plt
import cv2 as cv

from pyxvis.io import gdxraydb
from pyxvis.processing.images import fspecial, linimg, im_grad
from pyxvis.io.visualization import show_image_as_surface


image_set = gdxraydb.Baggages()
img = image_set.load_image(2, 1)
img = cv.resize(img, None, fx=0.25, fy=0.25, interpolation=cv.INTER_AREA)

hs = fspecial('sobel')      # Sobel kernel
hp = fspecial('prewitt')  # Prewitt kernel

hg = fspecial('gaussian', 9, 1.0)
hg = cv.filter2D(hg, cv.CV_64F, np.array([-1, 1]))

gs, __ = im_grad(img, hs)
gp, __ = im_grad(img, hp)
gg, __ = im_grad(img, hg)

gradients = np.hstack([linimg(gs), linimg(gp), linimg(gg)])  # Stack the results as a
      same image.

plt.figure(figsize=(12, 6))
plt.imshow(gradients, cmap='gray')
plt.show()

img_y = np.log(gg + 1)

show_image_as_surface(img_y[-5:5:-1, -5:5:-1], elev=80, azim=-185, fsize=(10, 10),
      colorbar=True)

fig, ax = plt.subplots(1, 1, figsize=(10, 10))
ax.imshow(img_y > 3, cmap='gray')
ax.axis('off')
plt.show()
```

The output of this code is shown in Figs. 4.16 and 4.17. The code uses command
im_grad of pyxvis Library.   □

4.4.2   Laplacian-of-Gaussian (LoG)

In the previous section, we learned that the edges of a function can be located by
detecting local maxima of the magnitudes of gradients. We know that the location
of the maximal values of the gradient coincides with zero-crossing of the second
derivative. In order to eliminate noisy zero-crossings, which do not correspond to
high gradient values, this method uses a Gaussian low-pass filter (see Fig. 4.18).
The method, known as Laplacian-of-Gaussian (LoG), is based on a kernel and a
zero-crossing algorithm [4]. LoG-kernel involves a Gaussian low-pass filter (4.10),
which is suitable for the pre-smoothing of the noisy X-ray images. LoG-kernel is
defined as the Laplacian of a 2D-Gaussian function:

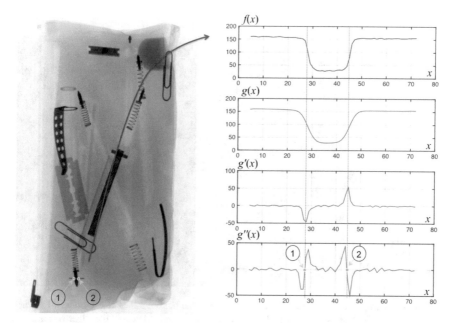

Fig. 4.18 Example of edge detection in 1D using LoG: The profile of the red line in an X-ray image is shown as $f(x)$. This function is filtered by a Gaussian low-pass filter obtaining $g(x)$. The gradient of $g(x)$, represented as $g'(x)$ shows the location of the maximal value (see dashed orange lines), that corresponds to the zero-crossing of the second derivative of $g(x)$. The edges '1' and '2' are then detected

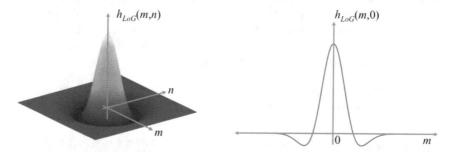

Fig. 4.19 LoG mask: (Left) representation of (4.19), (Right) profile for $n = 0$

$$h_{LoG}(m, n) = \frac{1}{2\pi\sigma^4} \cdot \left(2 - \frac{m^2 + n^2}{\sigma^2}\right) \cdot e^{-\frac{m^2+n^2}{2\sigma^2}}. \tag{4.19}$$

LoG-kernel is shown in Fig. 4.19. The parameter σ defines the width of the Gaussian function and, thus, the amount of smoothing and the edges detected (see Fig. 4.20). Using (4.8), we can calculate an image $\mathbf{Y}$ in which the edges of the original image are located by their zero-crossing. After zero-crossing, the detected edges $\mathbf{Z}$

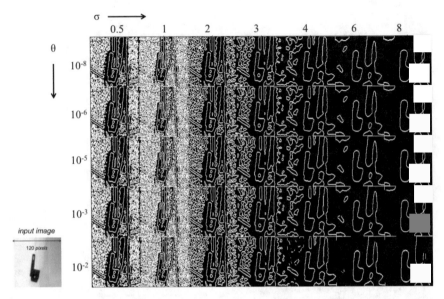

Fig. 4.20 Example of LoG edge detection of a slider (see bottom left of X-ray image of the pen case Fig. 4.18). Several values for σ and θ are presented. The smoothness of the edges is controlled by increasing σ. The reduction of noisy edges is controlled by increasing θ. [$\rightarrow$ Example 4.6 🐍]

correspond to the maximal (or minimal) values of the gradient image. In order to eliminate weak edges, a threshold θ is typically used. Thus, all edge pixels in **Z** that are not strong enough are ignored. The higher the threshold, the less edges will be detected. On the other hand, if $\theta = 0$, i.e., all zero-crossings are included, the edge image has closed and connected contours. As we will see in Sect. 4.5.2, this property is required when segmenting a region of the image.

🐍 Python Example 4.6: In this example, we show the edge detection of the object of a pen case (see Fig. 4.20) according to LoG algorithm explained in this Sect. 4.4.2:

Listing 4.6 : Edge detection using LoG

```python
import numpy as np
import matplotlib.pylab as plt
import cv2 as cv

from pyxvis.io import gdxraydb
from pyxvis.processing.images import Edge

image_set = gdxraydb.Baggages()

img = image_set.load_image(2, 1)
img = cv.resize(img, None, fx=0.5, fy=0.5, interpolation=cv.INTER_AREA)
img = img[595:715, 0:120]

plt.figure(figsize=(12, 6))
```

```
plt.imshow(img, cmap='gray')
plt.axis('off')
plt.show()

threshold = np.array([1e-8, 1e-6, 1e-5, 1e-3, 1e-2])  # Different threshold values
sigma = np.array([0.5, 1.0, 2.0, 3.0, 4.0, 6.0, 8.0]) # Different sigma values

rows = np.array([])
for t in threshold:
    cols = np.array([])
    for s in sigma:
        detector = Edge('log', t, s)
        detector.fit(img)
        cols = np.hstack([cols, detector.edges]) if cols.size else detector.edges
    rows = np.vstack([rows, cols]) if rows.size else cols

fig, ax = plt.subplots(1, 1, figsize=(12, 8))
ax.imshow(rows, cmap='gray');

# Figure configuration
from matplotlib.ticker import FixedLocator, FixedFormatter
ax.set_title('Sigma', y=1.05)
ax.set_ylabel('Threshold')
ax.tick_params(bottom=False, top=True, left=False, right=True)
ax.tick_params(labelbottom=False, labeltop=True, labelleft=True, labelright=False)
x_formatter = FixedFormatter(sigma)
y_formatter = FixedFormatter(threshold)
x_locator = FixedLocator(60 + 120 * np.arange(sigma.shape[0]))
y_locator = FixedLocator(60 + 120 * np.arange(threshold.shape[0]))
ax.xaxis.set_major_formatter(x_formatter)
ax.yaxis.set_major_formatter(y_formatter)
ax.xaxis.set_major_locator(x_locator)
ax.yaxis.set_major_locator(y_locator)
plt.show()
```

The output of this code—step by step—in Fig. 4.20. The code uses command Edge
of pyxvis Library.   ☐

4.4.3   Canny Edge Detector

Canny proposes a 2D linear mask for edge detection based on an optimization
approach [2], in which the following criteria are met:

- Good detection: The detection should respond to an edge (and not to noise).
- Good localization: The detected edge should be near the true edge.
- Single response: It should be one detected edge per true edge.

The optimal mask is similar to a derivative of a Gaussian. Thus, the idea is to
use this mask to find the local maxima of the gradient of the image. The practical implementation uses adaptive thresholding of the gradient (to detect strong and
weak edges) with hysteresis (weak edges are detected only if they are connected to
strong edges).

In Example 4.6, the code line `detector = Edge('log',t,s)` can be changed
by `E = Edge('canny',t,s)` to elucidate similarities and differences between
both edge detectors.

4.5   Segmentation

Image segmentation is defined as the process of subdividing an image into disjointed regions [3]. A region is defined as a set of connected pixels that correspond to a certain *object of interest*. Obviously, these regions of interest depend on the application. For instance, in the inspection of aluminum castings with X-ray images, the idea of segmentation is to find regions with defects. Here, the object of interest is the defects. An example is shown in Fig. 4.21, where the segmentation is the small spots that indicate defective areas.

Another example of segmentation in X-ray testing is weld inspection as illustrated in Fig. 4.22, where a weld seam with two regions is clearly distinguishable: the weld (*foreground*) and the base metal (*background*). In this example, the (first) object of interest is the weld because it is the region where defects can be present. The reader can clearly identify the defects in the weld (see small dark regions in the middle of the X-ray image). In this example, the defects, that have to be detected in a second segmentation stage, are our second object of interest. In this case, the background is the weld, and the foreground is the defects.

Segmentation is one of the most difficult processes in image processing. Clearly, there are some simple applications in which certain segmentation techniques are very effective (e.g., separation between weld and metal base as shown in Fig. 4.22), however, in many other applications segmentation is far from being solved as the appearance of the object of interest can become very intricate. This is the case of

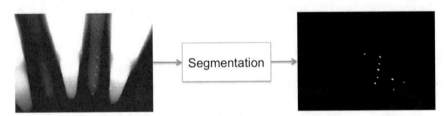

Fig. 4.21   Example of segmentation: detection of defects in an aluminum wheel (see details in Fig. 4.14). [→ Example 4.4 🐍]

Fig. 4.22   Segmentation of a weld. (Top) original X-ray image. (Bottom) segmentation. The first step in weld inspection is the segmentation of the weld, i.e., the region where the defects can be present (see segmentation in Fig. 4.27). The second step is the detection of defects. [→ Example 4.8 🐍]

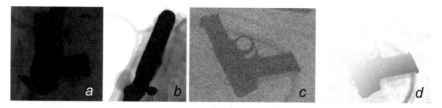

Fig. 4.23 Problems when detecting a gun. Detection can be a very complex task due to **a** occlusion, **b** self-occlusion, **c** noise, **d** wrong acquisition

baggage screening, where the segmentation of objects of interest inside a piece of luggage can be extremely difficult due to problems of (self-)occlusion, noise, and acquisition (see Fig. 4.23).

In image processing for X-ray testing, segmentation is used to detect (potential) regions that can be the objects of interest that we are looking for. As mentioned in previous examples, segmentation divides the X-ray image into two areas: foreground and background. Foreground means the pixels of the object(s) of interest. Background means the remaining pixels of the image. Usually, a *binary image* is the output of the segmentation process as we can see in Figs. 4.21 and 4.22, where a pixel equals to '1' (white) is foreground, whereas '0' (black) means background. We use the term 'potential' throughout to make it clear that a segmented region is not necessarily the final detected region. In many applications, the segmentation is just the first step of the whole detection process. In such cases, an additional step that analyzes the segmented region is required. This additional step can include multiple view analysis or a pattern recognition technique (see Fig. 1.22). The later extracts and classifies features of the segmented region in order to verify whether it corresponds to the object that we are detecting or it is a *false detection*.

Thus, segmentation basically acts as a focus of attention mechanism that filters the information that is fed to the following steps, as such a failure in the segmentation is catastrophic for the final performance. In this section, we will review some basic segmentation techniques that have been used in X-ray testing: thresholding (Sect. 4.5.1), region growing (Sect. 4.5.2), and maximally stable extremal regions (Sect. 4.5.3). Please note that more complex techniques based on computer vision algorithms will be addressed in the next sections.

4.5.1 Thresholding

In some X-ray images, we can observe that the background is significantly darker than the foreground. This is the case of an X-ray image of an apple placed on a uniform background as illustrated in Fig. 4.24. It is clear that the object of interest can be segmented using a very simple approach based on *thresholding*. In this section, we will explain a methodology based on two steps: (i) estimate of a global

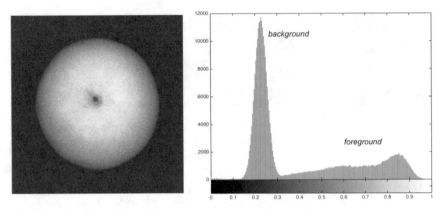

Fig. 4.24   X-ray image of an apple and its histogram

threshold using a statistical approach and (ii) a morphological operation in order to fill the possible holes presented in the segmented binary image. This method was originally presented for color food images [14], however, it can be easily adapted for X-ray images.

The X-ray image to be segmented is stored in matrix $\mathbf{I}$. In order to enhance the contrast of the image, a linear transformation can be performed (see Sect. 4.2.2). Additionally, a linear or non-linear filter can be used for noise removal (see Sect. 4.3). Here, after image enhancement and filtering, we obtain a new image $\mathbf{J}$, where $J_{max} = 1$ and $J_{max} = 0$. Image $\mathbf{J}$ has a bimodal histogram as shown in Fig. 4.24, where the left distribution corresponds to the background and the right to the food image. In this image, the first separation between foreground and background can be performed estimating a global threshold t. Thus, we define a binary image

$$K(i, j) = \begin{cases} 1 & \text{if} \quad J(i, j) > t \\ 0 & \text{else} \end{cases} \tag{4.20}$$

where '1' means foreground and '0' background, that define two classes of pixels in the image. Figure 4.25 illustrates different outputs depending on t. The problem is to determine the 'best' threshold t that separates the two modes of the histogram from each other. A good separation of the classes is obtained by ensuring (i) a small variation of the gray values in each class, and (ii) a large variation of the gray values in the image [6]. The first criterion is obtained by minimizing a weighted sum of the within-class variances (called *intraclass* variance $\sigma_W^2(t)$):

$$\sigma_W^2(t) = p_b(t)\sigma_b^2(t) + p_f(t)\sigma_f^2(t), \tag{4.21}$$

where the indices 'b' and 'f' denote respectively background and foreground classes, and p and σ^2 are respectively the probability and the variance for the indicated class. These values can be computed from the histogram.

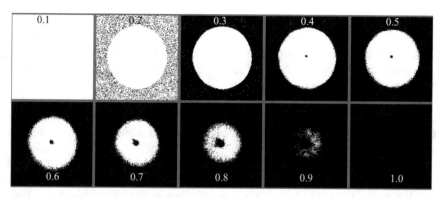

Fig. 4.25 Segmentation using threshold $t = 0.1, 0.2, \ldots 1.0$

The second criterion is obtained by maximizing the between-class variance (called *interclass* variance $\sigma_B^2(t)$):

$$\sigma_B^2(t) = p_b(\mu_b(t) - \mu)^2 + p_f(\mu_f(t) - \mu)^2, \tag{4.22}$$

where μ_b, μ_f, and μ indicate the mean value of the background, foreground, and the whole image respectively.

The best threshold t can be estimated by a sequential search through all possible values of t that minimizes $\sigma_W^2(t)$ (or maximizes $\sigma_B^2(t)$). Both criteria, however, lead to the same result because the sum $\sigma_W^2 + \sigma_B^2$ is a constant and corresponds to the variance of the whole image [6]. Skimage Library computes the global image threshold by minimizing the intraclass variance $\sigma_W^2(t)$. The threshold can be obtained with the function `threshold_otsu`. In our example, the obtained threshold is $t = 0.4824$, that is approximately 0.5 (see Fig. 4.25).

We can observe in Fig. 4.25 that the segmentation suffers from inaccuracy because there are many dark (bright) regions belonging to the foreground (background) that are below (above) the chosen threshold and therefore misclassified. For this reason, additional morphological processing must be obtained.

The morphological operation is performed in three steps as shown in Fig. 4.26: (i) remove small objects, (ii) close the binary image, and (iii) fill the holes.

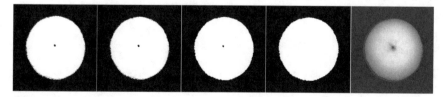

Fig. 4.26 Additional morphological operations. From left to right: **K**: binary image after thresholding, **A**: after removal of small objects, **C**: after closing process, **R**: after filling holes, and boundary superimposed onto the original image. [→ Example 4.7 🐍]

In the first step, we remove from binary image **K** obtained from (4.20) all connected regions that have fewer than n pixels (see image **A** in Fig. 4.26).This operation is necessary to eliminate those isolated pixels of the background that have a gray value greater than the selected threshold. Empirically, we set $n = NM/100$, where $N \times M$ is the number of pixels of the image.

The second step *closes* the image, i.e., the image is *dilated* and then *eroded*. The dilation is the process that incorporates into the foreground the background pixels that touch it. On the other hand, erosion is the process that eliminates all the boundary pixels of the foreground. The closing process (dilation followed by erosion) fills small holes and thins holes in the foreground, connecting nearby regions, and smoothing the boundaries of the foreground without changing the area significantly [3] (see image **C** in Fig. 4.26). This operation is very useful in objects that have spots in the boundary.

Finally, the last operation fills the holes in the closed image (see image **R** in Fig. 4.26). We use this operation to incorporate into the foreground all pixels '0' that are inside of the region. The whole algorithm is implemented in command seg_bimodal of pyxvis Library. In the implementation, as suggested in [14], an offset p that modifies the threshold is used because there are dark zones in the boundary that are not well included in the original segmented region.

Python Example 4.7: In this example, we show the segmentation of an X-ray image of an apple (see Fig. 4.24) according to the method explained in this Sect. 4.5.1:

Listing 4.7 : Apple segmentation using global thresholding

```python
import matplotlib.pylab as plt

from pyxvis.io import gdxraydb
from pyxvis.processing.segmentation import seg_bimodal
from pyxvis.io.visualization import binview


image_set = gdxraydb.Nature()
img = image_set.load_image(5, 9)

mask, contours = seg_bimodal(img)
seg = binview(img, mask, 'g')

fig, ax = plt.subplots(1, 2, figsize=(12, 6))
ax[0].imshow(img, cmap='gray')
for n, contour in enumerate(contours):
        ax[0].plot(contour[:, 1], contour[:, 0], color='r', linewidth=3)
ax[0].axis('off')
ax[1].imshow(seg)
ax[1].axis('off')
plt.show()
```

The output of this code—step by step—in Fig. 4.26. The code uses command seg_bimodal of pyxvis Library.   □

The above-mentioned methodology, based on a *global threshold*, does not segment appropriately when there is a large variation in the background or foreground intensity. For this reason, in certain cases, it is recommended to use an *adaptive threshold*. The idea is to divide the input image into partitions with some overlapping. Each partition is handled as a new image that is segmented by thresholding (using a global but an ad hoc threshold for each partition). The output image is a fusion of all segmented partitions, e.g., using logical OR operator. The next example shows an implementation that was used to segment the weld of Fig. 4.22. Since the weld area is horizontal, the proposed method uses vertical partitions that include background and foreground areas. The segmentation of each partition is performed by the same method used for the segmentation of the apple.

Python Example 4.8: This example shows the segmentation of a weld of Fig. 4.22 using adaptive thresholding. The approach is simple; the input image is divided into four partitions with an overlapping of 50%. Each partition is segmented using command seg_bimodal of pyxvis Library. The obtained binary images of the segmentation are superimposed using logical OR operator:

Listing 4.8 : Weld segmentation using adaptive thresholding

```python
import numpy as np
import matplotlib.pylab as plt

from skimage.measure import find_contours

from pyxvis.io import gdxraydb
from pyxvis.processing.segmentation import seg_bimodal
from pyxvis.io.visualization import binview


image_set = gdxraydb.Welds()
img = image_set.load_image(1, 1)

mask = np.zeros(img.shape, np.uint8)  # Create a uint8 mask image
max_width = img.shape[1]

d1 = int(np.round(max_width/4))
d2 = int(np.round(d1 * 1.5))

i1 = 0

while i1 < max_width:
    i2 = min(i1 + d2, max_width)  # second column of partition
    img_i = img[:, i1:i2]  # partition i
    bw_i, _ = seg_bimodal(img_i)  # segmentation of partition i
    roi = mask[:, i1:i2]
    overlap = np.bitwise_or(roi, bw_i)  # addition into whole segmentation
    mask[:, i1:i2] = overlap
    i1 = i1 + d1  # update of first column

seg = binview(img, mask, 'g', g=5)

contours = find_contours(np.float32(mask), 0.5)

fig, ax = plt.subplots(2, 1, figsize=(14, 5))
ax[0].imshow(img, cmap='gray');
for n, contour in enumerate(contours):
        ax[0].plot(contour[:, 1], contour[:, 0], color='r', linewidth=3)
```

```
ax[0].axis('off')
ax[1].imshow(seg)
ax[1].axis('off')
fig.tight_layout()
plt.show()
```

The output of this code—step by step—is shown in the last row of Fig. 4.27.   □

4.5.2   Region Growing

In region growing, we segment a region using an iterative approach. We start by choosing a seed pixel, as shown in Fig. 4.28. At this moment, our region is initialized and its size is one pixel only. We extract some feature of the region, e.g., the gray value. We extract the same feature of each neighboring pixel. In our example, there are four neighbors (up, down, right, and left), as we can see in third image of Fig. 4.28. We increase our region by adding similar neighboring pixels, i.e., those neighboring pixels that have a similar feature to the region. The whole process is continued, each added pixel is a new seed for the next iteration until no more neighboring pixels can be added.

In Fig. 4.28, we have a binary edge image. The feature that we use to establish the similarity is the value of the pixel. In our example, there are only two pixel values: '0' for the edge pixels, and '1' for the remaining pixels. That means that the value of the pixel of the seed is '1' and in each iteration, we can add only those neighboring pixels the value of which are '1'. As we can see, the red region grows up from 1

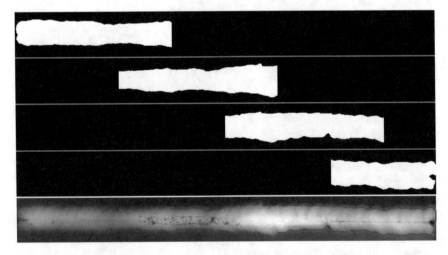

Fig. 4.27 Weld segmentation of Fig. 4.22 using adaptive thresholding of four partitions. The last image shows the segmentation after fusion the four individual segmentations using logical OR operator.) [→ Example 4.8 🐍]

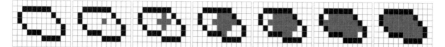

Fig. 4.28 Region growing: we start with a seed pixel that grows in each iteration in four directions until a boundary is found. The directions in this example are four: up, down, right, and left

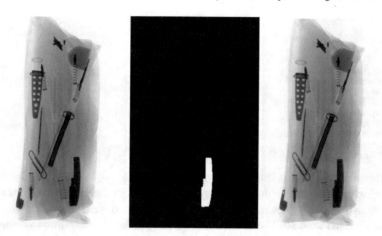

Fig. 4.29 Region growing in an X-ray image using a seed pixel in the object of interest. The region is well segmented as we can see in the binary image and in boundaries. [→ Example 4.9 🐍]

pixel to 5, 12, 16, 22, and finally, 24 pixels. The output is the red region of the last step.

Region growing can be used directly in X-ray images as illustrated in Fig. 4.29. We start with a seed pixel, and neighboring pixels are added if they are similar enough.

🐍 Python Example 4.9: In this example, we show the performance of region growing in the segmentation of an object in an X-ray image of a pen case (see Fig. 4.29). The seed is chosen at pixel (190,403). The seed grows by adding neighboring pixels with similar gray values. We use command region_growing of pyxvis Library. In this implementation, the similarity between region and neighboring pixels is established if $|\bar{R} - r_n| \le \theta$, where $\bar{R}$ is the average of the gray values of the region, r_n is the gray value of the neighboring pixel, and θ is a threshold. In this example, $\theta = 20$:

Listing 4.9 : Region Growing

```python
import matplotlib.pylab as plt
import cv2 as cv

from pyxvis.io import gdxraydb
from pyxvis.processing.segmentation import region_growing
from pyxvis.io.visualization import binview
```

```
image_set = gdxraydb.Baggages()

img = image_set.load_image(3, 4)
img = cv.resize(img, None, fx=0.35, fy=0.35, interpolation=cv.INTER_AREA)

th = 40  # threshold
si, sj = (403, 190)  # Seed

mask = region_growing(img, (si, sj), tolerance=th)

seg = binview(img, mask, 'g')

fig, ax = plt.subplots(1, 3, figsize=(14, 8))
ax[0].imshow(img, cmap='gray')
ax[0].plot(sj, si, 'r+')
ax[0].axis('off')
ax[1].imshow(mask, cmap='gray')
ax[1].axis('off')
ax[2].imshow(seg)
ax[2].axis('off')
plt.tight_layout()
plt.show()
```

The output of this code is shown in Fig. 4.29.   □

Region growing can be used in X-ray testing in defect detection (see, for example, interesting approaches in aluminum castings [10] and welds [11]). The method is illustrated in Fig. 4.30. The method uses an edge detection algorithm to obtain an edge image with closed and connected contours around the real defects. Thus, we use region growing to isolate each region enclosed by edges. The idea is to extract features from this isolated region (e.g., area, average of gray value, contrast, etc.) that can be used in a classification strategy. In our example, a region is segmented using a very simple classifier (the features of a segmented region must be in certain ranges, e.g., $A_{\min} \leq$ Area $\leq A_{\max}$). Obviously, more sophisticated features and classifiers can be used to improve the segmentation performance in more complex scenarios as we will see   in the following chapters.

Python Example 4.10: In this example, we show how to segment defects in aluminum castings using binary images of potential defects and some simple features that can be extracted from each potential region. In this example, we segment all those regions the area of which is between 200 and 2000 pixels, the average of the gray value is less than 150, and the contrast is greater than 1.1:

Listing 4.10 : Detection of defects in castings

```
import numpy as np
import matplotlib.pylab as plt

from skimage.segmentation import find_boundaries

from pyxvis.io import gdxraydb
from pyxvis.processing.segmentation import seg_log_feature
from pyxvis.io.visualization import binview


image_set = gdxraydb.Castings()
X = image_set.load_image(31, 19)
```

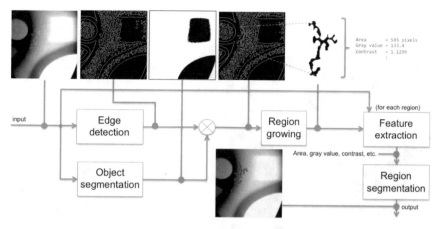

Fig. 4.30 Segmentation of defects in aluminum castings using region growing, edge detection, and some features. The size of the image in this example is 286×286 pixels [$\rightarrow$ Example 4.10 🐍]

```python
X = X[0:572:2, 0:572:2]  # Donwsampling the image

fig1, ax1 = plt.subplots(1, 1, figsize=(8, 8))
ax1.imshow(X, cmap='gray')
ax1.set_title('Input image')
ax1.axis('off')
plt.show()

R = X < 240

fig2, ax2 = plt.subplots(1, 1, figsize=(8,8))
ax2.imshow(R, cmap='gray')
ax2.set_title('Segmented object')
ax2.axis('off')
plt.show()

options = {
    'area': (30, 1500),  # Area range (area_min, area_max)
    'gray': (0, 150),  # Gray value range (gray_min, gray_max)
    'contrast': (1.08, 1.8),  # Contras range (cont_min, cont_max)
    'sigma': 2.5
}

Y, m = seg_log_feature(X, R, **options)

print(f'Found {m} regions.')

fig3, ax3 = plt.subplots(1, 1, figsize=(8, 8))
ax3.imshow(binview(X, find_boundaries(Y)), cmap='gray')
ax3.set_title('Segmented regions')
ax3.axis('off')
plt.show()
```

The output of this code is shown—step by step—in Fig. 4.30. We use command seg_log_feature of **pyxvis** Library.   □

This method is very effective for regions of interest that have gray values significantly different from the background (the reader, for instance, can try to segment

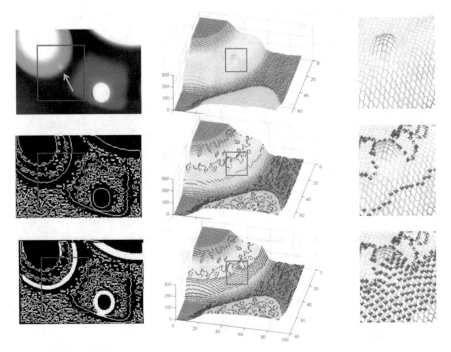

Fig. 4.31 X-ray image of an aluminum casting with a small defect at an edge (see defect pointed by green arrow). (First row) original image. (Second row) LoG. (Third row) LoG and high gradient pixels. (First column) image representation. (Second column) 3D representation of red square. (Third column) zoom of blue square. In this representation, the edge pixels are represented as red points superimposed onto the 3D surface. The output of this method is a binary image in which the real defects are closed by edges

the objects of the pen case of Fig. 4.29 using command seg_log_feature of pyxvis Library).

Nevertheless, the method may fail if the boundaries do not close a region of interest. This is the case in some defects of aluminum castings that are at an edge of a regular structure as illustrated in Fig. 4.31.[2] In this problem, we can see that the edges of LoG algorithm (and other edge detection algorithms like Sobel or Canny as well) cannot correctly find the defect's edge. Contrarily, it finds the regular structure's edge. To overcome this problem, we have to complete the remaining edges of these defects. A simple approach was suggested in [13] by thickening of the edges of the regular structure after LoG-edge detection: (i) The gradient of the original image is calculated. The gradient image is computed by taking the square root of the sum of the squares of the gradient in horizontal and in vertical directions. These are calculated by the convolution of the radioscopic image with the first derivative (in the corresponding direction) of the Gaussian low-pass filter used in the LoG fil-

---

[2] A video of this small defect can be watched at http://youtu.be/e3wDJhq2Tqg.

ter. (ii) High gradient pixels are detected by thresholding. (iii) The resulting image is added to the LoG-edge detection image. Afterwards, each closed region is segmented as a potential flaw. As can be observed the effectiveness of this method in Fig. 4.31, the defect on an edge of a regular structure could be satisfactorily closed. Thus, the method of Fig. 4.30 can be used.

4.5.3 Maximally Stable Extremal Regions (MSER)

In order to understand the MSER approach [8], the reader can imagine a simple video as follows. The video will have 256 frames. Frame t is defined as the binary image $\mathbf{I} < t$, where $\mathbf{I}$ is the input image we want to segment. If the binary image is black for '0' and white for '1', at the beginning our video will be very dark and at the end very bright. In the middle, we will have some regions depending on the threshold.[3] Thus, each region has an area $A(t)$, that depends on t. If the gray value of the region is very different from its background, the area of this region will be stable for some thresholds $t, t + 1 \ldots t + p$, i.e., $A(t) \approx A(t + 1) \cdots \approx A(t + p)$. The key idea of MSER is to segment those regions which fulfill:

$$\frac{\Delta A}{\Delta t} < \theta, \qquad (4.23)$$

where θ is a threshold. That means those regions whose sizes remain approximately stable by varying the segmentation threshold t are to be detected.

Python Example 4.11: In this example, we show the segmentation of an X-ray image of a pen case (see Fig. 4.15) according to MSER approach (see Sect. 4.5.3):

Listing 4.11 : Pencase segmentation using MSER algorithm

```python
import numpy as np
import matplotlib.pylab as plt
import cv2 as cv

from skimage.segmentation import find_boundaries
from skimage.morphology import binary_dilation

from pyxvis.io import gdxraydb
from pyxvis.processing.segmentation import seg_mser
from pyxvis.io.visualization import plot_bboxes


image_set = gdxraydb.Baggages()
img = image_set.load_image(2, 1)

fig1, ax1 = plt.subplots(1, 1, figsize=(10, 10))
ax1.imshow(img, cmap='gray')
ax1.set_title('Input image')
```

---

[3] The video can be found in http://youtu.be/tWdJ-NFE6vY.

```
ax1.axis('off')
plt.show()

mser_options = {
    'area': (60, 40000),  # Area of the ellipse (Max, Min)
    'min_div': 0.9,  # Minimal diversity
    'max_var': 0.2,  # Maximal variation
    'delta': 3,  # Delta
}

J, L, bboxes = seg_mser(img, **mser_options)

E = binary_dilation(find_boundaries(J, connectivity=1, mode='inner'), np.ones((3, 3)))

fig2, ax2 = plt.subplots(1, 1, figsize=(10, 10))
ax2.imshow(E, cmap='gray')
ax2.set_title('Edges')
ax2.axis('off')
plt.show()

fig3, ax3 = plt.subplots(1, 1, figsize=(10, 10))
ax3.imshow(L, cmap='gray')
ax3.set_title('Segmentation')
ax3.axis('off')
plt.show()

fig4, ax4 = plt.subplots(1, 1, figsize=(10, 10))
ax4.imshow(img, cmap='gray')
ax4 = plot_bboxes(bboxes, ax=ax4)
ax4.set_title('Bounding Boxes')
ax4.axis('off')
plt.show()
```

The output of this code—step by step—in Figs. 4.15 and 4.32. The code uses command seg_mser of pyxvis Library. This function uses OpenCV implementation of MSER.   □

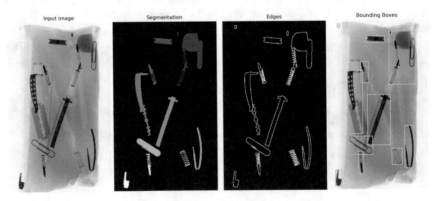

Fig. 4.32 Segmentation of objects in a pencase using MSER [→ Example 4.11 🖱]

4.6   Image Restoration

Image restoration involves recovering detail in severely blurred images. This process is more efficient when the causes of the imperfections are known a priori [5]. This knowledge may exist as an analytical model, or as a priori information in conjunction with knowledge (or assumptions) of the physical system that provided the imaging process in the first place. The purpose of restoration then is to estimate the best source image, given the blurred example and some a priori knowledge.

In this section, we concentrate on the particular case of blur caused by uniform linear motion, which may be introduced by relative motion between detector and object. Early work on restoring an image degraded by blurring calculated the deblurring function as an inverse filtering. The inverse filtering evaluation of the blurring function h (or point spread function PSF) in the frequency domain tends to be very sensitive to noise [5]. The cause of this sensitivity is the lowpass nature of the PSF: its frequency response $H(\omega)$ contains very small values, and small noise in the frequency regions where $1/H(\omega)$ is very large, maybe greatly emphasized. Sondhi [5] proposed a non-iterative algorithm to find a solution to the uniform-blurring case, but the computational load is extremely high in small motions. Another two non-iterative approaches are presented in [9]. In the first one, the matrix left division calculates the restored signal as a signal that has the fewest possible nonzero components. This solution differs strongly from the original signal because the original signal must not have necessarily many zero components. The second one, the Moore–Penrose pseudo-inverse of a matrix, finds a restored signal whose norm is smaller than any other solution. This solution is very good, but the estimation is based on Singular Value Decomposition (SVD), whose computation load is very high. In this section, we address the above problems and reduce the computational times significantly using a new technique that minimizes the norm between blurred and original.

A blurred X-ray image $g(x, y)$ that has been degraded by a motion in the vertical direction x and the horizontal direction y can be modeled by

$$g(x, y) = \frac{1}{T} \int_0^T f(x - x_t(t), y - y_t(t))dt, \qquad (4.24)$$

where f, T, $x_t(t)$ and $y_t(t)$ represent respectively the deterministic original X-ray image, the duration of the exposure and the time-varying component of motion in the x and y directions. In this case, the total exposure is obtained by integrating the instantaneous exposure over the time interval during which the shutter is open. By rotation of the camera or by using a transformation that rotates the blurred image, a new system of coordinates is chosen in which $x_t(t)$ is zero. Considering that the original image $f(x, y)$ undergoes uniform linear motion in the horizontal direction y only, at a rate given by $y_t(t) = ct/T$, let us write (4.24), with $u = y - ct/T$, as follows:

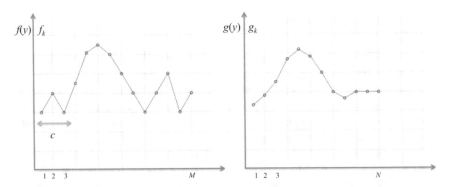

Fig. 4.33 Blurring process: (Left) original row f. (Right) Blurred row

$$g(y) = \frac{1}{T} \int_0^T f(y - ct/T)dt = \frac{1}{c} \int_{y-c}^y f(u)dt, \qquad (4.25)$$

or as a digital that has been discretized in spatial coordinates by taking N samples $\Delta y = Y/N$ units apart:

$$g_k = \frac{1}{n} \sum_{i=0}^{n-1} f_{k+i}, \qquad (4.26)$$

where

$$g_k = g\left(y_0 + (k-1)\frac{c}{n}\right), \quad f_k = f\left(y_0 + (k-1)\frac{c}{n} - c\right), \qquad (4.27)$$

with $n = c/\Delta y$. Figure 4.33 shows a row $\mathbf{f} = [f_1 \ ... \ f_M]^\mathsf{T}$ of an original image and its corresponding row $\mathbf{g} = [g_1 \ ... \ g_N]^\mathsf{T}$ of the blurred image for $n = 3$ pixels. Equation (4.26) describes an underdetermined system of N simultaneous equations (one for each element of vector $\mathbf{g}$) and $M = N + n - 1$ unknowns (one for each element of vector $\mathbf{f}$) with $M > N$. This process is carried out for each row of the image. The degradation of $\mathbf{f}$ can be modeled using a convolution of $\mathbf{f}$ with $\mathbf{h}$, where $\mathbf{h}$ is the PSF, a n-element vector defined as the impulse response of this linear system [5]. Thus, element g_i of vector $\mathbf{g}$ is calculated as a weighted sum of n elements of $\mathbf{f}$, i.e., $g_i = h_1 f_i + h_2 f_{i+1} + ... + h_n f_{i+n-1}$, for $i = 1, ..., N$. Using a circulant matrix, the convolution can be written as $\mathbf{Hf} = \mathbf{g}$:

$$\mathbf{g} = \mathbf{f} * \mathbf{h} = \begin{bmatrix} h_1 \ ... \ h_n \ 0 \ 0 \ 0 \ 0 \\ 0 \ h_1 \ ... \ h_n \ 0 \ 0 \ 0 \\ \vdots \qquad \vdots \\ 0 \ 0 \ ... \ 0 \ h_1 \ ... \ h_n \end{bmatrix} \begin{bmatrix} f_1 \\ f_2 \\ \vdots \\ f_M \end{bmatrix} = \begin{bmatrix} g_1 \\ g_2 \\ \vdots \\ g_N \end{bmatrix} \qquad (4.28)$$

Fig. 4.34 Degradation of an X-ray image of 2208 × 2688 pixels: Original image and degraded images with $n = 32$, 256, and 512 pixels

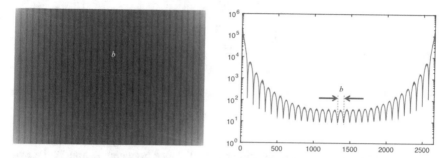

Fig. 4.35 Spectrum of a blurred image which was degraded by uniform linear motion with $n = 32$ pixels. (Left) 2D-Fourier Transformation of the original image of Fig. 4.34. Right Mean of the rows of the Fourier transformation. The size of the degraded image is 2208 × 2657 pixels. It can be demonstrated that b is approximately $2657/n$

An example of a degradation process is shown in Fig. 4.34. In this example, we can see how the objects cannot be recognized when the degradation is severe.

If the PSF is not exactly known, but if we know that it corresponds to a uniform linear motion, the parameter n can be estimated from the spectrum of the blurred image. An example is shown in Fig. 4.35. The 2D-Fourier Transformation of a blurred test is represented in Fig. 4.35a, in this case, a horizontal degradation took place with $n = 32$. The mean of its rows is illustrated in Fig. 4.35b. We can observe that the period of this function is inversely proportional to the length of the blurring process in pixels.

The problem of restoring an X-ray that has been blurred by uniform linear motion consists of solving the underdetermined system (4.28). The objective is to estimate an original row per row ($\mathbf{f}$), given each row of a blurred ($\mathbf{g}$) and a priori knowledge of the degradation phenomenon ($\mathbf{H}$). Since there is an infinite number of exact solutions for $\mathbf{f}$ in the sense that $\mathbf{Hf} - \mathbf{g} = \mathbf{0}$, an additional criterion that finds a sharp restored is required.

We observed that most solutions for $\mathbf{f}$ strongly oscillate. Figure 4.36 shows an example in which four different solutions for $\mathbf{f}$ are estimated, all solutions satisfy Eq. (4.28): $\mathbf{Hf} = \mathbf{g}$. Although these solutions are mathematically right, they do not correspond to the original signal. By the assumption that the components of the higher frequencies of $\mathbf{f}$ are not so significant in the wanted solution, these oscil-

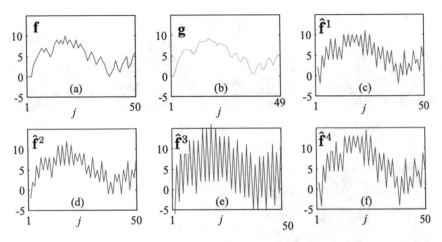

Fig. 4.36 Restoration of row **f**: **a** original row, **b** degraded row with $n = 2$, **c**, **d**, **e** and **f** four possible solution that satisfy $\mathbf{H}\hat{\mathbf{f}} = \mathbf{g}$

lations can be reduced by minimization of the distance between f_k and g_k, i.e., we take a vector as a sharp solution of $\mathbf{Hf} = \mathbf{g}$, so that this presents the smallest distance between original signal and blurred signal: we seek then to minimize the objective function

$$J(\mathbf{f}, \mathbf{g}) = \sum_{k=1}^{N} (f_k - g_k)^2 \rightarrow \min. \tag{4.29}$$

The application of criteria of the minimization of the norm between input and output (MINIO) does not mean that **f** is equal to **g** because this solution does not satisfy the system of equations (4.28) and the size of **f** and **g** are different. The solution also is defined as the vector in the solution space of the underdetermined system $\mathbf{Hf} = \mathbf{g}$ whose first N components has the minimum distance to the measured data, i.e., where the first N elements are of **f**. We can express vector $\hat{\mathbf{f}} = \mathbf{Pf}$, with **f** a $N \times M$ matrix which projects the vector **f** on the support of **g**:

$$\mathbf{P} = \begin{bmatrix} 1\ 0\ ...\ 0\ 0\ ...\ 0 \\ 0\ 1\ ...\ 0\ 0\ ...\ 0 \\ \quad\vdots\qquad\quad\vdots\ 0 \\ 0\ 0\ ...\ 1\ 0\ ...\ 0 \end{bmatrix}. \tag{4.30}$$

The original optimization problem is now:

$$\hat{\mathbf{f}} = \underset{\mathbf{f}}{\operatorname{argmin}} \parallel \mathbf{Pf} - \mathbf{g} \parallel^2 \tag{4.31}$$

subject to the constraint $\| \mathbf{Pf} - \mathbf{g} \|^2 = \mathbf{0}$. Applying the technique of *Lagrange multipliers* this problem can be alternatively formulated as an optimization problem without constraints:

$$V(\mathbf{f}) = \lambda \parallel \mathbf{Hf} - \mathbf{g} \parallel^2 + \parallel \mathbf{Pf} - \mathbf{g} \parallel^2 \rightarrow \min, \qquad (4.32)$$

if λ is large enough (e.g., $\lambda = 10^6$). The solution of this problem can be easily obtained by computing the partial derivative of criterion V with respect to the unknown $\mathbf{f}$:

$$\frac{\partial}{\partial \mathbf{f}} V(\mathbf{f}) = 2\lambda \mathbf{H}^\mathsf{T}(\mathbf{Hf} - \mathbf{g}) + 2\mathbf{P}^\mathsf{T}(\mathbf{Pf} - \mathbf{g}) = \mathbf{0}, \qquad (4.33)$$

then is

$$\hat{\mathbf{f}} = \left[\lambda \mathbf{H}^\mathsf{T}\mathbf{H} + \mathbf{P}^\mathsf{T}\mathbf{P} \right]^{-1} \left[\lambda \mathbf{H} + \mathbf{P} \right]^\mathsf{T} \mathbf{g}. \qquad (4.34)$$

This solution for the example of Fig. 4.36b is almost identical to the original sharp input signal of Fig. 4.36a. Figure 4.37 shows three different restoration examples.

Python Example 4.12: In this example, we simulate an X-ray image that has been degraded by a horizontal motion. The image is restored using MINIO algorithm (4.34):

Listing 4.12 : X-ray image restoration.

```python
import numpy as np
import matplotlib.pylab as plt
import cv2 as cv

from pyxvis.io import gdxraydb
from pyxvis.processing.images import res_minio

image_set = gdxraydb.Baggages()
img = image_set.load_image(46, 90)

n = 128
h = np.ones((1, n)) / n

img_g = cv.filter2D(img.astype('double'), cv.CV_64F, h)
fs = res_minio(img_g, h, method='minio')

fig, ax = plt.subplots(1, 3, figsize=(16, 8))
ax[0].imshow(img, cmap='gray')
ax[0].set_title('Original image')
ax[0].axis('off')
ax[1].imshow(img_g, cmap='gray')
ax[1].set_title('Degraded image')
ax[1].axis('off')
ax[2].imshow(fs, cmap='gray')
ax[2].set_title('Restored image')
ax[2].axis('off')
plt.show()
```

The output of this code is shown in the last row of Fig. 4.37. Details of the baggage are not discernible in the degraded image but are recovered in the restored image.

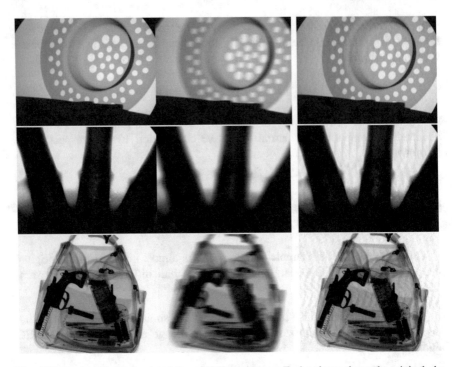

Fig. 4.37 Restoration in simulated degraded X-ray images. Each column shows the original, the degraded with n pixels, and the restored images. The size of the images are respectively 574 × 768, with $n = 30$; 574 × 768, with $n = 40$; and 2208 × 2688, with $n = 128$. [→ Example 4.12 🐍]

In this code, we use function res_minio of **pyxvis** Library. This function computes MINIO restoration algorithm as defined in (4.34).   □

The restoration quality is equally as good as the classical methods (see for example [5]), while the computation load is decreased considerably (see comparisons in [12]).

4.7   Summary

In this chapter, we covered the main techniques of image processing used in X-ray testing.

They are:

- Image preprocessing: Noise removal, contrast enhancement, and shading correction.
- Image Filtering: linear and non-linear filtering.
- Edge detection: Gradient estimation, Laplacian-of-Gaussian, and Canny.

- Image segmentation: Thresholding, region growing, and maximally stable extremal regions.
- Image restoration: Minimization of the norm between input and output.

The chapter provided a good overview, presenting several methodologies with examples using real and simulated X-ray images.

References

1. Boerner, H., Strecker, H.: Automated X-ray inspection of aluminum casting. IEEE Trans. Pattern Anal. Mach. Intell. **10**(1), 79–91 (1988)
2. Canny, J.: A computational approach to edge detection. IEEE Trans. Pattern Anal. Mach. Intell. **PAMI-8**(6), 679–698 (1986)
3. Castleman, K.: Digital Image Processing. Prentice-Hall, Englewood Cliffs (1996)
4. Faugeras, O.: Three-Dimensional Computer Vision: A Geometric Viewpoint. The MIT Press, Cambridge (1993)
5. Gonzalez, R., Woods, R.: Digital Image Processing, 3rd edn. Prentice Hall, Pearson (2008)
6. Haralick, R., Shapiro, L.: Computer and Robot Vision. Addison-Wesley Publishing Co., New York (1992)
7. Heinrich, W.: Automated inspection of castings using X-ray testing. Ph.D. thesis, Institute for Measurement and Automation, Faculty of Electrical Engineering, Technical University of Berlin (1988). (in German)
8. Matas, J., Chum, O., Urban, M., Pajdla, T.: Robust wide-baseline stereo from maximally stable extremal regions. Image Vis. Comput. **22**(10), 761–767 (2004)
9. MathWorks: Image Processing Toolbox for Use with MATLAB: User's Guide. The Math-Works Inc. (2014)
10. Mery, D.: Crossing line profile: a new approach to detecting defects in aluminium castings. In: Proceedings of the Scandinavian Conference on Image Analysis (SCIA 2003). Lecture Notes in Computer Science, vol. 2749, pp. 725–732 (2003)
11. Mery, D., Berti, M.A.: Automatic detection of welding defects using texture features. Insight-Non-Destruct. Testing Condit. Monitor. **45**(10), 676–681 (2003)
12. Mery, D., Filbert, D.: A fast non-iterative algorithm for the removal of blur caused by uniform linear motion in X-ray images. In: Proceedings of the 15th World Conference on Non-Destructive Testing (WCNDT–2000). Rome (2000)
13. Mery, D., Filbert, D.: Automated flaw detection in aluminum castings based on the tracking of potential defects in a radioscopic image sequence. IEEE Trans. Robot. Autom. **18**(6), 890–901 (2002)
14. Mery, D., Pedreschi, F.: Segmentation of colour food images using a robust algorithm. J. Food Eng. **66**(3), 353–360 (2004)

Chapter 5
X-ray Image Representation

Abstract In this chapter, we cover several topics that are used to represent an X-ray image (or a specific region of an X-ray image). This representation means that new features are extracted from the original image that can give us more information than the raw information expressed as a matrix of gray values. This kind of information is extracted as features or descriptors, i.e., a set of values, that can be used in pattern recognition problems such as object recognition, defect detection, etc. The chapter explains geometric and intensity features, and local descriptors and sparse representations that are very common in computer vision applications. It is worthwhile to mention, that the features mentioned in this chapter are called *handcrafted* features, in contrast to the *learned* features that are explained in Chap. 7 using deep learning techniques. Finally, the chapter addresses some feature selection techniques that can be used to chose which features are relevant in terms of extraction.

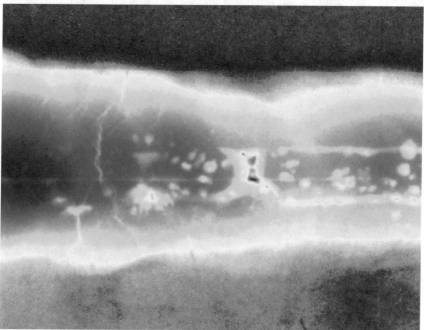

Cover image: *Welding defects (from X-ray image* W0001_0001, *well known as BAM5, colored with 'sinmap' colormap).*

D. Mery and C. Pieringer, *Computer Vision for X-Ray Testing*,
https://doi.org/10.1007/978-3-030-56769-9_5

169

5.1   Introduction

As we learned in the previous chapter, in image processing for X-ray testing, segmentation is used to detect (potential) regions that can be the objects of interest that
we are looking for (see Sect. 4.5). As segmented potential regions frequently set
off false detections, an analysis of the segmented regions can significantly improve
the effectiveness of detection. Measuring certain characteristics of the segmented
regions (*feature extraction*) can help us to distinguish the false detection, although
some of the features extracted are either irrelevant or are not correlated. Therefore,
a *feature selection* must be performed. Depending on the values returned for the
selected features, we can try to classify each segmented potential region in one of
the following two classes: *background* or *object of interest*.

In this chapter, we will explain several features that are normally used in image
analysis and computer vision for X-ray testing. In our description, features will be
divided into two groups: *geometric* and *intensity* features. Furthermore, we will
cover some local descriptors and sparse representations that can be used in many
X-ray testing applications. In this chapter, we shall concentrate on the extraction
and selection of features, whereas in the following chapter, we will discuss the classification problem itself.

We will use Fig. 5.1 as our example in the description of features. In our example, we use an X-ray image of a circular defect. The segmentation is a binary image
that gives information about the pixels that belongs to our object of interest (the
defect). Geometric features are extracted from this binary image. Moreover, intensity features are extracted from the intensity image considering the pixels of the
segmentation. Some intensity features consider only the gray values inside the segmented region, other ones take into account both gray values inside and outside the
region (e.g., contrast).

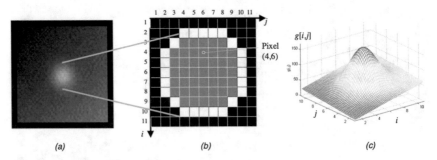

Fig. 5.1 Example of a region: **a** X-ray image, **b** segmented region (gray pixels), **c** 3D representation of the gray values

5.2   Geometric Features

These provide information on the location, size, and shape of the segmented region. Location and size features, such as center of mass, perimeter, Height, and width, are given in pixels. Shape features are usually coefficients without units. It is worth mentioning that we distinguish three different zones in the segmented image (see Fig. 5.1b), the segmented region (gray zone $\Re$), the boundary (white edge pixels ℓ), and the background (black zone).

5.2.1   Basic Geometric Features

In this section, we will summarize basic geometric features that can be easily extracted.

Height and Width
The eight and width of a region can be defined as:

$$h = i_{max} - i_{min} + 1 \qquad \text{and} \qquad w = j_{max} - j_{min} + 1, \qquad (5.1)$$

where i_{max} and i_{min} is the maximal and minimal value that takes coordinate i in the region respectively. The same is valid for j_{max} and j_{min} in j-direction. In our example of Fig. 5.1, $h = w = 7$ pixels.

Area and Perimeter
We define the area A of a region as the number of pixels that belong to the region. On the other hand, the perimeter L is the number of pixels that belong to the boundary. In the region of Fig. 5.1, the area and the perimeter are $A = 45$ and $L = 24$ pixels respectively. More accurate measurements for area and perimeter can also be estimated [6]: for instance, the boundary of the region can be fitted to a curve with known area and length (in our example the boundary can be fitted to a circle with radius $r = 4$ pixels, so $A = \pi r^2 = 50.26$ pixels and $L = 2\pi r = 25.13$ pixels), however, the computational time of such approaches can be extremely long if there are thousands of regions to be measured. Moreover, the shape of the region can be much more complex than a simple circle as shown in Fig. 4.30. We should remember, therefore, that the goal of feature extraction is not the accurate measurement, rather it is simply the extraction of features that can be used in a classification approach to separate our classes (objects of interest from background). Thus, it is not relevant that the measurement of the area of the region is just 45 pixels and not 50.26 pixels.

Center of Mass
This provides information about the location of the region. It is computed as the average of coordinate i and coordinate j in pixels that belong to region $\Re$:

$$\bar{i} = \frac{1}{A} \sum_{i \in \Re} i \qquad \bar{j} = \frac{1}{A} \sum_{j \in \Re} j, \qquad\qquad (5.2)$$

where A is the area of the region, i.e., the number of pixels of the region.

Roundness
Shape features are usually attributed coefficients without units. An example is roundness that is defined as

$$R = \frac{4 \cdot A \cdot \pi}{L^2} \qquad\qquad (5.3)$$

The roundness R is a value between 1 and 0. $R = 1$ means a circle, and $R = 0$ corresponds to a region without an area. In our example, $R = 4 \cdot 45 \cdot \pi / 24^2 = 0.98$.

Other Basic Features
There are some useful features that can be extracted employing pybalu Library and pyxvis Library:

- Danielson factor: A shape factor based on the distance transform of region $\Re$ [8].
- Euler Number: The number of objects in the region $\Re$ minus the number of holes in those objects [38].
- Equivalent Diameter: The diameter of a circle with the same area as the region $\Re$ [38].
- Major Axis Length and Minor Axis Length: The length (in pixels) of the major and minor axes of the ellipse that has the same normalized second central moments as region $\Re$ [38].
- Orientation: The angle (in degrees ranging from -90 to 90 degrees) between the x-axis and the major axis of the ellipse that has the same second moments as region $\Re$ [38].
- Solidity: The proportion of the pixels in the convex hull that are also in region $\Re$ [38].
- Extent: The ratio of pixels in region $\Re$ to pixels in the total bounding box [38].
- Eccentricity: The eccentricity of the ellipse that has the same second moments as region $\Re$ [38].
- Convex Area: area of the convex hull the region $\Re$ [38].
- Filled Area: area of the filled region $\Re$ [38].

All basic geometric features explained in this section can be extracted by function extract_features of pyxvis Library with parameters 'basicgeo' and bw=R, where R is the binary image from which the features are extracted. An example is shown in Table 5.1, where the basic 15 geometric features (divided by 1000) are presented for ten regions of Fig. 5.2: f_1: Center of mass in i direction. f_2: Center of mass in j direction. f_3: Height. f_4: Width. f_5: Area. f_6: Perimeter. f_7: Roundness. f_8: Danielsson factor. f_9: Euler Number. f_{10}: Equivalent Diameter. f_{11}: Major Axis Length. f_{12}: Minor Axis Length. f_{13}: Orientation. f_{14}: Solidity. f_{15}: Extent. f_{16}: Eccentricity. f_{17}: Convex Area. f_{18}: Filled Area.

Table 5.1 Basic geometric features of ten apples (see Fig. 5.2) [→ Example 5.1]

	1	2	3	4	5	6	7	8	9	10
f_1	0.2371	0.3011	0.3283	0.3230	0.4653	0.4721	0.4604	0.4936	0.5459	0.6325
f_2	1.4339	0.4666	0.2972	1.2602	0.4729	1.4571	1.6117	0.3225	1.3124	0.4238
f_3	0.1480	0.1670	0.1690	0.1480	0.1510	0.1580	0.1380	0.1490	0.1430	0.1740
f_4	0.1500	0.1640	0.1580	0.1370	0.1260	0.1530	0.1530	0.1520	0.1470	0.1710
f_5	15.9180	19.3980	19.1130	14.3240	15.0450	18.2380	16.9430	16.2320	15.4250	20.4780
f_6	0.4642	0.5278	0.5102	0.4975	0.4700	0.4998	0.4813	0.4888	0.4625	0.6552
f_7	0.0009	0.0009	0.0009	0.0007	0.0009	0.0009	0.0009	0.0009	0.0009	0.0006
f_8	0.0019	0.0019	0.0019	0.0022	0.0017	0.0018	0.0016	0.0019	0.0018	0.0024
f_9	0.0010	-0.0010	0.0010	0.0020	-0.0010	0.0010	0.0010	0.0010	0.0010	-0.0050
f_{10}	0.1424	0.1572	0.1560	0.1350	0.1384	0.1524	0.1469	0.1438	0.1401	0.1615
f_{11}	0.1516	0.1661	0.1657	0.1569	0.1536	0.1608	0.1559	0.1526	0.1496	0.1791
f_{12}	0.1353	0.1512	0.1490	0.1191	0.1257	0.1458	0.1400	0.1393	0.1327	0.1494
f_{13}	0.0514	0.0330	-0.0735	0.0621	0.0833	-0.0698	0.0095	-0.0536	0.0456	0.0407
f_{14}	0.0010	0.0010	0.0010	0.0009	0.0010	0.0010	0.0010	0.0010	0.0010	0.0010
f_{15}	0.0007	0.0007	0.0007	0.0007	0.0008	0.0008	0.0008	0.0007	0.0007	0.0007
f_{16}	0.0005	0.0004	0.0004	0.0007	0.0006	0.0004	0.0004	0.0004	0.0005	0.0006
f_{17}	16.3180	20.0000	19.7450	15.1820	15.5005	18.7300	17.4140	17.0750	15.8885	21.5085
f_{18}	15.9180	19.4000	19.1130	14.3240	15.0470	18.2380	16.9430	16.2330	15.4250	20.5360

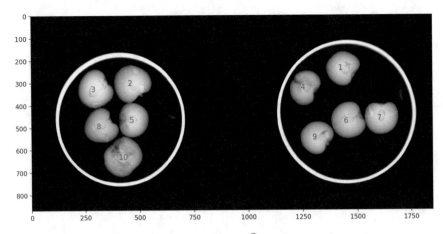

Fig. 5.2 X-ray image of ten apples. [→ Example 5.1]

Python Example 5.1: In this example, we show how to extract the basic geometric features of ten apples as segmented in Fig. 5.2. The segmentation in this
example is performed by thresholding the X-ray image and by selecting those segmented objects that present an appropriate size (area).

Listing 5.1 : Basic geometric features

```python
import numpy as np
import matplotlib.pyplot as plt
from skimage.measure import label
from pyxvis.features.extraction import extract_features

# Input Image
fig    = plt.figure()
ax     = fig.add_subplot(111)
img    = plt.imread('../images/N0001_0004b.png')
implot = plt.imshow(img,cmap='gray')

# Segmentation
R   = img>0.27              # thresholding of light objects
L   = label(R)              # labeling of objects
n   = np.max(L)             # number of detected objects
T   = np.zeros((n,18))      # features of each object will stored in a row

# Analysis of each segmented object
t    = 0 # count of recognized fruits
for i in range(n):
    R = (L == i)*1                            # binary image of object i
    f = extract_features('basicgeo',bw=R)     # feature extraction for object i
    area = f[4]
    # recognition of fruits according to the size
    if area>14000 and area<21000:
        T[t,:] = f                            # storing the features of the fruit t
        t = t+1
        # labeling each recognized fruit in the plot
        ax.text(f[1]-20, f[0]+10, str(t), fontsize=12,color='Red')

# Display and save results
```

```
plt.show()
F = T[0:t,:]
print('Basic Geo—Features:')
print(F)
np.save('GeoFeatures.npy',F)          # save features
```

The output of this code is shown in Fig. 5.2 and Table 5.1. The reader can observe the use of function extract_features to extract the basic geometric features.   □

5.2.2 Elliptical Features

Elliptical features can be used to extract information about location, size, and shape of a region. They are extracted from a fitted ellipse to the boundary of the region [14]. From this ellipse, we can extract the center, the length of the axes, the orientation, and the eccentricity.

The pixels of the boundary are defined as (x_i, y_i) for $i = 1 \ldots L$. It is well known that an ellipse is defined as

$$ax^2 + bxy + cy^2 + dx + ey + f = 0, \tag{5.4}$$

that can be written as $\mathbf{a}^\mathsf{T}\mathbf{x} = 0$, where $\mathbf{a} = [a\ b\ c\ d\ e\ f]^\mathsf{T}$ is a vector that includes the parameters of the ellipse and $\mathbf{x} = [x^2\ xy\ y^2\ x\ y\ 1]^\mathsf{T}$ is a vector that includes the coordinates of a point (x, y) that lies on the ellipse.

If our region is elliptical, then for each point (x_i, y_i), we have $\mathbf{a}^\mathsf{T}\mathbf{x}_i = 0$ with $\mathbf{x}_i = [x_i^2\ x_i y_i\ y_i^2\ x_i\ y_i\ 1]^\mathsf{T}$. Nevertheless, in practice, the regions are not perfectly elliptical, not only because real regions have different shapes but also there is a discretization error when forming a digital image. For this reason, we look for a vector $\mathbf{a}$ so that $\mathbf{a}^\mathsf{T}\mathbf{x}_i \to \min$ for every point $i = 1 \ldots L$. That is, we can formulate the estimation of the parameters of the ellipse as an optimization problem as follows:

$$\| \mathbf{X}\mathbf{a} \| \to \min, \tag{5.5}$$

where $\mathbf{X}$ is matrix with L rows whose ith row is $\mathbf{x}_i^\mathsf{T}$. Usually, a solution can be found by minimizing (5.5) subject to $\| \mathbf{a} \| = 1$. In this case, $\mathbf{a}$ is the last column of matrix $\mathbf{V}$, where $\mathbf{X} = \mathbf{U}\mathbf{S}\mathbf{V}^\mathsf{T}$ is the singular value decomposition (SVD) of $\mathbf{X}$ [21].

The elliptical features can be extracted by writing (5.4) as follows:

$$\left(\frac{x - x_0}{a_e}\right)^2 + \left(\frac{y - y_0}{b_e}\right)^2 = 1, \tag{5.6}$$

where

$$a_e = \frac{1}{\sqrt{s\ a_p}}, \, b_e = \frac{1}{\sqrt{s\ b_p}} \tag{5.7}$$

with

$$s = \frac{1}{v - f} \qquad v = \mathbf{t}^\mathsf{T} \mathbf{T} \mathbf{t}$$

$$\mathbf{T} = \begin{bmatrix} a & b/2 \\ b/2 & c \end{bmatrix} \qquad \mathbf{t} = \begin{bmatrix} x_0 \\ y_0 \end{bmatrix} = \frac{1}{2} \mathbf{T}^{-1} \begin{bmatrix} d \\ e \end{bmatrix}$$

$$a_p = a \cos^2(\alpha) + b \cos(\alpha) \sin(\alpha) + c \sin^2(\alpha)$$
$$b_p = a \sin^2(\alpha) - b \cos(\alpha) \sin(\alpha) + c \cos^2(\alpha)$$

and

$$\alpha = \frac{1}{2} \arctan \left(\frac{b}{a - c} \right). \tag{5.8}$$

The axes of the ellipse are defined by a_e and b_e, the center of the ellipse is located on (x_0, y_0) and the orientation is α. Thus, the eccentricity is defined by

$$e_x = \frac{\min(a_e, b_e)}{\max(a_e, b_e)}. \tag{5.9}$$

For circular shapes, the eccentricity as the roundness (5.3), takes values between 0 and 1, where 1 means a perfect circle.

Python Example 5.2: In this example, we show how to extract elliptical features of a shape. We test this approach on an X-ray of a cherry with an elliptical shape as shown in Fig. 5.3.

> **Listing 5.2 : Elliptical boundary of a fruit**

```python
import matplotlib.pyplot as plt

from pyxvis.processing.segmentation import seg_bimodal
from pyxvis.features.extraction import extract_features
from pyxvis.io.plots import plot_ellipses_image


img    = plt.imread('../images/N0006_0003b.png') # input image with a fruit
R,_,   = seg_bimodal(img)                         # segmentation
fxell  = extract_features('ellipse',bw=R)         # extraction of elliptical features
print('Elliptical Features:')                     # show results
print(fxell)                                       # print elliptical features
plot_ellipses_image(img,fxell)                     # draw ellipse onto image
```

The output of this code is shown in Fig. 5.3. The elliptical features are extracted by function extract_features of pyxvis Library with parameters 'ellipse' and bw=R, where R is the binary image from which the features are extracted. Additionally,

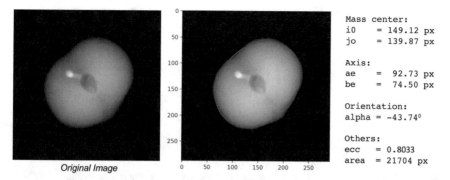

Original Image

Mass center:
i0   = 149.12 px
jo   = 139.87 px

Axis:
ae   = 92.73 px
be   = 74.50 px

Orientation:
alpha = -43.74⁰

Others:
ecc   = 0.8033
area = 21704 px

Fig. 5.3 Elliptical features of a cherry. In this example, the extracted features are the coordinates of the center of the ellipse ($i_0 = 149.12$, $j_0 = 139.87$), the estimated length of each axis are 92.73 and 74.50 pixels, the orientation (with respect to vertical axis in a counterclockwise direction) is -43.74^0, the eccentricity is 0.8033, and the area is 21704 pixels. [→ Example 5.2 🐍]

the ellipse can be superimposed onto the original X-ray image using function plot_ellipses_image of pyxvis Library to see the estimated ellipse. ☐

5.2.3 Fourier Descriptors

Shape information—invariant to scale, orientation and position—can be measured using *Fourier descriptors* [5, 52, 65]. The coordinates of the pixels of the boundary are arranged as a complex number $i_k + j \cdot j_k$, with $j = \sqrt{-1}$ and $k = 0, ..., L - 1$, where L is the perimeter of the region, and pixel k and $k + 1$ are connected. The complex boundary function can be considered as a periodical signal of period L. The discrete Fourier transformation [4] gives a characterization of the shape of the region. The Fourier coefficients are defined by

$$F_n = \sum_{k=0}^{L-1}(i_k + j \cdot j_k)e^{-j\frac{2\pi kn}{L}} \qquad \text{for} n = 0, ..., L - 1. \qquad (5.10)$$

The Fourier descriptors correspond to the coefficients F_n for $n > 0$. The Fourier coefficient F_0 is not used because it gives information about the location of the region. The magnitude and phase of Fourier descriptors give information about orientation and symmetry of the region. In general, only the magnitude $|F_n|$ is used. Fourier descriptors are invariant under rotation. The Fourier descriptors of our example in Fig. 5.1a are illustrated in Fig. 5.4. The first pixel of the periodic function is $(i_0, j_0) = (6, 10)$. In case the region is a perfect circle, $|F_n| = 0$ for $1 < n < L$ because (i_k, j_k) represent a perfect sinusoid. In our example, the region is not a

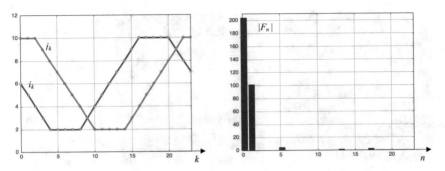

Fig. 5.4 Coordinates of the boundary of region of Fig. 5.1 and the Fourier descriptors

perfect circle, however, as we can see the Fourier descriptors are very small for $2 < n < L$.

Fourier descriptors can be extracted using function extract_features of pyxvis Library with parameters 'fourier' and bw=R, where R is the binary image from which the features are extracted.

5.2.4   Invariant Moments

The statistical moments are defined by

$$m_{rs} = \sum_{i,j \in \Re} i^r j^s \qquad \text{for } r, s \in \mathbb{N}, \tag{5.11}$$

where $\Re$ is the set of pixels that belong to the region (see gray pixels in Fig. 5.1b). In this example, pixel $(i = 4, j = 6) \in \Re$. The parameter $r + s$ corresponds to the order of the moment. The reader can demonstrate that the zeroth moments m_{00} is equal to the area A of the region. Moreover, the center of mass of the region is easily defined by

$$\bar{i} = \frac{m_{10}}{m_{00}} \qquad \bar{j} = \frac{m_{01}}{m_{00}}. \tag{5.12}$$

The reader can compare this definition with (5.2). The coordinates of the center of mass can be computed using function extract_features of pyxvis Library with parameters 'centroid' and bw=R, where R is the binary image from which the centroid is computed.

The center of mass and statistical moments of higher order, however, are not invariant to the location of the region. This can be useful for detecting objects that must be in certain locations. Nevertheless, when objects of interest may be everywhere in the image we must use features that are invariant to the position. Using the center of mass, the central moments are defined. They are invariant to the position:

$$\mu_{rs} = \sum_{i,j \in \mathfrak{R}} (i - \bar{i})^r (j - \bar{j})^s \qquad \text{for } r, s \in \mathbb{N}. \tag{5.13}$$

Other known moments that can be used are the well-known Hu-moments [22, 57]. These were developed using the central moments as follows:

$$
\begin{aligned}
\phi_1 &= \eta_{20} + \eta_{02} \\
\phi_2 &= (\eta_{20} - \eta_{02})^2 + 4\eta_{11}^2 \\
\phi_3 &= (\eta_{30} - 3\eta_{12})^2 + (3\eta_{21} - \eta_{03})^2 \\
\phi_4 &= (\eta_{30} + \eta_{12})^2 + (\eta_{21} + \eta_{03})^2 \\
\phi_5 &= (\eta_{30} - 3\eta_{12})(\eta_{30} + \eta_s 12)[(\eta_{30} + \eta_{12})^2 - 3(\eta_{21} + \eta_{03})^2] + \\
&\quad (3\eta_{21} - \eta_{03})(\eta_{21} + \eta_{03})[3(\eta_{30} + \eta_{12})^2 - (\eta_{21} + \eta_{03})^2] \\
\phi_6 &= (\eta_{20} - \eta_{02})[(\eta_{30} + \eta_{12})^2 - (\eta_{21} + \eta_{03})^2] + \\
&\quad 4\eta_{11}(\eta_{30} + \eta_{12})(\eta_{21} + \eta_{03}) \\
\phi_7 &= (3\eta_{21} - \eta_{03})(\eta_{30} + \eta_{12})[(\eta_{30} + \eta_{12})^2 - 3(\eta_{21} + \eta_{03})^2] - \\
&\quad (\eta_{30} - 3\eta_{12})(\eta_{21} + \eta_{03})[3(\eta_{30} + \eta_{12})^2 - (\eta_{21} + \eta_{03})^2]
\end{aligned}
\tag{5.14}
$$

with

$$\eta_{rs} = \frac{\mu_{rs}}{\mu_{00}^t} \qquad t = \frac{r+s}{2} + 1.$$

Hu-moments are invariant to translation, rotation, and scale. That means that regions that have the same shape, but have a different size, location, and orientation, will have similar Hu-moments.

In addition, there are similar invariant features, called Gupta moments, that are derived from the pixels of the boundary (instead of the region) [19]. They are invariant to translation, rotation, and scale.

Sometimes, it is necessary to have features that are invariant to affine transformation as well (see Sect. 3.2.2). For this reason Flusser moments, i.e., features invariant to translation, rotation, scale, and affine transformations were derived from second- and third-order central moments [15, 56]:

$$
\begin{aligned}
I_1 &= (\mu_{20}\mu_{02} - \mu_{11}^2)/\mu_{00}^4 \\
I_2 &= (\mu_{30}^2\mu_{03}^2 - 6\mu_{30}\mu_{21}\mu_{12}\mu_{03} + 4\mu_{30}\mu_{12}^3 + 4\mu_{21}^3\mu_{03} - 3\mu_{21}^2\mu_{12}^2)/\mu_{00}^{10} \\
I_3 &= (\mu_{20}(\mu_{21}\mu_{03} - \mu_{12}^2) - \mu_{11}(\mu_{30}\mu_{03} - \mu_{21}\mu_{12}) + \mu_{02}(\mu_{30}\mu_{12} - \mu_{21}^2))/\mu_{00}^7 \\
I_4 &= (\mu_{20}^3\mu_{03}^2 - 6\mu_{20}^2\mu_{11}\mu_{12}\mu_{03} - 6\mu_{20}^2\mu_{02}\mu_{21}\mu_{03} + 9\mu_{20}^2\mu_{02}\mu_{12}^2 \\
&\quad + 12\mu_{20}\mu_{11}^2\mu_{21}\mu_{03} + 6\mu_{20}\mu_{11}\mu_{02}\mu_{30}\mu_{03} - 18\mu_{20}\mu_{11}\mu_{02}\mu_{21}\mu_{12} \\
&\quad - 8\mu_{11}^3\mu_{30}\mu_{03} - 6\mu_{20}\mu_{02}^2\mu_{30}\mu_{12} + 9\mu_{20}\mu_{02}^2\mu_{21} \\
&\quad + 12\mu_{11}^2\mu_{02}\mu_{30}\mu_{12} - 6\mu_{11}\mu_{02}^2\mu_{30}\mu_{21} + \mu_{02}^3\mu_{30}^2)/\mu_{00}^{11}.
\end{aligned}
\tag{5.15}
$$

Python Example 5.3: In this example, we show how to measure invariant moments that can be used as a shape feature of objects of interest. We tested this approach on an X-ray containing ten apples. We superimpose onto this image four

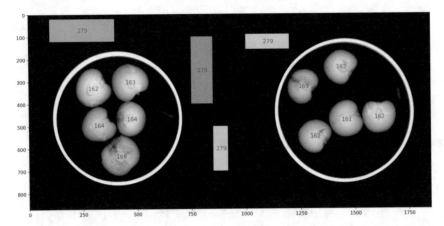

Fig. 5.5 First Hu-moment (ϕ_1) of apples and rectangles. Since ϕ_1 for apples is approximately 163 and for these rectangles is 279, it is evident that this feature can be used to discriminate them from each other. [$\rightarrow$ Example 5.3 🐍]

rectangles the size of which is $a \times b$ pixels (where $b = 3a$). The rectangles are located in horizontal and vertical directions as shown in Fig. 5.5. Thus, we can simulate an input X-ray image containing apples and rectangles. The idea is to separate them. We see that the first Hu-moment can be used to effectively discriminate apples from rectangles.

Listing 5.3 : Detection using invariant moments

```python
import matplotlib.pyplot as plt
import numpy as np
from skimage.measure import label
from pyxvis.features.extraction import extract_features

fig     = plt.figure()
ax      = fig.add_subplot(111)
img     = plt.imread('../images/N0001_0004b.png')
img[100:399,750:849]  = 0.5
img[500:699,850:916]  = 0.75
img[20:119,100:399]   = 0.6
img[90:156,1000:1199] = 0.75
implot = plt.imshow(img,cmap='gray')
R       = img>0.27    # segmentation
L       = label(R)    # labeling
n       = np.max(L)   # number of segmented objects
t       = 0
T       = np.zeros((n,7))
for i in range(n):
    R = (L == i)*1                      # binary image of object i
    fx = ['basicgeo','hugeo']
    f = extract_features(fx,bw=R)       # feature extraction
    area = f[4]
    # recognition of fruits according to the size
    if area>10000 and area<31000:
        h       = f[18:]                # hu moments
        T[t,:] = h
        t       = t+1
```

```
        x       = round(1000*h[0])       # first hu moment
        ax.text(f[1]-20, f[0]+10, str(int(x)), fontsize=12,color='Red')
plt.show()
F = T[0:t,:]
print('Hu Features:')
print(F)
np.save('HuFeatures.npy',F)              # save features
```

The output of this code is shown in Fig. 5.5. In this example, the features (basic geometric features for centroid and area, and Hu-moments) are computed by function extract_features of pyxvis Library with parameters ['basicgeo','hugeo'] and bw=R, where R is the binary image from which the features are extracted. The output of this function is a vector f computed by concatenation of two vectors, one for the basic geometric features (of 18 elements) and one for the Hu-moments (of 7 elements). Thus, the first Hu-moment is stored in f[18]. The reader can test Flusser and Gupta moments using functions parameters 'flusser' and 'gupta' respectively in function extract_features.   □

5.3   Intensity Features

These provide information about the intensity of a region. For gray value images, e.g.., X-ray images, there is only one intensity channel. The following features are computed using the gray values in the image, where $x(i, j)$ denotes the gray value of pixel (i, j).

5.3.1   Basic Intensity Features

In this section, we summarize basic intensity features that can be easily extracted.

Mean gray value
The mean gray value of the region is computed as

$$G = \frac{1}{A} \sum_{i,j \in \Re} x(i, j),$$ (5.16)

where $\Re$ is the set of pixels of the region and A the area. A 3D representation of the gray values of the region and its neighborhood of our example is shown in Fig. 5.1. In this example, $G = 121.90$ ($G = 0$ means 100% black and $G = 255$ corresponds to 100% white).

Mean Gradient in the Boundary
This feature gives information about the change of the gray values in the boundary of the region. It is computed as

$$C = \frac{1}{L} \sum_{i,j \in \ell} x'(i, j), \tag{5.17}$$

where $x'(i, j)$ means the gradient of the gray value function in pixel (i, j) (see Sect. 4.4.1) and ℓ the set of pixels that belong to the boundary of the region. The number of pixels of this set corresponds to L, the perimeter of the region. Using a Gaussian gradient operator in our example in Fig. 5.1, we obtain $C = 35.47$.

Mean Second Derivative
This feature is computed as

$$D = \frac{1}{A} \sum_{i,j \in \Re} x''(i, j), \tag{5.18}$$

where $x''(i, j)$ denotes the second derivate of the gray value function in pixel (i, j). The Laplacian-of-Gauss (LoG) operator can be used to calculate the second derivate of the image. If $D > 0$, we have a region that is darker than its neighborhood as shown in Fig. 4.18.

Other Basic Features
A simple texture feature is the local variance [24]. This is given by:

$$\sigma_g^2 = \frac{1}{4hb + 2h + 2b} \sum_{i=1}^{2h+1} \sum_{j=1}^{2b+1}, (g(i, j) - \bar{g})^2 \tag{5.19}$$

where $\bar{g}$ denotes the mean gray value in the zone.

Other basic intensity features such as kurtosis and skewness can be computed as (5.19). All intensity geometric features explained in this section can be extracted by command `basic_int_features` of pybalu library. An example is shown in Table 5.2, where the basic six intensity features are presented for ten regions of Fig. 5.2: f_1: Intensity mean. f_2: Intensity standard deviation. f_3: Intensity kurtosis. f_4: Intensity skewness. f_5: Mean Laplacian. f_6: Mean boundary gradient.

Table 5.2 Basic intensity features of apples of Fig. 5.2 [→ Example 5.4 🐍]

	1	2	3	4	5	6	7	8	9	10
f_1	0.5974	0.6213	0.6600	0.5416	0.6064	0.5999	0.6294	0.5704	0.5892	0.5321
f_2	0.1482	0.1651	0.1744	0.1283	0.1671	0.1492	0.1563	0.1426	0.1474	0.1175
f_3	2.1264	2.0493	2.2211	2.1782	1.8771	2.0972	2.2371	1.9955	2.0586	2.4228
f_4	−0.5083	−0.3613	−0.4826	−0.4042	−0.3009	−0.4695	−0.5713	−0.3677	−0.4328	−0.3964
f_5	−0.0011	−0.0013	−0.0012	−0.0013	−0.0012	−0.0011	−0.0013	−0.0011	−0.0013	−0.0012
f_6	0.0360	0.0365	0.0397	0.0313	0.0345	0.0356	0.0438	0.0329	0.0355	0.0307

Python Example 5.4: In this example, we show how to extract basic intensity features of ten apples as segmented in Fig. 5.2.

Listing 5.4 : Basic intensity features

```python
import numpy as np
import matplotlib.pyplot as plt
from skimage.measure import label
from pyxvis.features.extraction import extract_features

fig    = plt.figure()
ax     = fig.add_subplot(111)
img    = plt.imread('../images/N0001_0004b.png')
implot = plt.imshow(img,cmap='gray')
R      = img>0.27        # segmentation
L      = label(R)        # labeling
n      = np.max(L)       # number of segmented regions
t      = 0
T      = np.zeros((n,6))
for i in range(n):
    R = (L == i)*1     # binary image of object i
    f = extract_features('basicgeo',bw=R)
    area = f[4]
    # recognition of fruits according to the size
    if area>14000 and area<21000:
        # extract int features only in the segmented region
        h       = extract_features('basicint',img=img,bw=R)
        T[t,:] = h
        t = t+1
        ax.text(f[1]-20, f[0]+10, str(t), fontsize=12,color='Red')
plt.show()
F = T[0:t,:]
print('Basic Int-Features:')
print(F)
np.save('IntFeatures.npy',F) # save features
```

The output of this code is shown in Fig. 5.2 and Table 5.2. The basic geometric features are extracted by function extract_features of pyxvis Library with parameters 'basicint', img=img, and bw=R, where img is the original X-ray image and R is the binary image that indicates the pixels where the intensity features are extracted.

□

5.3.2   Contrast

The contrast gives a measure of the difference in the gray value between region and its neighborhood. The smaller the gray value difference, the smaller the contrast. In this work, region and neighborhood define a zone. The zone is considered as a window of the image:

$$g(i, j) = x(i + i_r, j + j_r) \tag{5.20}$$

for $i = 1, ..., 2h + 1$ and $j = 1, ..., 2w + 1$, where h and w are the height and width as expressed in (5.1). The offsets i_r and j_r are defined as $i_r = \bar{i} - h - 1$ y $j_r =$

$j - b - 1$, where $(\bar{i}, \bar{j})$ denotes the center of mass of the region as computed in (5.12).

Contrast is a very important feature in fault detection, as the differences in the gray values are good for distinguishing a region from its neighborhood. The smaller the gray value difference, the smaller the contrast. In order to visualize the contrast, we can use a 3D representation with three coordinates (x, y, z), where (x, y) are used to represent the location of a pixel (i, j), and z is used for the representation of the gray value. An example is illustrated in Fig. 5.1c that shows the 3D representation of Fig. 5.1a. The reader can observe in this example a high-contrast region.

There are many definitions of contrast. A common definition of contrast is given using texture features (as explained in Sect. 5.3.5). Other simple definitions of contrast are given in [26, 56]:

$$ K_1 = \frac{G - G_e}{G_e}, \qquad K_2 = \frac{G - G_e}{G + G_e} \qquad y \qquad K_3 = \ln(G/G_e), \tag{5.21} $$

where G an G_e denote the mean gray value in the region and in the neighborhood respectively.

Two further definitions of contrast are given in [41] where new contrast features are suggested. According to Fig. 5.6, these new features can be calculated in four steps: *(i)* we take a profile in i direction and in j direction centered in the mass center of the region (see P_1 and P_2 respectively); *(ii)* we calculate the ramps R_1 and R_2 that are estimated as a first-order function that contains the first and last points of P_1 and P_2; *(iii)* new profiles without background are computed as $Q_1 = P_1 - R_1$ and $Q_2 = P_2 - R_2$ (they are stored together as $Q = [Q_1 \; Q_2]$); *(iv)* the new contrast features are given by

$$ K_\sigma = \sigma_Q \qquad \text{and} \qquad K = \ln(Q_{max} - Q_{min}). \tag{5.22} $$

Another definition of contrast can be found in [28], where the contrast is given by the mean of absolute differences between pixel values and mean of adjacent (e.g., eight adjacent pixels):

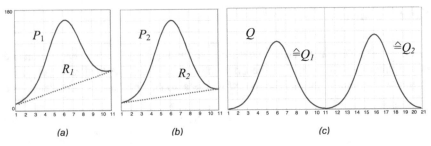

Fig. 5.6 Computation of Q for contrast features for region of Fig. 5.1: **a** Profile in i direction, **b** profile in j direction, **c** fusion of profiles: $Q = [Q_1 \; Q_2]$

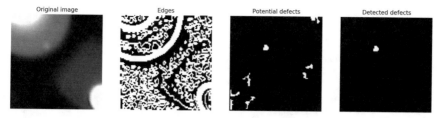

Fig. 5.7 Detection of small defects in apples using area contrast features: input image, edge detection, labeled regions, and detection. [→ Example 5.5 🐍]

$$K_c = \frac{1}{A_T} \sum_{(i,j)\in\mathbb{T}} |g(i, j) - \mu_{A(i,j)}, |.$$
(5.23)

where A_T is the area of the region and its neighborhood and $\mu_{A(i,j)}$ is the mean value of pixels locations adjacent of pixel (i, j).

Python Example 5.5: In this example, we show how to detect small defects in an X-ray image of a casting (see Fig. 5.7) using area and contrast features. We follow the general block diagram of Fig. 4.30. Here, area and contrast features are extracted for each region as defined by enclosed edges. The detection is performed if the size of the region is between some thresholds and the contrast is high enough.

Listing 5.5 : Defects detections using area and contrast features

```python
import numpy as np
import matplotlib.pyplot as plt
from pyxvis.processing.images import gradlog
from skimage.measure import label
from pyxvis.features.extraction import extract_features

img     = plt.imread('../images/small_wheel.png')  # input image with a defect
(N,M)   = img.shape
e       = gradlog(img,1.25,4/250)
L       = label(~e)          # labeling of objects
n       = np.max(L)          # number of detected objects

K1      = np.zeros((N,M), dtype=bool)
K2      = np.zeros((N,M), dtype=bool)
# Analysis of each segmented object
for i in range(n):
    R = (L == i)                             # binary image of object i
    f = extract_features('basicgeo',bw=R*1)  # feature extraction for object i
    area = f[4]
    # recognition of potential defects according to the size
    if area>20 and area<40:
        K1 = np.bitwise_or(K1,R)
        i0 = int(round(f[0]))
        j0 = int(round(f[1]))
        h  = int(round(f[2]/2))
        w  = int(round(f[3]/2))
        i1 = max(i0-h,0)
        j1 = max(j0-w,0)
```

```
        i2 = min(i0+h,N-1)
        j2 = min(j0+w,M-1)
        I  = img[i1:i2,j1:j2]
        bw = R[i1:i2,j1:j2]
        x  = extract_features('contrast',img=I,bw=bw)
        if x[3]>1.5:
            print('contrast features:')
            print(x)
            print('area = '+str(area)+' pixels')
            K2 = np.bitwise_or(K2,R)

fig, ax = plt.subplots(1, 4, figsize=(16, 8))
ax[0].imshow(img, cmap='gray')
ax[0].set_title('Original image')
ax[0].axis('off')
ax[1].imshow(e, cmap='gray')
ax[1].set_title('Edges')
ax[1].axis('off')
ax[2].imshow(K1, cmap='gray')
ax[2].set_title('Potential defects')
ax[2].axis('off')
ax[3].imshow(K2, cmap='gray')
ax[3].set_title('Detected defects')
ax[3].axis('off')
plt.show()
```

The output of this code is shown in Fig. 5.7. In this example, the contrast features are extracted using function extract_features of pyxvis Librarywith parameters img=I and bw=bw, i.e., for the grayscale image I with a potential defect located at the pixels equals to '1' of binary image bw. In this example, we use feature K_σ from (5.22) that is stored in variable x[3].   □

5.3.3   Crossing Line Profiles

An approach based on *crossing line profiles* (CLP) was originally developed to detect aluminum casting defects [39], however, it can be used to detect spots in general, or regions that have some gray value difference with their neighborhood. As the contrast between a defect and a defect-free neighborhoods is distinctive, the detection is usually performed by thresholding this feature (as we already learned in Sect. 5.3.2). Nevertheless, this measurement suffers from accuracy error when the neighborhood is not homogeneous, for example, when a defect is at an edge of a regular structure of the test object (see Fig. 4.31). For this reason, many approaches use a priori information about the location of regular structures of the test piece. CLP is able to detect those defects without a priori knowledge using *crossing line profiles*, i.e., the gray level profiles along straight lines crossing each segmented potential region in the middle. The profile that contains the most similar gray levels in the extremes is selected. Hence, the homogeneity of the neighborhood is ensured. Features from the selected profile are extracted.

In this approach, we follow a simple automated segmentation approach based on Fig. 4.30 and Fig. 4.31. The steps of detection based on CLP are shown in Fig. 5.8. First, a LoG kernel and a zero-crossing algorithm are used to detect the edges of the

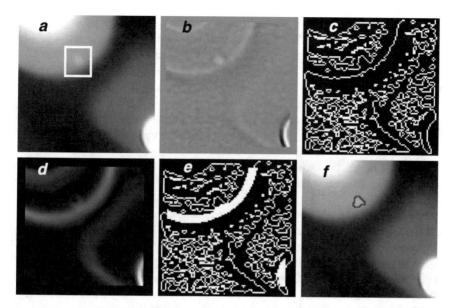

Fig. 5.8 Detection of flaws: **a** radioscopic image with a small flaw at an edge of a regular structure, **b** Laplacian-filtered image with $\sigma = 1.25$ pixels (kernel size = 11×11), **c** zero crossing image, **d** gradient image, **e** edge detection after adding high gradient pixels, and **f** detected flaw using feature F_1 extracted from a crossing line profile. [→ Example,5.6]

X-ray images. The LoG-operator involves a Gaussian low-pass filter which is a good choice for the pre-smoothing of our noisy images that are obtained without frame averaging. The resulting binary edge image should produce at real defects closed and connected contours which demarcate *regions*. However, a region of interest may not be perfectly enclosed if it is located at an edge of a regular structure as shown in Fig. 5.8c. In order to complete the remaining edges of these defects, a thickening of the edges of the regular structure is performed as follows: (a) the gradient of the original image is calculated (see Fig. 5.8d); (b) by thresholding the gradient image at a high gray level a new binary image is obtained; and (c) the resulting image is added to the zero-crossing image (see Fig. 5.8e). Afterwards, each closed region is segmented as a potential flaw. For details, see a description of the method in [40].

This is a very simple detector of potential regions with a large number of false detections flagged erroneously. However, the advantages are as follows: *(i)* it is a single detector (it is the same detector for each image), *(ii)* it is able to identify potential defects independent of the placement and the structure of the specimen, i.e., without a priori information of the design structure of the test piece, and *(iii)* the detection rate of real flaws is very high (approximately 90%). In order to reduce the number of the false positives, the segmented regions must be measured and classified.

A segmented potential region is defined as a region enclosed by edges of the binary image obtained in the edge detection (see connected black pixels in Fig. 5.8e). For each segmented region, a window g is defined from the X-ray image x as: $g(i, j) = x(i + i_r, j + j_r)$ for $i = 1 \ldots 2h + 1$, and $j = 1 \ldots 2w + 1$, where h and w are the height and width of the region as defined in (5.1). The offsets i_r and j_r are defined as $i_r = \bar{i} - h - 1$ and $j_r = \bar{j} - w - 1$, where $(\bar{i}, \bar{j})$ denotes the coordinates of the center of mass of the region (5.12), rounded to the nearest integers. Hence, g is a window of size $(2h + 1) \times (2w + 1)$, in which the middle pixel corresponds to the center of mass of the segmented potential flaw, i.e., $g(h + 1, w + 1) = x(\bar{i}, \bar{j})$.

Now, we define the crossing line profile P_θ as the gray-level function along a straight line of window g through the middle pixel $(h + 1, w + 1)$ forming an angle θ with i-axis. In Sect. 5.3.2, P_0 and $P_{\pi/2}$ were analyzed together in order to obtain two features, K and K_σ, that give a measurement of the difference between maximum and minimum, and the standard deviation of both crossing line profiles. However, the analysis does not take into account that the profiles could include a non-homogeneous area. For example, if a non-defect region is segmented at an edge of a regular structure, it could be that P_0 (or $P_{\pi/2}$) includes a significant gray-level change of the regular structure. In this case, the variation of the profile will be large and therefore the region will be erroneously classified as defect.

In order to avoid this problem, we suggest an individual analysis of eight crossing line profiles P_θ, at $\theta = k\pi/8$, for $k = 0, \ldots, 7$, as illustrated in Fig. 5.9. In this analysis, the crossing line profile that contains the most similar gray levels in the extremes is selected. Hence, the attempt is made to ensure the homogeneity of the neighborhood filtering out those profiles that present a high gray-level change in the edge of the regular structure. In the example of Fig. 5.9, the selected profile is obtained for $k = 5$, where the gray values of the extremes are both approximately equal to 150. We can observe that the selected crossing line is approximately perpendicular to the direction of the gradient of the X-ray image without defect. This coincides with one of the criteria used by approaches with a priori knowledge: the selected pixels of the defect-free area are located perpendicular to the direction of the gradient of the piece's contour [42].

Before the features are extracted, a pre-processing of the selected crossing line profile is performed as follows: (1) The selected profile is resized to size $n = 32$ using the nearest neighbor interpolation. The resized profile will be denoted by P. (2) In order to obtain a defect profile without the background of the regular structure, P is linearly transformed by $Q_i = mP_i + b$, for $i = 1, \ldots, n$, where m and b are so chosen that $Q_1 = Q_n = 0$.

Finally, the proposed features are extracted from the normalized profile Q. They are defined as follows:

$$
\begin{aligned}
\bar{Q} &= \text{mean}(Q) \\
\sigma_Q &= \text{std}(Q) \\
\Delta_Q &= \max(Q) - \min(Q) \\
F_i &= \sum_{k=0}^{n-1} Q_{k+1} e^{-j\frac{2\pi ki}{n}} \quad \text{for } i = 1, \ldots, 4.
\end{aligned}
\tag{5.24}
$$

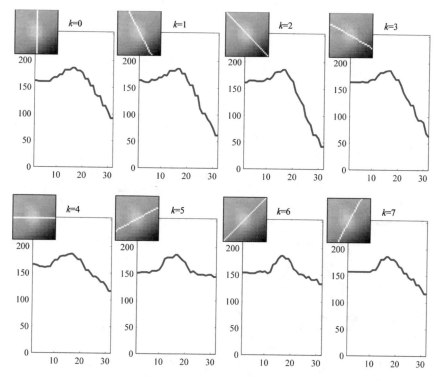

Fig. 5.9 Crossing line profiles for the window shown in Fig. 5.8a. [→ Example 5.6 🐍]

That is $\bar{Q}$: mean of Q; σ_Q: standard deviation of Q; Δ_Q: difference between maximum and minimum of Q; and F_i: magnitude of the ith harmonic of the discrete Fourier transform of Q for $i = 1, ..4$.

🐍   **Python Example 5.6:** In this example, we show how to detect a very small casting defect that is located at the edge of a regular structure as illustrated in Fig. 5.8 using area and CLP features. We follow the general block diagram of Fig. 4.30. That is area and contrast features are extracted for each region defined by enclosed edges. The detection is performed in two steps: *(i)* we detect *potential defects*, i.e., regions that are enclosed by edges that have a size between some thresholds, and *(ii)* we select from the potential defects those regions that have a CLP feature high enough. CLP feature is extracted from a window of the image that contains the potential defect in the middle.

Listing 5.6 : Defects detections using area and CLP features

```python
import numpy as np
import matplotlib.pyplot as plt

from pyxvis.processing.images import gradlog
from pyxvis.features.extraction import extract_features
```

```
from skimage.measure import label

img     = plt.imread('../images/small_wheel.png')  # input image with a defect
(N,M)   = img.shape
e       = gradlog(img,1.25,4/250)
L       = label(~e)          # labeling of objects
n       = np.max(L)          # number of detected objects

K1      = np.zeros((N,M), dtype=bool)
K2      = np.zeros((N,M), dtype=bool)

# Analysis of each segmented object
for i in range(n):
    R = L == i                            # binary image of object i
    f = extract_features('basicgeo', bw=R*1)  # feature extraction for object i
    area = f[4]

    # recognition of potential defects according to the size
    if area > 10 and area < 40:
        K1 = np.bitwise_or(K1,R)
        i0 = int(round(f[0]))
        j0 = int(round(f[1]))
        h  = int(round(f[2]/2))
        w  = int(round(f[3]/2))
        i1 = max(i0-h,0)
        j1 = max(j0-w,0)
        i2 = min(i0+h,N-1)
        j2 = min(j0+w,M-1)
        I  = img[i1:i2,j1:j2]
        x  = extract_features('clp', img=I)
        if x[5]>0.4:
            K2 = np.bitwise_or(K2,R)

fig, ax = plt.subplots(1, 4, figsize=(16, 8))
ax[0].imshow(img, cmap='gray')
ax[0].set_title('Original image')
ax[0].axis('off')
ax[1].imshow(e, cmap='gray')
ax[1].set_title('Edges')
ax[1].axis('off')
ax[2].imshow(K1, cmap='gray')
ax[2].set_title('Potential defects')
ax[2].axis('off')
ax[3].imshow(K2, cmap='gray')
ax[3].set_title('Detected defects')
ax[3].axis('off')
plt.show()
```

The output of this code is shown in Figs. 5.8 and 5.9. In this example, the edges are detected using command gradlog of pyxvis Library, that computes the logical OR of edge detection using LoG and edge detection by thresholding the gradient. The contrast features are extracted using command extract_features of pyxvis Library with parameters 'clp', img=I, where I is an image that contains the potential defects in the middle. In this example, we use features F_1 from (5.24). □

CLP features were tested on detecting casting defects. In this experiment, 50 X-ray images of aluminum wheels were analyzed. In the segmentation, approximately 23,000 potential flaws were obtained, in which there were 60 real defects. Some of these were existing blow holes. The other defects were produced by drilling small holes in positions of the casting which were known to be difficult to detect. In the

Fig. 5.10 Class distribution of CLP feature F_1 in detection of casting defects

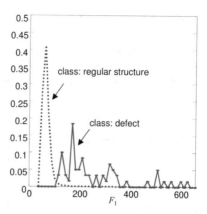

performance analysis, the best result was achieved by our feature F_1 (5.24). The class distribution between class 'defect' and 'non-defect' (or regular structure) is illustrated in Fig. 5.10. The reader can observe the effectiveness of the separation clearly. For more details, see [39].

5.3.4   Intensity Moments

In intensity moments, we use statistical moments (5.11) including gray value information [56]:

$$m'_{rs} = \sum_{i,j \in \Re} i^r j^s x(i, j) \qquad \text{for } r, s \in \mathbb{N}. \tag{5.25}$$

The summation is computed over the pixels (i, j) of the region $\Re$ only. Thus, it is possible to compute Hu, Flusser, and Gupta moments, as explained in Sect. 5.2.4 using the gray value information of the region. Hu-moments with intensity information can be computed by function extract_features of pyxvis Library with parameters 'huint', img=img, and bw=R, where img is an X-ray image and R is the binary image that indicates the pixels where the intensity moments are extracted.

5.3.5   Statistical Textures

These features provide information about the distribution of the gray values in the image. In this work, however, we restrict the computation of the texture features for a zone only defined as region and neighborhood (see Eq. 5.20).

Statistical texture features can be computed using the co-occurrence matrix $\mathbf{P}_{kl}$ [20]. The element $P_{kl}(i, j)$ of this matrix for a zone is the number of times, divided by N_T, that gray levels i and j occur in two pixels separated by that distance and direction given by the vector (k, l), where N_T is the number of pixels pairs contributing to build matrix $\mathbf{P}_{kl}$. In order to decrease the size $N_x \times N_x$ of the co-occurrence matrix, the grayscale is often reduced to 8 gray levels. From the co-occurrence matrix, several texture features can be computed. Haralick in [20] proposes (here $p(i, j) := P_{kl}(i, j)$):

Angular second moment:	$f_1 = \sum_{i=1}^{N_x} \sum_{j=1}^{N_x} [p(i, j)]^2$
Contrast:	$f_2 = \sum_{n=0}^{N_x-1} n^2 \sum_{i=1}^{N_x} \sum_{j=1}^{N_x} p(i, j) \text{ for } |i - j| = n$
Correlation:	$f_3 = \frac{1}{\sigma_x \sigma_y} \sum_{i=1}^{N_x} \sum_{j=1}^{N_x} \left[ij \cdot p(i, j) - \mu_x \mu_y \right]^2$
Sum of squares:	$f_4 = \sum_{i=1}^{N_x} \sum_{j=1}^{N_x} (i - j)^2 p(i, j)$
Inverse difference moment:	$f_5 = \sum_{i=1}^{N_x} \sum_{j=1}^{N_x} \frac{p(i,j)}{1+(i-j)^2}$
Sum average:	$f_6 = \sum_{i=2}^{2N_x} i \cdot p_{x+y}(i)$
Sum variance:	$f_7 = \sum_{i=2}^{2N_x} (i - f_8) \cdot p_{x+y}(i)$
Sum entropy:	$f_8 = -\sum_{i=2}^{2N_x} p_{x+y}(i) \cdot \log(p_{x+y}(i))$
Entropy:	$f_9 = -\sum_{i=1}^{N_x} \sum_{j=1}^{N_x} p(i, j) \log(p(i, j))$
Difference variance:	$f_{10} = \text{var}(\mathbf{p}_{x+y})$
Difference entropy:	$f_{11} = -\sum_{i=0}^{N_x-1} p_{x-y}(i) \cdot \log(p_{x-y}(i))$
Information measures of correlation 1:	$f_{12} = \frac{f_9 - HXY1}{\max(HX,HY)}$
Information measures of correlation 2:	$f_{13} = \sqrt{1 - \exp(-2(HXY2 - HXY))}$
Maximal correlation coefficient:	$f_{14} = \sqrt{\lambda_2}$

$$(5.26)$$

where μ_x, μ_y, σ_x, and σ_y are the means and standard deviations of p_x and p_y respectively with

$$\begin{aligned}
p_x &= \sum_{j=1}^{N_x} p(i, j) \\
p_y &= \sum_{i=1}^{N_x} p(i, j) \\
p_{x+y}(k) &= \sum_{i=1}^{N_x} \sum_{j=1 | i+j=k}^{N_x} p(i, j) \text{ for } k = 2, 3, ...2N_x \\
p_{x-y}(k) &= \sum_{i=1}^{N_x} \sum_{j=1 | i-j|=k}^{N_x} p(i, j) \text{ for } k = 0, 1, ...N_x - 1
\end{aligned},$$

and

$$\begin{aligned}
HX &= -\sum_{i=1}^{N_x} p_x(i) \log(p_x(i)) \\
HY &= -\sum_{j=1}^{N_x} p_y(j) \log(p_y(j)) \\
HXY1 &= -\sum_{i=1}^{N_x} \sum_{j=1}^{N_x} p(i, j) \log(p_x(i)p_y(j)) \\
HXY2 &= -\sum_{i=1}^{N_x} \sum_{j=1}^{N_x} p_x(i)p_y(j) \log(p_x(i)p_y(j))
\end{aligned}.$$

In f_14, λ_2 is the second largest eigenvalue of Q defined by

$$Q(i, j) = \sum_{k=1}^{N_x} \frac{p(i,k)p(j,k)}{p_x(i)p_y(k)}$$

The texture features are extracted for four directions ($0°–180°$, $45°–225°$, $90°–270°$, and $135°–315°$) in different distances $d = \max(k, l)$. That is, for a given distance d, we have four possible co-occurrence matrices: P_{0d}, P_{dd}, P_{d0}, and P_{-dd}. For example, for $d = 1$, we have $(k, l) = (0,1)$; $(1,1)$; $(1,0)$; and $(-1,1)$. After Haralick, 14 texture features using each co-occurrence matrix are computed (5.26), and the mean and range for each feature are calculated, i.e., we obtain $14 \times 2 = 28$ texture features for each distance d. The features will be denoted as $\bar{f}_i$ for the mean and f_i^Δ for the range, for $i = 1\ldots 14$.

The statistical textures based on Haralick can be computed by function extract_features of pyxvis Library with parameters 'haralick-d' and img=img, where img is the input X-ray image and d is the distance d defined above. In pyxvis Library, Haralick features are averaged for all four directions $(0, d)$, (d, d), $(d, 0)$, and $(-d, d)$, so we obtain a sort of rotation invariant feature.

5.3.6 Gabor

The Gabor functions are Gaussian shaped band-pass filters, with dyadic treatment of the radial spatial frequency range and multiple orientations, which represent an appropriate choice for tasks requiring simultaneous measurement in both space and frequency domains. The Gabor functions are a complete (but a nonorthogonal) basis set given by

$$f(x, y) = \frac{1}{2\pi\sigma_x\sigma_y} \exp\left(-\frac{1}{2}\left(\frac{x^2}{\sigma_x^2} + \frac{y^2}{\sigma_y^2}\right),\right) \quad (5.27)$$

where σ_x and σ_y denote the Gaussian envelope along the x and y-axes, and u_0 defines the radial frequency of the Gabor function. Examples of Gabor functions are illustrated in Fig. 5.11. In this case, a class of self-similar functions are generated by rotation and dilation of $f(x, y)$.

Each Gabor filter has a real and an imaginary component that are stored in $M \times M$ masks, called $\mathbf{R}_{pq}$ and $\mathbf{I}_{pq}$ respectively, where $p = 1\ldots S$, denotes the scale, and $q = 1\ldots L$, denotes the orientation (for details see [30]). Usually, $S = 8$ scales, and $L = 8$ orientations as shown in Fig. 5.11, with $M = 27$.

The Gabor filters are applied to each segmented window $\mathbf{W}$, that contains the segmented region and its surrounding (see Fig. 5.1). The filtered windows $\mathbf{G}_{pq}$ are computed using the 2D convolution (4.9) of the window $\mathbf{W}$ of the X-ray image with the Gabor masks as follows:

$$\mathbf{G}_{pq} = \left[(\mathbf{W} * \mathbf{R}_{pq})^2 + (\mathbf{W} * \mathbf{I}_{pq})^2\right]^{1/2}. \quad (5.28)$$

Fig. 5.11 Example of Gabor functions in spatial domain: (Top) imaginary components of self-similar filter bank by using $p = 1 \ldots 8$ scales and $q = 1 \ldots 8$ orientations, (Bottom) 3D representations of two Gabor functions of (a)

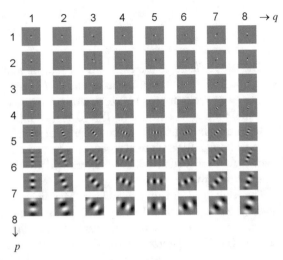

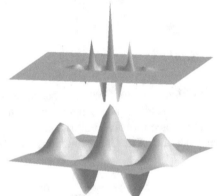

The Gabor features, denoted by g_{pq}, are defined as the average output of $\mathbf{G}_{pq}$, i.e., it yields $S \times L$ Gabor features for each segmented window:

$$g_{pq} = \frac{1}{n_w n_w} \sum_{i=1}^{n_w} \sum_{j=1}^{m_w} G_{pq}(i, j), \tag{5.29}$$

where the size of the filtered windows $\mathbf{G}_{pq}$ is $n_w \times m_w$.

Three additional Gabor features can be extracted: *(i)* maximum of all Gabor features: $g_{\max} = \max(\mathbf{g})$, *(ii)* minimu of all Gabor features: $g_{\min} = \min(\mathbf{g})$, and *(iii)* range of all Gabor features: $g_\Delta = g_{\max} - g_{\min}$. These features are very useful because they are rotation invariant.

The Gabor features can be computed by function extract_features of pyxvis Library with parameters 'gabor' and img=img, where img is the input X-ray image. Additionaly, we can compute rotation invariant Gabor features by averaging all Gabor features of the same scale, this is obtained by parameter `'gabor-ri'`. An example of the use of these features is given in Example 5.9.

5.3.7   Filter Banks

Filter banks can be used to extract texture information [53]. They are used in image transformations like Discrete Fourier Transform (DFT) (magnitude and phase), Discrete Cosine Transform (DCT) [16], and wavelets as Gabor features based on 2D Gabor functions (see Sect. 5.3.6).

For an image $\mathbf{X}$ of $N \times N$ pixels, the DFT in 2D is defined as follows:

$$F(m, n) = \sum_{i=1}^{M} \sum_{k=1}^{N} X(i, k) e^{-2\pi j \left(\frac{(m-1)(i-1)}{N} + \frac{(n-1)(k-1)}{N} \right)}, \tag{5.30}$$

where $j = \sqrt{-1}$. $F(m, n)$ is a complex number. That means magnitude and phase can be used as features. Fourier features can be computed by function extract_features of pyxvis Library with parameters `'fourier'`.

DCT in 2D is defined as

$$D(m, n) = \alpha_m \alpha_n \sum_{i=1}^{N} \sum_{k=1}^{N} X(i, k) \cos\left(\frac{\pi(2i - 1)(m - 1)}{2N} \right) \cos\left(\frac{\pi(2k - 1)(n - 1)}{2N} \right), \tag{5.31}$$

where $\alpha_1 = 1/\sqrt{N}$ and $\alpha_m = \sqrt{2/N}$, for $m = 2 \dots N$. DCT features are real numbers instead of complex number such as Fourier features. DCT features can be computed by function extract_features of pyxvis Library with parameters `'dct'`.

It is worth mentioning that these features are not rotation invariant, however, we can extract rotation-invariant features if we use maximum, minimum, and a range of them as we did for the Gabor features in Sect. 5.3.6. An example of the use of these features is given in Example 5.9.

5.4   Descriptors

Descriptors have been very relevant on computer vision applications [43]. This is because they are able to provide highly distinctive features, and can be used in applications such as multiple view analysis, in object recognition, texture recognition, and others. In this section, we provide some descriptors that are very useful in X-ray testing.

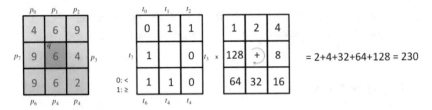

Fig. 5.12 LBP coding: a central pixel q the gray value of which is 6 has eight neighbors with gray values $p_0 = 4$, $p_1 = 6$, $p_2 = 9$, $p_3 = 4$, $p_4 = 2$, $p_5 = 6$, $p_6 = 9$, and $p_7 = 9$. A new mask with 8 bits is built, where $t_i = 1$ if $p_i \geq q$, otherwise $t_i = 0$. The LBP code is computed as $\sum_i t_i 2^i$, in this example the code is 230

5.4.1 Local Binary Patterns

LBP, *Local Binary Patterns* was proposed as a texture feature [46]. The idea is to extract texture information from occurrence histogram of local binary patterns computed from the relationship between each pixel intensity value with its eight neighbors. The LBP features are the frequencies of each one of the histogram bins. LBP is computed in three steps: *(i)* coding, *(ii)* mapping, and *(iii)* histogram.

Coding
Each pixel (i, j) of the input image has a set of neighbors. Typically, the set of eight neighbors defined by the 8-connected pixels is used. However, more neighbors for different distances can be defined as well. For eight connected pixels, the locations are $(i - 1, j - 1)$; $(i - 1, j)$; $(i - 1, j + 1)$; $(i, j + 1)$; $(i + 1, j + 1)$; $(i + 1, j + 1)$; $(i + 1, j)$, and $(i + 1, j - 1)$ respectively as shown in Fig. 5.12. The central pixel has a gray value q, and the neighbors have gray values p_i, for $i = 0 \ldots 7$. The code is computed by

$$y = \sum_{i=0}^{7} t_i 2^i, \tag{5.32}$$

where $t_i = 1$ if $p_i \geq q$, otherwise $t_i = 0$. That means, a pixel q with its neighbors can be coded as a number $y \in \{0 \ldots 255\}$. The code can be represented as a string of bits as shown in Fig. 5.12.

Mapping
We can observe that the code generated by the previous step can be categorized according to the number of changes (from '1' to '0', or from '0' to '1') in a cycle. For instance, in the example of Fig. 5.12, where the code is 01100111, we define a cycle with eight transitions as follows: $0 \rightarrow 1 \rightarrow 1 \rightarrow 0 \rightarrow 0 \rightarrow 1 \rightarrow 1 \rightarrow 1 \rightarrow 0$ (the last bit is the repetition of the first one because it is a cycle). The number of changes is $U = 4$. Thus, we can have codes with $U = 0, 2, 4, 6$, and 8 as illustrated in Fig. 5.13. After the authors, there are *uniform* and *non-uniform* patterns. The first ones ($U = 0$ and 2) correspond to textures with a low number of changes, the last ones ($U > 2$) can be interpreted as noise because there are many changes in the gray

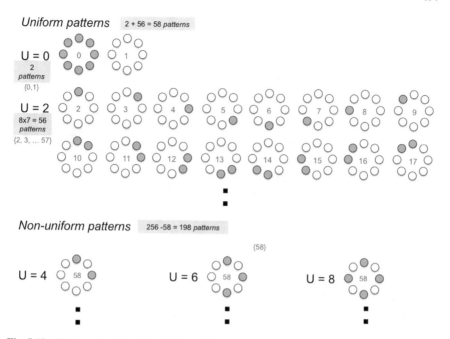

Fig. 5.13 LBP mapping for eight neighbors. Each small circle represents a bit t_i of the code, green means '1', white means '0'. U is the number of changes from '1' to '0', or from '0' to '1' in one cycle. For a small number of changes, i.e., $U = 0$ and 2, the codes represent *uniform patterns*, for $U > 2$ the patterns are *non-uniform*

values. There are 58 uniform patterns and 198 non-uniform patterns. Each uniform code is mapped as a number from 0 to 57 as illustrated in Fig. 5.13, whereas all non-uniform codes are mapped as number 58. This descriptor is known as LBP-u2. LBP-u2 mapping corresponds to a mapping that varies with the orientation of the image, i.e., it is not rotation-invariant. In order to build a rotation-invariant LBP descriptor, all patterns that have the same structure but with different rotations are mapped as a unique number. For instance, all patterns of the second row of Fig. 5.13 are mapped with the same number. The same is valid for the third row. In this mapping, we have 36 different numbers. This descriptor is known as LBP-ri.

Histogram

The process of coding and mapping is performed at each pixel of the input image. Thus, each pixel is converted into a number from 0 to $M - 1$, with a mapping of M numbers. Afterwards, a histogram of M bins of this image is computed. The LBP descriptor of the image is this histogram.

LBP is very robust in terms of grayscale and rotation variations [46]. An example is shown in Fig. 5.14. Other LBP features like *semantic* LBP (sLBP) [44] can be used in order to bring together similar bins. LBP is implemented in function extract_features of pyxvis Library with parameters 'lbp' (for uniform LBP features)

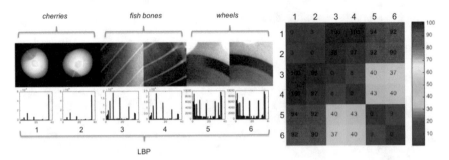

Fig. 5.14 Comparison of six textures using LBP-ri descriptor. It is clear that descriptors of the same texture are very similar, and descriptors from different textures are very different. A measurement of the Euclidean distance between all six descriptors is shown in the right color matrix

or parameter 'lbp-ri' (for rotation invariant LBP features). An example of the use of these features is given in Example 5.9.

The reader can find a descriptor with similar properties in [47], where LPQ (from *local phase quantization*) is proposed.

5.4.2 Binarized Statistical Image Features (BSIF)

BSIF, *binarized statistical image features*, was proposed as a texture descriptor [27]. As LBP and LPQ, it computes a binary code for each pixel of the input image. Thus, a histogram that encodes texture information is built by counting the frequency of each code.

In BSIF, the input image is filtered using a set of linear filters. The linear filters are learned from a training set of natural image patches ensuring statistical independence of the filter responses. BSIF computes the bits of the binary code by thresholding the response of the linear filters.

Therefore, instead of manually predefined sets of filters (like LBP or LPQ), BSIF uses filters based on statistics of natural images. After the authors, this improves its modeling capacity and the accuracy in texture recognition.[1] The reader can use this function to obtain similar results to those obtained by LBP in Fig. 5.14.

---

[1] BSIF has been originally implemented in Matlab [27]. A Python implementation of BSIF is given at https://github.com/CVRL/OpenSourceIrisPAD that was used for iris recognition [11].

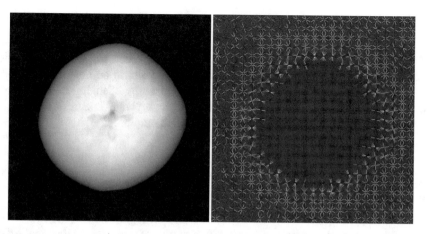

Fig. 5.15 Computation of HOG descriptors of an X-ray image of a fruit. The descriptors give information about shape and appearance

5.4.3   Histogram of Oriented Gradients

HOG, *histogram of oriented gradients*, was originally proposed as a descriptor that is able to detect pedestrians [7], however, the powerful of this descriptor can be used in many computer vision problems that require local object appearance and shape information. The key idea of HOG is to compute the distribution of intensity gradients in uniformly spaced cells arranged in a grid manner. A cell is typically defined as a squared region of the image.

In HOG, the gradient of the input image in both directions G_i and G_j is computed (see Sect. 4.4.1). Thus, for each pixel, we have the magnitude $G(i, j)$ and the angle $A(i, j)$ using (4.14) and (4.15) respectively.[2] In order to compute the cell histogram, we define n bins, where bin k corresponds to the orientation between θ_k and θ_{k+1}, with $\theta_{k+1} = \theta_k + \Delta\theta$, for $k = 1 \ldots n$. For example, for $n = 9$ bins, we could define $\Delta\theta = 360^0/9 = 40^0$ and $\theta_1 = -\Delta\theta/2 = -20^0$, so the first bin will be for orientations from -20^0 to $+20^0$, the second from $+20^0$ to $+60^0$, and so on. Therefore, a pixel (i, j) of the cell whose orientation is $\theta_k < A(i, j) \leq \theta_{k+1}$, registers a weighted vote in bin k based on its gradient value $G(i, j)$. This operation is repeated for every pixel of the cell (see Fig. 5.15).

In order to improve the performance of HOG descriptor and make it robust against changes in illumination and contrast, the authors used a dense grid of cells and an overlapping local contrast normalization [7]. That means normalized cells are grouped together into connected blocks. Then, the descriptor is a concatenation of the normalized cell histograms of all blocks. HOG is implemented in function extract_features of pyxvis Library with parameter 'hog'. An example of the use of these features is given in Example 5.9.

---

[2]Sometimes the magnitude of the angle is used.

5.4.4 Scale-Invariant Feature Transform (SIFT)

The *Scale-invariant feature transform*, SIFT, was proposed in [32] to detect and describe *keypoints*. A keypoint is a distinguishable point in an image, i.e., it represents a salient image region that can be recognized by changing its viewpoint, orientation, scale, etc. In SIFT methodology, each keypoint is described using a 128-element vector called *SIFT-descriptor*. SIFT-descriptor is

- Scale invariant
- Rotation invariant
- Illumination invariant
- Viewpoint invariant

SIFT-descriptor can be used as a 'signature' and it is highly distinctive, i.e., SIFT-descriptors of corresponding points (in different images) are very similar, and SIFT-descriptors of different points are very different. SIFT has two main stages: *(i)* keypoint detection and *(i)* keypoint description. In the following, these stages are presented in further details.

Keypoint Detection
Keypoints are then taken as maxima/minima of the Difference of Gaussians (DoG) that occur at multiple scales. Keypoints can be detected in four steps (see Fig. 5.16):

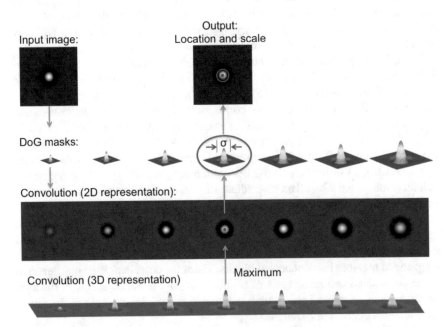

Fig. 5.16 Detection of a keypoint in a synthetic image. The image is convolved with several DoG masks. The maximal response defines the location (x, y) of the keypoint. The used mask for the convolution defines the scale σ

1. We define two Gaussian masks: $G(x, y, \sigma)$ and $G(x, y, k\sigma)$ from (4.10) at scales σ and $k\sigma$.
2. The input image $I(x, y)$ is convolved with both Gaussian filters obtaining $L(x, y, \sigma)$ and $L(x, y, k\sigma)$ respectively.
3. The Difference of Gaussians (DoG) is computed as:

$$D(x, y, \sigma) = L(x, y, \sigma) - L(x, y, k\sigma). \qquad (5.33)$$

4. Keypoints are found as maxima of $|D(x, y, \sigma)|$ that can occur at different values of σ. We compare each pixel in the DoG images to its 26 neighbors (8 at the same scale and 9 from the next scale and 9 from the previous scales). If the pixel value is the maximum or minimum among all compared pixels, it is selected as a candidate keypoint.

Keypoint Description

For each keypoint, we need a description. A keypoint is defined by its location (x, y) and its scale σ. The descriptor is computed in seven steps (see Fig. 5.17):

1. We define a window of size 1.5σ centered in (x, y).
2. The window is rotated $-\theta$, where θ is the orientation of the gradient in (x, y).
3. The rotated window is divided into $4 \times 4 = 16$ regular cells distributed in a grid manner.
4. For each cell, the histogram of gradients is computed using 8 bins.
5. All 16 histograms with 8 bins are concatenated, i.e., we obtain a descriptor of $16 \times 8 = 128$ elements.
6. Finally, the descriptor is normalized to unit length.

We can observe that SIFT descriptor is invariant to scale because the size of the window in step 1 depends on scale factor σ. SIFT descriptor is invariant to rotation because the window is rotated according to the orientation of the gradient (see step 2). Thus, if an image is resized and rotated, it will have the same window after these two steps. The SIFT descriptor is invariant to illumination because the descriptor is normalized to unit length. SIFT has been proven to be robust against perspective distortions and viewpoint changes when the rotation of the 3D object is less than 30 degrees rotation. An example of this can be found in Fig. 5.18.

Python Example 5.7: In this example, we find matching points in two views. SIFT keypoints are estimated in each view, and those with the most similar descriptors are matched.

Listing 5.7 : Matching points using SIFT

```python
import numpy as np
import matplotlib.pylab as plt

from pyxvis.io import gdxraydb
from pyxvis.features.descriptors import compute_descriptors, match_descriptors
from pyxvis.io.visualization import plot_matches
```

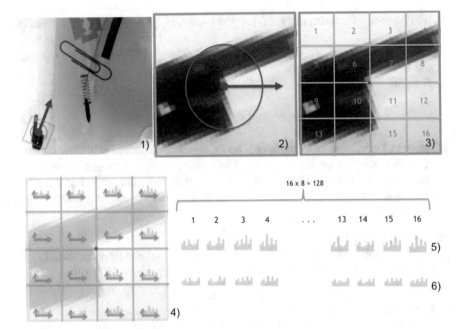

Fig. 5.17 Keypoint description (see explanation of six steps in text). This example corresponds to one of keypoints Fig. 5.18 (see object at bottom-left side of the X-ray image)

```
image_set = gdxraydb.Baggages()
I1 = image_set.load_image(2, 1)   # Image 1
I2 = image_set.load_image(2, 2)   # Image 2

fig1, (ax1, ax2) = plt.subplots(1, 2, figsize=(7, 7))
ax1.imshow(I1, cmap='gray')
ax1.axis('off')
ax2.imshow(I2, cmap='gray')
ax2.axis('off')
fig1.tight_layout()
plt.show()

kp1, desc1 = compute_descriptors(I1, 'sift') # SIFT descriptor for image 1
kp2, desc2 = compute_descriptors(I2, 'sift') # SIFT descriptor for image 2

matches = match_descriptors(desc1, desc2, matcher='flann', max_ratio=0.7) # Matching
    points using KDTREE

# Display results of matched points
fig, ax = plt.subplots(1, 1, figsize=(10, 10))
plot_matches(ax, I1, I2, kp1, kp2, matches, keypoints_color='lawngreen')
ax.axis('off')
plt.show()
```

The output of this code is shown in Fig. 5.18. In this example, the SIFT descriptors are detected using class compute_descriptors of pyxvis Library with parameter 'sift'. The implementation is based on OpenCV.

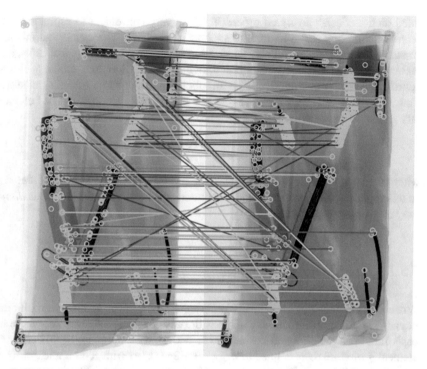

Fig. 5.18 Matching points of two different views of the same object. The object was rotated 10° around its horizontal axis from first to second image. The SIFT approach is able to find key points (green small circles) and their matchings (lines between keypoints of different images). [→ Example 5.7 🐍]

The reader can find descriptors with similar properties in SURF: *Speeded Up Robust Feature* [2], BRIEF: *Binary robust independent elementary features* [3], BRISK: *Binary Robust Invariant Scalable Keypoints* [31] among others.

5.5 Sparse Representations

In recent years, sparse representation has been widely used in signal processing [55], neuroscience [50], statistics [9], sensors [63], and computer vision [61, 64]. In many computer vision applications, under assumption that natural images can be represented using sparse decomposition [49] state-of-the-art results have been significantly improved. In these applications, the performance can be improved by learning non-parametric dictionaries for the sparse representation (instead of using fixed dictionaries).[3] In signal processing, it is very convenient to estimate a new representation of a signal in order to analyze it efficiently. The idea is that this representation captures a useful characterization of the signal for analytical tasks, e.g., feature extraction for pattern recognition, frequency spectrum for denoising, etc. An appropriate representation, due to its simplicity, is obtained by a linear transform. Thus, a signal $\mathbf{x} \in \mathbb{R}^n$ can be expressed as a linear combination of a set of elementary signals $\mathbf{D} = [\mathbf{d}_1 \ \mathbf{d}_2 \ \ldots \ \mathbf{d}_K] \in \mathbb{R}^{n \times K}$ as follows:

$$\mathbf{x} = \mathbf{D}\mathbf{z}, \tag{5.34}$$

where the vector $\mathbf{z} \in \mathbb{R}^K$ corresponds to the representation coefficients of signal $\mathbf{x}$. In this representation, matrix $\mathbf{D}$ and its columns $\mathbf{d}_k$ are commonly known as *dictionary* and *atoms* respectively.

5.5.1   Traditional Dictionaries

When every signal can be uniquely represented by a linear combination, the dictionary $\mathbf{D}$ corresponds to a *basis*. This is the case of DFT, for example, where the basis functions are sine and cosine waves with unity amplitude. In this case, the element j of atom $\mathbf{d}_k$ is defined as $d_{jk} = \exp(2\pi ijk/n)$ with $K = n$ and $i = \sqrt{-1}$ [16]. It is well known that for some applications, e.g., signal filtering, instead of processing the signal $\mathbf{x}$, it can be more convenient to process the signal in frequency domain $\mathbf{z}$ because it can be used to separate low and high frequencies effectively. Nevertheless, the Fourier basis is very inefficient when representing, for example, a discontinuity, because its representation coefficients are over all frequencies and the analysis becomes difficult or even impossible. Other *predefined* bases, i.e., where the atoms are *fixed*, are DCT and wavelets (e.g., Gabor) among others [16]. In many applications, since these dictionaries are fixed, they cannot represent more complex and high-dimensional signals satisfactorily [54].

In order to avoid the mentioned problem with fixed dictionaries, another way to represent a signal is using a *learned* dictionary, i.e., a dictionary that is estimated from representative signal examples. This is the case of Principal Component Analysis (PCA), or Karhunen–Loève Transform (KLT) [25], where the dictionary $\mathbf{D}$ is computed using the first K eigenvectors of the eigenvalue decomposition of the covariance matrix Σ, which is usually estimated from a set of zero-means signal examples $\mathbf{X} = \{\mathbf{x}_i\}_{i=1}^N$. The basis here represents K orthogonal functions (with $K \leq n$) that transforms $\mathbf{X}$ into a set of linearly uncorrelated signals $\mathbf{Z} = \{\mathbf{z}_i\}_{i=1}^N$ called the K *principal components*. This relationship is expressed as $\mathbf{X} = \mathbf{D}\mathbf{Z}$. In this case, KLT represents a signal more efficiently than DFT because the dictionary is not fixed and it is *learned* from signal examples [54].

---

[3] A good library for sparse representation in computer vision is SPAMS (SPArse Modelling Software) [34], see Matlab, Python and R implementations on http://spams-devel.gforge.inria.fr.

The mentioned dictionaries are *orthogonal*, i.e., each atom $\mathbf{d}_i$ is orthogonal to atom $\mathbf{d}_j$ in $\mathbb{R}^n$ space $\forall i \neq j$. Therefore, a signal $\mathbf{x}$ is represented as a sum of orthogonal vectors $z_i \mathbf{d}_i$. In addition, most of these dictionaries are *orthonormal*, with $\|\mathbf{d}_i\| = 1$ and $\mathbf{D}^T\mathbf{D} = \mathbf{I}$, where $\mathbf{I}$ is the identity matrix. Hence, it is very simple to calculate $\mathbf{Z} = \mathbf{D}^T\mathbf{X}$.

5.5.2   Sparse Dictionaries

Due to their mathematical simplicity, the orthogonal dictionaries dominated this kind of analysis for years. Nevertheless, there is no reason to accept as true that the number of atoms, required to characterize a set of signals, must be smaller than the dimension of the signal. Moreover, why should the atoms of the dictionary be orthogonal? The limited effectiveness of these dictionaries led to the development of newer dictionaries that can represent a wider range of signal phenomena, namely the *overcomplete* ones that have more atoms than the dimension of the signal ($K > n$) with no necessarily orthogonal atoms [58]. A seminal work in learning overcomplete dictionaries for image representation was presented by Olshausen and Field [48, 49]. They estimated—from small image patches of natural images—a *sparse* representation which was extremely similar to the mammalian simple-cell receptive fields (at that time, this phenomenon could only be described using Gabor filters). The key idea for representing natural signals is that although the number of possible atoms in the overcomplete dictionary is huge, the number of those atoms required to represent a signal is much smaller, i.e., the signals are sparse in the set of all possible atoms [58].

Sparse coding models a signal as a linear combination (5.34), or approximate, $\mathbf{x} \approx \mathbf{D}\mathbf{z}$, using a *sparse* linear combination of atoms from a learned dictionary, i.e., only a few atoms from $\mathbf{D}$ are allowed to be used in the linear combination (most coefficients of $\mathbf{z}$ are zero) and the atoms are not fixed (the dictionary is adapted to fit a given set of signal examples). In this case, the basis is not orthogonal.

Thus, from a representative set of signals $\mathbf{X} = \{\mathbf{x}_i\}_{i=1}^N$, the idea is *i)* to learn a dictionary $\mathbf{D} = \{\mathbf{d}_k\}_{k=1}^K$ and *ii)* to estimate the corresponding sparse representations $\mathbf{Z} = \{\mathbf{z}_i\}_{i=1}^N$ of the original signals $\mathbf{X}$.

In K-means algorithm—a very well-known algorithm used in clustering—the sparsity is *extreme* because for the representation of $\mathbf{x}$ only one atom of $\mathbf{D}$ is allowed, and the corresponding coefficient of $\mathbf{z}$ is 1. In this case, the dictionary and coefficients are estimated by:

$$\mathbf{D}^*, \mathbf{Z}^* = \underset{\mathbf{D},\mathbf{Z}}{\text{argmin}} \, \|\mathbf{X} - \mathbf{D}\mathbf{Z}\|_F^2 \quad \text{subject to } \forall i, \mathbf{z}_i = \mathbf{e}_k \text{ for some} k, \qquad (5.35)$$

where $\mathbf{e}_k$ is a vector from the trivial basis, with all zero entries except a one in kth position. In this equation, the Frobenius norm is used defined as $\|\mathbf{A}\|_F^2 = \sum_{ij} a_{ij}^2$. In clustering problems, the atom $\mathbf{d}_k$ is the centroid of samples $\mathbf{x}_i$ that fulfill $\mathbf{z}_i = \mathbf{e}_k$.

Thus, a signal $\mathbf{x}$ belongs to cluster k if it is closer to centroid k than any other centroids (in this case, its representation is $\mathbf{z} = \mathbf{e}_k$ and the corresponding atom is $\mathbf{d}_k$).

Sparsisty in general, can be expressed as follows:

$$\mathbf{D}^*, \mathbf{Z}^* = \underset{\mathbf{D},\mathbf{Z}}{\operatorname{argmin}} \, ||\mathbf{X} - \mathbf{DZ}||_F^2 \quad \text{subject to } ||\mathbf{x}_i||_0 \leq T, \tag{5.36}$$

where $||\mathbf{x}_i||_0$ is the ℓ^0 norm, counting the nonzero entries of $\mathbf{x}_i$. The goal is to express a new signal $\mathbf{x}$ as a linear combination of a small number of signals take from the dictionary. This optimization problem can be expressed as follows:

$$\mathbf{z}^* = \underset{\mathbf{z}}{\operatorname{argmin}} (||\mathbf{x} - \mathbf{D}^*\mathbf{z}||_2^2 + \lambda||z||_1) \tag{5.37}$$

It can be demonstrated that the solution of the ℓ^0 minimization problem (5.36) is equivalent to the solution of the ℓ^1 minimization problem [10]:

$$\underset{\mathbf{D},\mathbf{Z}}{\operatorname{argmin}} \, ||\mathbf{X} - \mathbf{DZ}||_F^2 \quad \text{subject to } ||\mathbf{z}_i||_1 \leq T \tag{5.38}$$

Thus, on the one hand, the *dictionary learning problem* is as follows: given a set of training signals $\mathbf{X} = \{\mathbf{x}_i\}_{i=1}^N$, find the dictionary $\mathbf{D}$ (and a set of representation coefficients $\mathbf{Z} = \{\mathbf{z}_i\}_{i=1}^N$) that represents at best each signal using the sparsity constraint (5.38), where no more than T atoms are allowed in each decomposition $\mathbf{z}_i$. On the other hand, the *sparse coding problem* can be stated as follows: given a signal $\mathbf{x}$ and a learned dictionary $\mathbf{D}$, find $\mathbf{z}$, the representation of signal $\mathbf{x}$, as follows::

$$\underset{\mathbf{z}}{\operatorname{argmin}} \, ||\mathbf{z}||_0 \quad \text{subject to } ||\mathbf{x} - \mathbf{Dz}||_2 < \epsilon, \tag{5.39}$$

where ϵ is the error tolerance.

5.5.3 Dictionary Learning

There are three categories of algorithms used to learn dictionaries [58]: *(i)* probabilistic methods, *(ii)* methods based on clustering, and *(iii)* methods with a particular construction. Probabilistic methods are based on a maximum likelihood approach, i.e., given the generative model (5.34), the objective is to maximize the likelihood that the representative samples have efficient, sparse representations in a redundant dictionary given by $\mathbf{D}$ [17, 29, 49, 62]. In clustering-based methods, the representative samples are grouped into patterns such that their distance to a given atom is minimal. Afterwards, the atoms are updated such that the overall distance in the group of patterns is minimal. This schema follows a K-means algorithm. In order to generalize the K-means algorithm, the 'K-SVD' algorithm was developed [1]. The method has two steps: a) it uses orthogonal matching pursuit (OMP) algorithm for the sparse approximation,[4] b) the columns of the dictionary are sequentially updated using SVD decomposition to minimize the approximation error. It is reported that dictionaries learned with K-SVD show excellent performance in image denoising [13, 35] among other applications. Finally, dictionaries with specific structures use (instead of general forms of atoms) a set of *parametric functions* that can describe the atoms shortly, i.e., the generating functions and the parameters build the dictionary functions. Thus, the problem is reduced to learning the parameters for one or more generating functions (see for example, [18, 33]). In Sect. 6.2.9, we will see how to use sparse representations for a classification task.

5.6   Feature Selection

Which features are relevant? or which features should be extracted? Such questions arise because there is a huge number of features that can be extracted and unfortunately, we don't know which or which of them are really necessary. First, we should not forget the reason why we extract features... so at least we could answer the question: why are they really necessary? As we explained in the introduction of this chapter (see Sect. 5.1), our task is to recognize or detect our *objects of interest*, and we need to differentiate them from the *background*. For example, in X-ray images of aluminum castings, we can have several *potential defects* that were detected using some segmentation approaches (see Sect. 4.5). As the segmentation is far from perfect, the potential defects consist of not only 'defects', but 'regular structures' as well. Our object of interest in this example is the defects, whereas the background corresponds to the regular structure of the aluminum casting. From the X-ray images, we can extract features that describe the potential defects (e.g., area, width, height, location, contrast, statistical textures, etc.). In order to recognize the defects, we have to analyze the extracted features of the available potential defects and select those features that are able to properly separate the defects from the regular structures. In this example, we could expect a good separability of both classes by selecting the contrast (see Sect. 5.3.2) because it gives a measure of the difference in the gray value between the segmented region and its neighborhood.

5.6.1   Basics

In general, if we have two classes (ω_1 for 'object of interest' and ω_0 for 'background') and we want to analyze the performance of extracted feature x, e.g., contrast, we can investigate the frequency distribution for each class as illustrated in his-

---

[4]OMP is a greedy algorithm that iteratively selects locally optimal basis vectors [59].

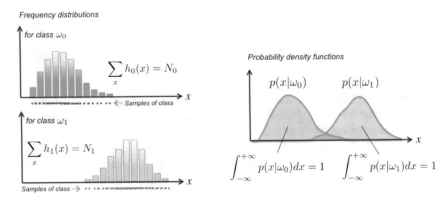

Fig. 5.19 A good class distribution for feature x and two classes ω_0 and ω_1

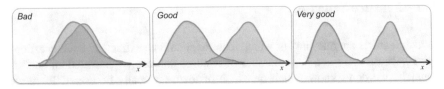

Fig. 5.20 Class distribution for three different features. It is clear that the best separability is achieved by the last features

tograms of Fig. 5.19. In this case, for frequency distribution of class ω_k, we only take into account the samples that belong to the kth class. In this *supervised* approach, the label d_i of ith sample must be available, for $i = 1 \ldots N$ for N samples. That means, someone, for example, an expert, must annotate the label of each sample of the dataset. Thus, if the ith sample belongs to class ω_k, then $d_i = k$. For N samples, we will have a vector **d** with N elements.

The available data should be representative enough, that means on the one hand that N_k, the number of samples of class ω_k, must be large enough, and on the other hand, for each class, the samples of the dataset must include the full range of variations that exist in the class itself. In our example, if x is the contrast of potential defects, we compute the frequency distribution of class ω_1 and the frequency distribution of class ω_0 by considering only the samples of 'defects' and 'regular structures' respectively. In addition, we can estimate the probability density functions from each frequency distribution known as $p(x|\omega_k)$, i.e., the probability of x given class ω_k. As we can see in Fig. 5.19, feature x is able to properly separate both classes because it takes low values for class ω_0 and high values for class ω_1, however, there is some degree of overlapping.

In feature selection, we have to decide just which features (extracted from our potential objects of interest) are relevant to the classification. By analyzing each extracted feature, three general scenarios are possible (see Fig. 5.20): a bad, a good, and a very good separability. In the first scenario, the confusion between both classes

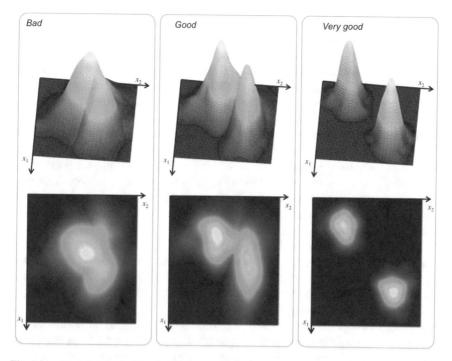

Fig. 5.21 Class distribution for three different pairs of features (x_1, x_2). As in Fig. 5.20, it is clear that the best separability is achieved by the last feature. The figure shows two types of visualization of the feature space of two features: a 3D representation and a top view using a colormap. A third type of visualization for this data is available in Fig. 5.23

is so high that it is impossible to separate the classes satisfactorily, i.e., a classifier cannot distinguish either of the classes. In the second scenario, a good separation is possible with some overlapping of the classes, i.e., a classifier will not recognize both classes perfectly, however, in many cases, this scenario can be acceptable. In the third scenario, the separability is very good, and a classifier could identify both classes in approximately 100% of the cases. If all extracted features are in the first scenario, there is no classifier that can separate both classes, i.e., new features are required. On the other hand, if we have a feature of the third scenario, the recognition can be easily performed by thresholding. In this case, no sophisticated classifiers are required. Unfortunately, the third scenario seldom occurs and we have to deal with some degree of overlapping.

In order to overcome the overlapping problem, more than one feature can be selected, however, the same three scenarios are also possible (see Fig. 5.21 for two features).

In this section, we will review some known techniques that can be used in feature selection. The reasons why feature selection is necessary are as follows:

1. It is possible that some extracted features are not *discriminative* enough, i.e., there is no information in these features for separating the classes. An example of this case is illustrated in the first scenario in Fig. 5.20. This may occur for example when we consider the mean gray value (5.16) of potential defects when detecting defects in welds. The (absolute) gray value of some defects can be very similar to the gray value of some regions of the background. In this example, we need rather a relative gray value such as a contrast (5.21).

2. Some extracted features with good separability could be redundant, i.e., they are somehow correlated. An example of this case is shown in the third scenario of Fig. 5.21 because x_1 are highly correlated with x_2. In this example, the separability by using (x_1, x_2) is very similar to the separability by using x_1 only. This may occur for example when we use two contrasts (5.21) to discriminate defects from background, maybe one contrast is enough and the second one does not increase the separability at all because it is redundant.

3. In order to simplify the testing stage, it is much better to extract a low number of features. In the training stage, we are allowed to investigate a huge number of features (in order to select some of them), however, in the testing stage, it is recommended to use a reduced subset of these. Thus, the computational time of the testing stage will be significantly reduced.

4. In order to avoid the *curse of the dimensionality*, it is highly recommended to train a classifier with a low number of features. When we increase the number of selected features, the volume of our feature space increases exponentially. Thus, in order to be statistical significant, we need to collect exponentially larger amounts of samples. This is not possible with a limited number of samples, for this reason, the performance of the classifier tends to become reduced as the number of features increases [23].

5. Last but not least, in order to avoid *false correlations* some features should not have been extracted at all and must be filtered out in this step... just in case they were extracted. This is a very common mistake and it must be avoided before a classifier is trained. An example maybe by trying to recognize a threat object (e.g., a knife) in baggage screening using features that are not rotation invariant. Imagine that we extract all elliptical features (see Sect. 5.2.2) of potential knives. The orientation α of the fitted ellipse is extracted as well (5.8). It is possible, that in our training dataset the orientation of the potential knives is always very vertical, as in the series https://www.dropbox.com/sh/pmgoyrstox6x6jk/AAAiQ0QgmkVPXr0sLqEsTdyRa?dl=0B0008 of GDXray+ (see Fig. 2.10). That means, the extracted feature α could have a distribution like a scenario two or three of Fig. 5.20. The *separability* of this feature could lead to misinterpretation because we could think that we found an extraordinary good feature that can separate knives from background, however, we are saying that a knife must be always vertical if we want to recognize it! It is clear that the orientation should not have been extracted in order to avoid a false correlation. Another typical mistake occurs when considering the location (5.2) as feature in defect recognition. In our training data, it is possible that all defects are located in one part of the image, however, in real life, they can be everywhere. Obviously, there

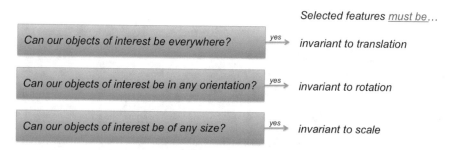

Fig. 5.22 In order to avoid false correlations, we can follow these steps when extracting features. In these cases, extracted features must be manually eliminated

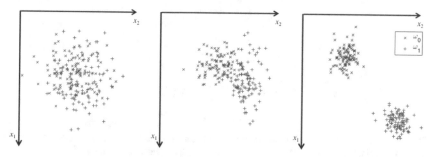

Fig. 5.23 In this visualization each, sample is represented as a point in the feature space of two dimensions (x_1, x_2). The figure shows the visualization for the three examples of Fig. 5.21

is no algorithm that detects this error. When we design an automated system, we have to be very careful in order to select manually those features that could lead to false correlations. A guide to avoid this problem is suggested in Fig. 5.22.

Formally, the extracted features of a sample can be represented as a row vector $\mathbf{x}$ of m elements, where m is the number of extracted features. Thus, a sample can be viewed as a point $\mathbf{x} = [x_1 \ldots x_m]$ in the feature space of m dimensions (see Fig. 5.19 for one dimension and Fig. 5.23 for two dimensions). The feature vector of all samples can be stored in matrix $\mathbf{X}$ of size $N \times m$, where N is the number of samples, i.e., $N = \sum_k N_k$, and N_k is the number of samples of class ω_k. The jth column of $\mathbf{X}$, called $\mathbf{x}_j$, consists of the values that take feature x_i in all samples. In addition, element x_{ij} means the feature x_j of ith sample. The features are usually normalized as

$$\tilde{x}_{ij} = \frac{x_{ij} - \mu_j}{\sigma_j} \tag{5.40}$$

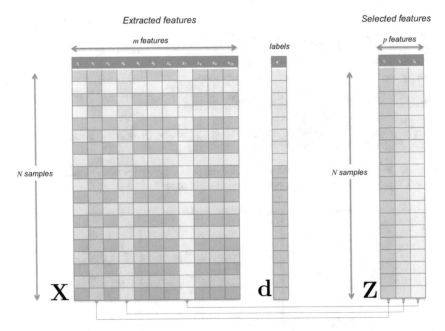

Fig. 5.24 Feature selection: there are m extracted features, from them p are selected. As we can see in the labels, the first samples belong to one class and the last one to another

for $i = 1, ..., N$ and $j = 1, ..., m$, where μ_j and σ_j are the mean and standard deviation respectively of the $\mathbf{x}_j$. The normalized features have zero mean and a standard deviation equal to one.[5]

A very good practice is to eliminate *(i)* those features that are very constant, i.e., $\sigma_j < \theta_1$, where θ_1 is some threshold, e.g., 10^{-8}, and *(ii)* those features that are very correlated, i.e., if two of any extracted features ($\mathbf{x}_i$ and $\mathbf{x}_j$) are highly correlated (if $|\text{cov}(\mathbf{x}_i, \mathbf{x}_j)|/(\sigma_i \sigma_j) > \theta_2$) one of them is eliminated. We can set θ_2 to 0.99, for example. The feature 'cleaning' is implemented in function `clean` of pybalu library.

The key idea of the feature selection is to select a subset of p features ($p \leq m$) that leads to the smallest classification error. The selected p features are arranged in a new row vector of p elements $\mathbf{z} = [z_1 \ldots z_p]$. The selected feature vector of all samples can be stored in matrix $\mathbf{Z}$ of size $N \times p$. This process is illustrated in Fig. 5.24 for $m = 10$ and $p = 3$. The p selected features are columns $s_1, s_2 \ldots s_p$ of $\mathbf{X}$, that means column j of $\mathbf{Z}$ is equal to column s_j of $\mathbf{X}$, $\mathbf{z}_j = \mathbf{x}_{s_j}$, for $j = 1 \ldots p$.

For a given set of selected features $\mathbf{s} = (s_1, s_2 \ldots s_p)$, we need some measurement of *separability* that can be used to assess the performance of the selection, i.e., for our three scenarios (see Figs. 5.20 and 5.21), this measurement should be low, high and very high respectively. We define the separability J as a function of $\mathbf{Z}$

---

[5]In pybalu library, (5.40) is implemented in `normalize` of pybalu library.

(selected features) and $\mathbf{d}$ (labels of the samples). Since $\mathbf{Z}$ corresponds to the selected columns of $\mathbf{X}$ that are defined by $\mathbf{s}$, we can write the separability as $J(\mathbf{X}, \mathbf{s}, \mathbf{d})$.

The problem of feature selection can be stated as follows, given the extracted features for N samples $(\mathbf{X})$ and the labels of each sample $(\mathbf{d})$, find a set of features (indexed by $\mathbf{s} = (s_1, s_2 \ldots s_p))$ that maximizes the separability $(J(\mathbf{X}, \mathbf{s}, \mathbf{d}))$. This is an optimization problem

$$\hat{\mathbf{s}} = \operatorname*{argmax}_{\mathbf{s} \subseteq \mathbf{Q}} J(\mathbf{X}, \mathbf{s}, \mathbf{d}), \qquad \text{s.t.} |\mathbf{s}| = p, \tag{5.41}$$

where $\mathbf{Q} = (1, 2, \ldots m)$ is the set of all possible indices that can take $\mathbf{s}$.

There are many approaches that can be used to measure the separability. A very common one is based on Fisher criterion that ensures: (i) a small intraclass variation and (ii) a large interclass variation in the space of the selected features.

For the first condition, the interclass covariance (known also as between-class covariance matrix) is used:

$$\mathbf{C}_b = \sum_k p_k (\bar{\mathbf{z}}_k - \bar{\mathbf{z}})(\bar{\mathbf{z}}_k - \bar{\mathbf{z}})^\mathsf{T}, \tag{5.42}$$

where p_k denotes the a priori probability of the kth class, $\bar{\mathbf{z}}_k$ and $\bar{\mathbf{z}}$ are the mean value of the kth class and the mean value of the selected features.

For the second condition, the intraclass covariance (known also as within-class covariance matrix) is used:

$$\mathbf{C}_w = \sum_{k=1}^{K} p_k \mathbf{C}_k, \tag{5.43}$$

where the covariance matrix of the kth class is given by:

$$\mathbf{C}_k = \frac{1}{N_k - 1} \sum_{j=1}^{N_k} (\mathbf{z}_{kj} - \bar{\mathbf{z}}_k)(\mathbf{z}_{kj} - \bar{\mathbf{z}}_k)^\mathsf{T}, \tag{5.44}$$

with $\mathbf{z}_{kj}$ the jth selected feature vector of the kth class, N_k is the number of samples in the kth class. Selection performance can be evaluated using the spur criterion for the selected features $\mathbf{z}$:

$$J = \operatorname{spur}\left(\mathbf{C}_w^{-1} \mathbf{C}_b\right). \tag{5.45}$$

where 'spur' means the sum of the diagonal. The larger the objective function J, the higher the selection performance. For the examples of Fig. 5.23, this function takes the values 0.1, 2.1, and 27.8 respectively. The objective function defined in (5.45) can be used directly in (5.41).

Another approach that can be used to measure the separability is to compute the accuracy of a classifier with the selected features. In this approach, we divide $\mathbf{Z}$ into two subsets of samples: training and testing datasets. A classifier is designed using the training set, and afterwards is tested using the testing set. The separability J is defined as the accuracy evaluated on the testing set, i.e., the ratio of samples that were correctly classified to the total number of samples.[6]

The features can be selected using several state-of-art algorithms reported in the literature. In the following, some selection algorithms are presented.

5.6.2   Exhaustive Search

The selection of the features is performed by evaluating (5.41) for all possible combination of p features of $\mathbf{X}$. The combination that achieves the highest value for J is selected. This approach ensures that global maximum of J is attained, however, it requires $n = m!/(p!(m - p)!)$ evaluations of J. The number n can be prohibited for large m and p values. For instance, if we have $m = 100$ extracted features and we want to select $p = 10$ features, then 1.73×10^{13} evaluations of J are required using exhaustive search. This function is implemented in command `exsearch` of pybalu library.

5.6.3   Branch and Bound

In branch and bound, the global maximum of J is ensured also [45]. Given that J is a monotonically increasing function, i.e., $J(\mathbf{z}_1) < J(\mathbf{z}_1, \mathbf{z}_2) < \ldots J(\mathbf{z}_1, \ldots \mathbf{z}_p)$, we can considerably reduce the number of evaluations of J. In branch and bound technique, we use a tree representation, where the root corresponds to the set of all features, and a node of the tree corresponds to a combination of features. The children's nodes are subsets of their parents. Nodes in the kth level represent combinations of $m - k$ features. We start by evaluating J at the main node ($k = 0$) with all features. This will be our *bound*, the current maximum. The key idea of the algorithm is to evaluate those children nodes that have a separability J higher than the bound. If that is the case, then we update the bound. Consequently, nodes whose separability J is lower than the bound will not be evaluated.

5.6.4   Sequential Forward Selection

This method selects the best single feature and then adds one feature at a time that, in combination with the selected features, maximizes the separability. The iteration is stopped once the selected subset reaches p features. This method requires $n = pm - p(p - 1)/2$ evaluations. For instance, if we have $m = 100$ extracted fea-

---

[6]Classifiers and accuracy estimation are covered in Chap. 6.

tures and we want to select $p = 10$ features, then 955 evaluations of J are required using SFS, this is a very low number in comparison with the number of evaluations required for exhaustive search. This method is implemented in command `sfs` of pybalu library, command `SequentialFeatureSelector` and mlxtend Library and fsel of pyxvis Library.

Python Example 5.8: The basic syntax of how to use feature selection algorithms in pyxvis Libraryis given in this code. An example that uses these commands is shown in Example 5.9.

Listing 5.8 :   Basic syntax of feature selection with pyxvis Library.

```
# [INPUT]   X  : matrix of training features, one sample per row
#           d  : vector of training labels
#           Xt : matrix of testing features, one sample per row
#           dt : vector of testing labels
#           s  : string with the name of the model
#           p  : number of features to be selected
# [OUTPUT]  q  : indices of selected features (columns)
#           X  : new matrix of training features
#           Xt : new matrix of testing features

from pyxvis.features.selection import fse_model, fsel
from pyxvis.io.data import load_features
from sklearn.neighbors import KNeighborsClassifier as KNN
from pyxvis.io.plots import print_confusion

# Definition of input variables
(X,d,Xt,dt) = load_features('../data/F40/F40')
s = 'lda'
p = 5

# Feature selection
(name,params) = fse_model(s)
q = fsel([name,params],X,d,p,cv = 5, show = 1)
print(str(len(q)) +' from ' + str(X.shape[1]) +' features selected.')

# New training and testing data
X = X[:,q]
Xt = Xt[:,q]

# Classification and Evaluation
clf = KNN(n_neighbors=5)
clf.fit(X, d)
ds = clf.predict(Xt)
print_confusion(dt,ds)
```

The use of fsel is the following:

1. We load in `name,params` the name and the parameters of the objective function for the separability using function fse_model with the string s.[7]
2. We use function fsel with the name and parameters of the model [name,params]; the original features stored in matrix X and the labels stored in vector d (in this case, the size of matrix X is $N \times m$, and the size of vector d is $N \times 1$ as illustrated in Fig. 5.24); and the number of features to be selected p. In addition, we can use

---

[7]The available names of models are `'LR'` (logistic regression), `'Ridge'` (Ridge function), `'LDA'` (linear discriminant analysis), `'QDA'` (quadratic discriminant analysis), `'SVM-LIN'` (SVM classifier with linear kernel), `'SVM-RBF'` (SVM classifier with RBF kernel).

a number of folds for the measurement of the objective function using cross-validation,[8] e.g., cv=5, and we can indicate if the curve of the performance is plotted with show=1, as illustrated in Fig. 5.24b.

3. The indices of the selected features (numbers of the columns of matrix X are stored in vector q of $p \times 1$ elements. We use it to build a new matrix of the samples with the selected columns using X[:,q].   □

Python Example 5.9: In this example, we extract intensity features of small cropped X-ray images (100×100 pixels) of salmon filets. The cropped images are in series https://www.dropbox.com/sh/d95h06ykl0w7xa5/AAAtq1ZtSWKEFqfhbqy2Cgbwa?dl=0N0002 of GDXray+. There are 100 cropped images with fishbones and 100 with no fishbones.[9] The idea is to select those features that can be relevant for the separation between both classes 'fishbones' and 'background' (labels 1 and 0 respectively). Using the selected features, we could detect small regions with fishbones in an X-ray image of a salmon filet. In this series, the labels (0 or 1) of the cropped images are available for this supervision task. We initially extract several intensity features (more than 80) and their corresponding labels using function extract_features_labels of pyxvis Library. Additionally, high correlated or constant features are eliminated as well using clean function of pybalu Library. Both functions clean and normalize of pybalu are merged together into function clean_norm of pyxvis Library (the indices of the selected 'cleaned' and the parameters a,b for linear scaling of the features can be used in function clean_norm_transform for the testing features). We select 15 features using SFS, we compute the 6 principal components of them using PCA, and finally 3 from them using exhaustive search. The computational time of the feature selection step is short because we are dealing with a small number of features and samples.

> Listing 5.9 : Feature selection with SFS.

```
import numpy as np
from pybalu.feature_selection import exsearch
from pybalu.feature_transformation import pca
from pybalu.feature_analysis import jfisher
from pyxvis.features.extraction import extract_features_labels
from pyxvis.features.selection import fsel, fse_model, clean_norm, clean_norm_transform
from pyxvis.io.plots import plot_features3, print_confusion
from sklearn.neighbors import KNeighborsClassifier as KNN

# Training-Data
path        = '../images/fishbones/'
fx          = ['basicint','gabor-ri','lbp-ri','haralick-2','fourier','dct','hog']
X,d         = extract_features_labels(fx,path+'train','jpg')
X,sclean,a,b = clean_norm(X)
(name,params) = fse_model('QDA')
```

---

[8]Cross-validation is explained in Sect. 6.3.2.

[9]In this example, we used an augmented version of this subset that is available in the folder images/fishbones. The original subset has 80 samples per class for training and 20 samples per class per testing. In the training stage of our example, the 80 samples per class are augmented to 320 per class by rotating them in 0^0, 90^0, 180^0, and 270^0. The testing samples of our examples correspond to the 20 samples per class of the original dataset (with no augmentation).

```
ssfs           = fsel([name,params],X,d,15,cv = 5, show = 1)
X              = X[:,ssfs]
Ypca,_,A,Mx,_  = pca(X, n_components=6)
X              = np.concatenate((X,Ypca),axis=1)
sf             = exsearch(X, d, n_features=3 ,method="fisher",show=True)
X              = X[:,sf]
print('Jfisher = ' + str(jfisher(X,d)))
plot_features3(X,d,'Fishbones')

# Testing-Data
Xt,dt          = extract_features_labels(fx,path+'test','jpg')
Xt             = clean_norm_transform(Xt,sclean,a,b)
Xt             = Xt[:,ssfs]
Ytpca          = np.matmul(Xt - Mx, A)
Xt             = np.concatenate((Xt,Ytpca),axis=1)
Xt             = Xt[:,sf]

# Classification and Evaluation
clf            = KNN(n_neighbors=5)
clf.fit(X, d)
ds             = clf.predict(Xt)
print_confusion(dt,ds)
```

The output of this code is shown in Fig. 5.25. In this example, the features were extracted using commands extract_features_labels, and the features were selected using fsel of pyxvis Library. This function requires a function that gives a score of the separability. In our case, we use function QDA of sklearn library. The use of fsel is explained in details in Listing 5.8.

In this example, we use from pybalu library following functions: pca for computing of PCA, exsearch for exhaustive search, AND jfisher for computing the Fisher objective function.   □

5.6.5   Sequential Backward Selection

This method selects all features and then eliminates one feature at a time that maximizes the separability. The iteration is stopped once the selected subset reaches p features. This method requires $n = (m - p + 1)m - (m - p)(m - p + 1)/2$ evaluations. For instance, if we have $m = 100$ extracted features and we want to select $p = 10$ features, then 5005 evaluations of J are required using SBS.

5.6.6   Ranking by Class Separability Criteria

Features are ranked using an independent evaluation criterion to assess the significance of every feature for separating two labeled groups. The absolute value two-sample t-Student test with pooled variance estimate is used as an evaluation criterion [37].

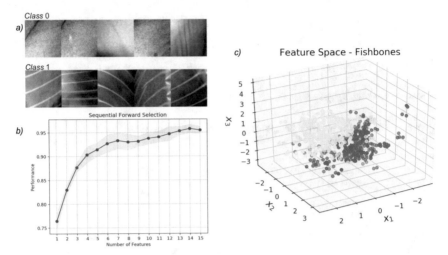

Fig. 5.25 Feature selection using SFS and exhaustive search: **a** Some cropped X-ray images of both classes 'background' and 'fishbones'. In this example, there are $m = 88$ extracted features, i.e., the extracted features are stored in matrix **X**. **b** Sequential forward selection. There are 15 selected features and the performance in the classification for each step. **c** Using PCA and exhaustive search, we select three features that are represented in the 3D feature space using command plot_features3 of **pyxvis** Library. We can see that the separability is 'good' and correspond to our second scenario. [→ Example 5.9] A classification example using a similar strategy on this dataset can be found in Example 6.13

5.6.7   Forward Orthogonal Search

In FOS, features are selected one at a time, by estimating the capability of each specified candidate feature subset to represent the overall features in the measurement feature space using a squared correlation function to measure the dependency between features [60].

5.6.8   Least Square Estimation

In LSE, features are selected one at a time, evaluating the capacity of the select feature subsets to reproduce sample projections on principal axis usingPCA [36].

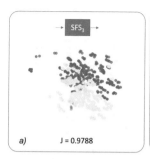
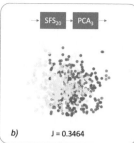
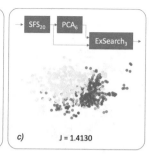

Fig. 5.26 Separability of three different feature selection methods for fishbone detection (see Example 5.9 for details). Each visualization is a 3D plot of the distribution of the two classes, the axes and the grid are not represented for the sake of simplicity. **a** The best three features after SFS. **b** The three principal components, obtained by PCA, of the 20 features selected by SFS. **c** The three best features, computed by an exhaustive search, of the concatenation of the 20 features selected by SFS and the 6 principal components of them. This plot is the same as Fig. 5.25c with the best separability J after Fisher criterion (5.45)

5.6.9 Combination with Principal Components

The first p principal components of the large set of features $\mathbf{X}$ (or a pre-selected subset of features using one of the mentioned approaches) are appended as new columns (features) of $\mathbf{X}$. Thus, we have a new set of features $\mathbf{X}_{new} = [\mathbf{X} \; pca(\mathbf{X}, p)]$. Afterwards, a feature selection algorithm (like SFS or exhaustive search) is computed on $\mathbf{X}_{new}$. As result, the selected features can be some original features and some principal components [12]. An example is shown in Fig. 5.26. In this example, this method achieved the best separability with only three features, however, it is worth mentioning that using this method the computational time is increased significantly in the testing stage. The reason is not because that we have to compute the PCA transformation, but because we have to extract all features required by PCA.

5.6.10 Feature Selection Based in Mutual Information

In mRMR, the features are selected based on two criteria: minimal redundancy in order to remove redundant variables; and maximal relevance in order to select the relevant features that are able to separate the classes [51].

5.7 A Final Example

In this example, we show how to extract and select features for a three-class problem. We want to separate handguns, shuriken, and razor blades (see some samples

Fig. 5.27 Some objects used in example of Sect. 5.7: a handgun, a shuriken, and a razor blade, from GDXray+ series https://www.dropbox.com/sh/4patsy7zf7bnwhf/AADtQ5bmbC3Il0hjI9-jN9Zra?dl=0в0049, https://www.dropbox.com/sh/hcjoso1t1urku4y/AAD31u3LbIRSQC4jonqAqasJa?dl=0в0050, and https://www.dropbox.com/sh/hffzvlzxcmhucab/AACsckxAV0QXb9xmaW5ENm_4a?dl=0в0051 respectively

in Fig. 5.27). We extract geometric features that are invariant to rotation, translation, and scale. The separation is easy because the shapes are very different. Probably, this particular example does not have any application in real life, but it shows how we can use pyxvis Library easily to extract and select features for a classification task.

The features can be extracted using a simple Python code (as shown in Example 5.10). With these commands, it is really simple to design a program that is able to extract and select many features. The general strategy follows the schema presented in Fig. 5.28.

Python Example 5.10: This example shows a simple code that is used to extract and select features geometric features. The task is to separate handguns, shuriken, and razor blades (Fig. 5.27) according to their shapes. For this end, we use isolated threat objects that are segmented using seg_bimodal segmentation approach of pyxvis Library (see Sect. 4.5.1). The reader can easily adapt this code to similar recognition problems. In this example, 20 features are selected (we use PCA of three principal components for visualization purposes only).

Listing 5.10 : Feature selection with SBS.

```python
import numpy as np
import matplotlib.pyplot as plt
from skimage.measure import label
from pyxvis.features.extraction import extract_features

# Input Image
fig    = plt.figure()
ax     = fig.add_subplot(111)
img    = plt.imread('../images/N0001_0004b.png')
implot = plt.imshow(img,cmap='gray')

# Segmentation
R      = img>0.27          # thresholding of light objects
```

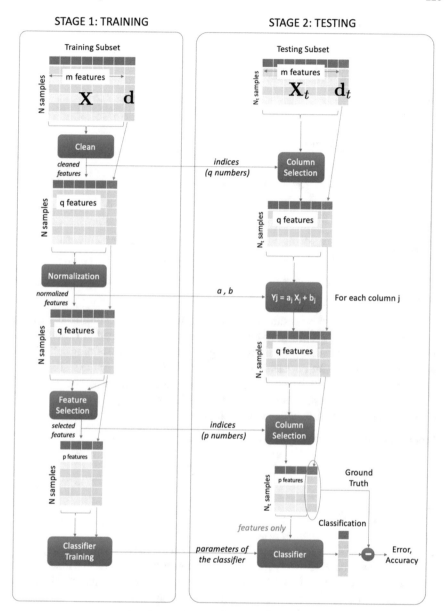

Fig. 5.28 Supervised pattern recognition schema. In the training stage, features are extracted and selected. In addition, a classifier is designed (see Chap. 6). In the testing stage, selected features are extracted and the test samples are classified

```
L      = label(R)          # labeling of objects
n      = np.max(L)         # number of detected objects
T      = np.zeros((n,18))  # features of each object will stored in a row

# Analysis of each segmented object
t       = 0 # count of recognized fruits
for i in range(n):
    R = (L == i)*1                          # binary image of object i
    f = extract_features('basicgeo',bw=R)   # feature extraction for object i
    area = f[4]
    # recognition of fruits according to the size
    if area>14000 and area<21000:
        T[t,:] = f                          # storing the features of the fruit t
        t = t+1
        # labeling each recognized fruit in the plot
        ax.text(f[1]-20, f[0]+10, str(t), fontsize=12,color='Red')

# Display and save results
plt.show()
F = T[0:t,:]
print('Basic Geo-Features:')
print(F)
np.save('GeoFeatures.npy',F)               # save features
```

The output of this code is shown in Fig. 5.29. In this example, we use several powerful functions of pyxvis Library:

- extract_features_labels: Feature extraction and their corresponding labels from a set of images. The list of the features to be extracted are defined in variable `fx`.
- clean_norm: For cleaning and normalizing the features of the training set.
- clean_norm_transform: For cleaning and normalizing features of the testing set.
- fse_model: For definition of objective function of the separability.
- fse_sbs: For sequential backward selection of features.
- print_confusion: To display the confusion matrix.

From Fig. 5.29 it is very simple to design a classification strategy (e.g., using KNN).

□

5.8   Summary

In this Chapter, we covered several topics that are used to represent an X-ray image (or a specific region of an X-ray image). This representation means that new features are extracted from the original image and that they can give us more information than the raw information expressed as a matrix of gray values.

In the first part of this chapter, we learned about geometric and intensity features. We reviewed basic geometric features (such as area and perimeter among others), elliptical features, Fourier descriptors, and invariant moments. Further, we addressed basic intensity features, several definitions of contrast, crossing line profiles (CLP), intensity moments, statistical textures, Gabor, and filter banks (such as Fourier and DCT).

In the second part of this chapter, we gave an overview of certain descriptors that are widely used in computer vision and can be a powerful tool in X-ray testing. We

Feature Space - PCA - Threat Objects

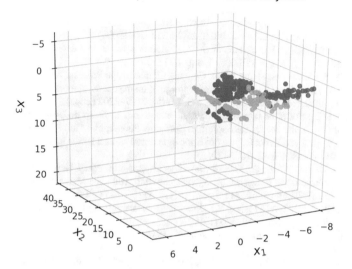

Fig. 5.29 Separation of three classes: (0) Handguns, (1) Shuriken, (2) Razor blades. The performance of the separation (accuracy) in this example is 97.0% using a simple KNN classifier. [→ Example 5.10 🐍]

covered local binary patterns (LBP), binarized statistical image features (BSIF), histogram of oriented gradients (HOG), and scale-invariant feature transform (SIFT).

In the third part of this chapter, we studied sparse representations. They have been widely used in computer vision. In X-ray testing, they can be used in problems of object recognition as we will see in the next chapter.

In the fourth part of this chapter, we presented different feature selection techniques that can be used to chose which features are relevant for a classification problem. Some of the techniques are sequential feature selection, branch and bound, and feature selection based on mutual information.

Finally, we gave a simple code as an example that can be used to extract and select features for a classification problem.

References

1. Aharon, M., Elad, M., Bruckstein, A.: K-SVD: An algorithm for designing overcomplete dictionaries for sparse representation. IEEE Trans. Signal Process. **54**(11), 4311–4322 (2006)
2. Bay, H., Tuytelaars, T., Van Gool, L.: Surf: speeded up robust features. In: 9th European Conference on Computer Vision (ECCV2006). Graz Austria (2006)
3. Calonder, M., Lepetit, V., Strecha, C., Fua, P.: BRIEF: Binary robust independent elementary features. In: Computer Vision–ECCV 2010, pp. 778–792. Springer (2010)
4. Castleman, K.: Digital Image Processing. Prentice-Hall, Englewood Cliffs (1996)

5. Chellappa, R., Bagdazian, R.: Fourier coding of image boundaries. IEEE Trans. Pattern Anal. Mach. Intell. **PAMI-6**(1), 102–105 (1984)
6. Coeurjolly, D., Klette, R.: A comparative evaluation of length estimators of digital curves. IEEE Trans. Pattern Anal. Mach. Intell. **26**(2), 252–258 (2004)
7. Dalal, N., Triggs, B.: Histograms of oriented gradients for human detection. Conf. Comput. Vis. Pattern Recognit. (CVPR2005) **1**, 886–893 (2005)
8. Danielsson, P.E.: A new shape factor. Comput. Graph. Image Process. **7**, 292–299 (1978)
9. Donoho, D., Elad, M.: Optimally sparse representation in general (nonorthogonal) dictionaries via ℓ_1 minimization. Proc. Natl. Acad. Sci. **100**(5), 2197–2202 (2003)
10. Donoho, D.L.: For most large underdetermined systems of linear equations the minimal ℓ_1-norm solution is also the sparsest solution. Commun. Pure Appl. Math. **59**(6), 797–829 (2006)
11. Doyle, J.S., Bowyer, K.W.: Robust detection of textured contact lenses in iris recognition using bsif. IEEE Access **3**, 1672–1683 (2015)
12. Duda, R., Hart, P., Stork, D.: Pattern Classification, 2nd edn. Wiley, New York (2001)
13. Elad, M., Aharon, M.: Image denoising via sparse and redundant representations over learned dictionaries. IEEE Trans. Image Process. **15**(12), 3736–3745 (2006)
14. Fitzgibbon, A., Pilu, M., Fisher, R.: Direct least square fitting ellipses. IEEE Trans. Pattern Anal. Mach. Intel. **21**(5), 476–480 (1999)
15. Flusser, J., Suk, T.: Pattern recognition by affine moment invariants. Pattern Recognit. **26**(1), 167–174 (1993)
16. Gonzalez, R., Woods, R.: Digital Image Processing, 3rd edn. Prentice Hall, Pearson (2008)
17. Gorodnitsky, I., Rao, B.: Sparse signal reconstruction from limited data using focuss: a reweighted minimum norm algorithm. IEEE Trans. Signal Process. **45**(3), 600–616 (1997)
18. Gribonval, R., Nielsen, M.: Sparse representations in unions of bases. IEEE Trans. Inf. Theory **49**(12), 3320–3325 (2003)
19. Gupta, L., Srinath, M.D.: Contour sequence moments for the classification of closed planar shapes. Pattern Recognit. **20**(3), 267–272 (1987)
20. Haralick, R., Shanmugam, K., Dinstein, I.: Textural features for image classification. IEEE Trans. Syst. Man Cybern. **SMC-3**(6), 610–621 (1973)
21. Hartley, R.I., Zisserman, A.: Multiple View Geometry in Computer Vision, 2nd edn. Cambridge University Press, Cambridge (2003)
22. Hu, M.K.: Visual pattern recognition by moment invariants. IRE Trans. Info. Theory **IT**(8), 179–187 (1962)
23. Hughes, G.: On the mean accuracy of statistical pattern recognizers. IEEE Trans. Inf. Theory **14**(1), 55–63 (1968)
24. Jähne, B.: Digitale Bildverarbeitung, 2nd edn. Springer, Berlin (1995)
25. Joliffe, I.: Principal Component Analysis. Springer, New York (1986)
26. Kamm, K.F.: Grundlagen der Röntgenabbildung. In: Ewen, K. (ed.) Moderne Bildgebung: Physik, Gerätetechnik, Bildbearbeitung und -kommunikation, Strahlenschutz, Qualitätskontrolle, pp. 45–62. Georg Thieme Verlag, Stuttgart, New York (1998)
27. Kannala, J., Rahtu, E.: BSIF: Binarized statistical image features. In: 2012 21st International Conference on Pattern Recognition (ICPR), pp. 1363–1366. IEEE (2012)
28. Klette, R.: Concise Computer Vision: An Introduction into Theory and Algorithms. Springer Science & Business Media (2014)
29. Kreutz-Delgado, K., Murray, J., Rao, B., Engan, K., Lee, T., Sejnowski, T.: Dictionary learning algorithms for sparse representation. Neural Comput. **15**(2), 349–396 (2003)
30. Kumar, A., Pang, G.: Defect detection in textured materials using gabor filters. IEEE Trans. Ind. Appl. **38**(2), 425–440 (2002)
31. Leutenegger, S., Chli, M., Siegwart, R.Y.: BRISK: Binary robust invariant scalable keypoints. In: 2011 IEEE International Conference on Computer Vision (ICCV), pp. 2548–2555. IEEE (2011)
32. Lowe, D.: Distinctive image features from scale-invariant keypoints. Int. J. Comput. Vis. **60**(2), 91–110 (2004)

33. Mailhé, B., Lesage, S., Gribonval, R., Bimbot, F., Vandergheynst, P., et al.: Shift-invariant dictionary learning for sparse representations: extending k-svd. Proc. Eur. Signal Process. Conf. **4** (2008)
34. Mairal, J., Bach, F., Ponce, J.: Sparse modeling for image and vision processing. Found. Trends® Comput. Graph. Vis. **8**(2-3), 85–283 (2014)
35. Mairal, J., Elad, M., Sapiro, G.: Sparse representation for color image restoration. IEEE Trans. Image Process. **17**(1), 53–69 (2008)
36. Mao, K.: Identifying critical variables of principal components for unsupervised feature selection. IEEE Trans. Syst. Man Cybern. Part B: Cybern.**35**(2), 339–344 (2005)
37. MathWorks: Matlab Toolbox of Bioinformatics: User's Guide. Mathworks Inc. (2007)
38. MathWorks: Image Processing Toolbox for Use with MATLAB: User's Guide. The MathWorks Inc. (2014)
39. Mery, D.: Crossing line profile: a new approach to detecting defects in aluminium castings. Proceedings of the Scandinavian Conference on Image Analysis (SCIA 2003), Lecture Notes in Computer Science vol. 2749, pp. 725–732 (2003)
40. Mery, D., Filbert, D.: Automated flaw detection in aluminum castings based on the tracking of potential defects in a radioscopic image sequence. IEEE Trans. Robot. Autom. **18**(6), 890–901 (2002)
41. Mery, D., Filbert, D.: Classification of potential defects in automated inspection of aluminium castings using statistical pattern recognition. In: 8th European Conference on Non-Destructive Testing (ECNDT 2002), pp. 1–10. Barcelona (2002)
42. Mery, D., Filbert, D., Jaeger, T.: Image processing for fault detection in aluminum castings. In: MacKenzie, D., Totten, G. (eds.) Analytical Characterization of Aluminum and Its Alloys. Marcel Dekker, New York (2003)
43. Mikolajczyk, K., Schmid, C.: A performance evaluation of local descriptors. IEEE Trans. Pattern Anal. Mach. Intel. **27**(10), 1615–1630 (2005)
44. Mu, Y., Yan, S., Liu, Y., Huang, T., Zhou, B.: Discriminative local binary patterns for human detection in personal album. In: IEEE Conference on Computer Vision and Pattern Recognition (CVPR 2008), pp. 1–8 (2008)
45. Narendra, P.M., Fukunaga, K.: A branch and bound algorithm for feature subset selection. IEEE Trans. Comput. **C-26**(9), 917–922 (1977)
46. Ojala, T., Pietikainen, M., Maenpaa, T.: Multiresolution gray-scale and rotation invariant texture classification with local binary patterns. IEEE Trans. Pattern Anal. Mach. Intel. **24**(7), 971–987 (2002)
47. Ojansivu, V., Heikkilä, J.: Blur insensitive texture classification using local phase quantization. In: Image and signal processing, pp. 236–243. Springer (2008)
48. Olshausen, B., Field, D.: Emergence of simple-cell receptive field properties by learning a sparse code for natural images. Nature **381**(6583), 607–609 (1996)
49. Olshausen, B., Field, D.: Sparse coding with an overcomplete basis set: a strategy employed by v1? Vis. Res. **37**(23), 3311–3325 (1997)
50. Olshausen, B., Field, D.: Sparse coding of sensory inputs. Curr. Opin. Neurobiol. **14**(4), 481–487 (2004)
51. Peng, H., Long, F., Ding, C.: Feature selection based on mutual information criteria of max-dependency, max-relevance, and min-redundancy. IEEE Trans. Pattern Anal. Mach. Intel. **27**(8), 1226–1238 (2005)
52. Persoon, E., Fu, K.: Shape discrimination using Fourier descriptors. IEEE Trans. Syst. Man Cybern. **SMC-7**(3), 170–179 (1977)
53. Randen, T., Husoy, J.: Filtering for texture classification: a comparative study. IEEE Trans. Pattern Anal. Mach. Intell. **21**(4), 291–310 (1999)
54. Rubinstein, R., Bruckstein, A., Elad, M.: Dictionaries for sparse representation modeling. Proc. IEEE **98**(6), 1045–1057 (2010)
55. Rubinstein, R., Zibulevsky, M., Elad, M.: Double sparsity: learning sparse dictionaries for sparse signal approximation. IEEE Trans.Signal Process. **58**(3), 1553–1564 (2010)

56. Sonka, M., Hlavac, V., Boyle, R.: Image Processing, Analysis, and Machine Vision, 2nd edn. PWS Publishing, Pacific Grove (1998)
57. Teh, C., Chin, R.: On digital approximation of moment invariants. Comput. Vis. Graph. Image Process. 33(3), 318–326 (1986)
58. Tosic, I., Frossard, P.: Dictionary learning. Signal Process. Mag. IEEE 28(2), 27–38 (2011)
59. Tropp, J.: Greed is good: algorithmic results for sparse approximation. IEEE Trans. Inf. Theory 50(10), 2231–2242 (2004)
60. Wei, H.L., Billings, S.: Feature subset selection and ranking for data dimensionality reduction. IEEE Trans. Pattern Anal. Mach. Intell. 29(1), 162–166 (2007)
61. Wright, J., Ma, Y., Mairal, J., Sapiro, G., Huang, T., Yan, S.: Sparse representation for computer vision and pattern recognition. Proc. IEEE 98(6), 1031–1044 (2010)
62. Yaghoobi, M., Blumensath, T., Davies, M.: Dictionary learning for sparse approximations with the majorization method. IEEE Trans. Signal Process. 57(6), 2178–2191 (2009)
63. Yang, A., Gastpar, M., Bajcsy, R., Sastry, S.: Distributed sensor perception via sparse representation. Proc. IEEE 98(6), 1077–1088 (2010)
64. Yang, J., Yu, K., Gong, Y., Huang, T.: Linear spatial pyramid matching using sparse coding for image classification. In: IEEE Conference on Computer Vision and Pattern Recognition, 2009. CVPR 2009, pp. 1794–1801 (2009)
65. Zahn, C., Roskies, R.: Fourier descriptors for plane closed curves. IEEE Trans. Comput. C-21(3), 269–281 (1971)

Chapter 6
Classification in X-Ray Testing

•

Abstract In this chapter, we will cover known classifiers that can be used in X-ray testing. Several examples will be presented using Python. The reader can easily modify the proposed implementations in order to test different classification strategies. We will then present how to estimate the accuracy of a classifier using hold-out, cross-validation and leave-one-out. Finally, we will present an example that involves all steps of a pattern recognition problem, i.e., feature extraction, feature selection, classifier's design, and evaluation. We will thus propose a general framework to design a computer vision system in order to select—automatically—from a large set of features and a bank of classifiers, those features and classifiers that can achieve the highest performance.

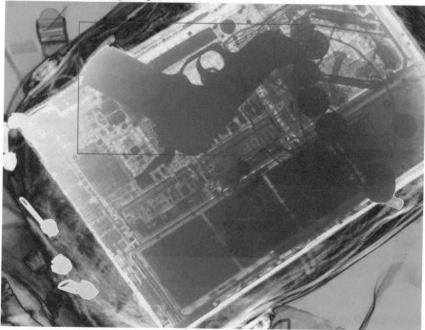

Ideal detection of a handgun superimposed onto a laptop (X-ray image `B0019_0001` *colored with 'sinmap').*

© Springer Nature Switzerland AG 2021
D. Mery and C. Pieringer, *Computer Vision for X-Ray Testing*,
https://doi.org/10.1007/978-3-030-56769-9_6

6.1   Introduction

Considerable research efforts in computer vision applied to industrial applications have been developed in recent decades. Many of them have been concentrated on using or developing tailored methods based on visual features that are able to solve a specific task. Nevertheless, today's computer capabilities are giving us new ways to solve complex computer vision problems. In particular, a new paradigm on machine learning techniques has emerged posing the task of recognizing visual patterns as a search problem based on training data and a hypothesis space composed of visual features and suitable classifiers. Furthermore, now we are able to extract, process, and test in the same time more image features and classifiers than before. In our book, we propose a general framework that designs a computer vision system automatically, i.e., it finds—without human interaction—the features and the classifiers for a given application avoiding the classical trial and error framework commonly used by human designers. The key idea of the proposed framework is to design a computer vision system as shown in Fig. 6.1 in order to select—automatically—from a large set of features and a bank of classifiers, those features and classifiers that achieve the highest performance.

Whereas Chap. 5 covered feature extraction and selection, the focus of this chapter will be the classification. Once the proper features are selected, a classi-

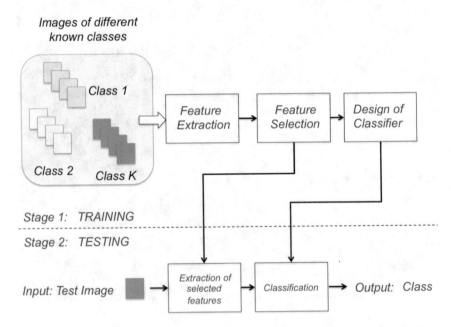

Fig. 6.1 Supervised pattern recognition schema. In the training stage, features are extracted and selected (see Chap. 5 and details in Fig. 5.28). In addition, a classifier is designed. In the testing stage, selected features are extracted and the test image is classified

fier can be designed. Typically, the classifier assigns a feature vector **x** with n features $(x_1 \ldots x_n)$ to one class. In case of defects detection, for example, there are two classes: *flaws* or *no-flaws*. In case of baggage screening, there can be more classes: *knives, handguns, razor blades*, etc. In pattern recognition, classification can be performed using the concept of similarity: patterns that are *similar* are assigned to the same class [12]. Although this approach is very simple, a good metric defining the similarity must be established. Using representative samples, we can make a supervised classification finding a discriminant function $h(\mathbf{x})$ that provides us information on how similar a feature vector **x** is to a class representation.

In this chapter, we will cover many known classifiers (such as linear discriminant analysis, Bayes, support vector machines, neural networks among others). Several examples will be presented using Python. The reader can easily modify the proposed implementations in order to test different classification strategies. Afterwards, we present how to estimate the accuracy of a classifier using hold-out, cross-validation, and leave-one-out. The well-known confusion matrix and receiver-operation-characteristic curve will be outlined as well. We will explain by detailing the advantages and disadvantages of each one. Finally, we will present an example that involves all steps of a pattern recognition problem, i.e., feature extraction, feature selection, classifier's design, and evaluation.

6.2   Classifiers

In this section, the most relevant classifiers are explained with several examples. Before we start with the explanation of the classifiers, let us review the syntax of some basic functions of pyxvis Library. The implementation of this functions is based on sklearn library.

Python Example 6.1 The basic syntax of how to use classification algorithms in pyxvis Library is given in this code. Examples that use these commands are shown in this section (e.g., see Example 6.2).

Listing 6.1 : Basic syntax of classification with pyxvis Library.

```
# [INPUT]  X  : training features (matrix of N x p elements)
#          d  : vector of training labels (vector of N elements)
#          Xt : testing features (matrix of Nt x p elements)
#          dt : vector of training labels (vector of Nt elements)
#          s  : string with the name of the model
# [OUTPUT] ds : classification (vector of Nt elements)
#          clf: trained classifier

from pyxvis.io.data import load_features
from pyxvis.learning.classifiers import clf_model, define_classifier
from pyxvis.learning.classifiers import train_classifier, test_classifier
from pyxvis.io.plots import print_confusion

# Definition of input variables
(X,d,Xt,dt) = load_features('../data/G3/G3')
```

```
s                = 'knn5'

# Training and Testing
(name,params) = clf_model(s)                 # function name and parameters
clf           = define_classifier([name,params]) # classifier definition
clf           = train_classifier(clf,X,d)    # classifier training
ds            = test_classifier(clf,Xt)      # clasification on testing

# Evaluation of performance
print_confusion(dt,ds)
```

The training and testing stages of a classification process is given in following four steps (see Fig. 6.18):

1. We load in `name,params` the name and the parameters of the classifier using function clf_model with the string s [1]
2. We define a classifier using function define_classifier with the name and parameters of the model [name,params]. The defined classifier is stored in clf.
3. Classifier clf is trained using training data (X,d) with function train_classifier. The defined classifier is stored in clf.
4. Trained classifier clf is tested on testing data (Xt) using function test_classifier. The classification, i.e., labels of the testing samples, are stored in vector ds. To evaluate the effectiveness of the classifier, we can count the number of coincidences between dt (real labels of testing data) and ds (classification using trained classifier). [2]   □

6.2.1 Minimal Distance

The simplest classifier is probably based on the concept of 'minimal distance'. In this classifier, each class is represented by its center of mass that can be viewed as a template [10]. Thus, a mean value $\bar{\mathbf{x}}_k$ of each class is calculated on the training data:

$$\bar{\mathbf{x}}_k = \frac{1}{N_k} \sum_{i=1}^{N_k} \mathbf{x}_{jk}, \tag{6.1}$$

where $\mathbf{x}_{jk}$ is the jth sample of class ω_k of the training data, and N_k is the number of samples of the kth class. A test sample $\mathbf{x}$ is assigned to class ω_k if the Euclidean distance $\| \mathbf{x} - \bar{\mathbf{x}}_k \|$ is minimal. Formerly,

$$h_{\text{dmin}}(\mathbf{x}) = \underset{k}{\operatorname{argmin}} \{\| \mathbf{x} - \bar{\mathbf{x}}_k \|\} . \tag{6.2}$$

---

[1]The available names of models are: 'LR' (logistic regression), 'dmin' (Minimal Distance), 'LDA' (linear discriminant analysis), 'QDA' (quadratic discriminant analysis), 'KNN' (nearest neighbors), 'RF' (random forest), 'NN' (neural network), 'AdaBoost' (AdaBoost), 'SVM-LIN' (SVM classifier with linear kernel), 'SVM-RBF' (SVM classifier with RBF kernel).

[2]Usually, for this end we can use the *accuracy* metric explained in Sect. 6.3.

A useful formulation is defining the distance function $d_{\mathrm{dmin}}(\mathbf{x}, k) = \| \mathbf{x} - \bar{\mathbf{x}}_k \|$. Thus, we can write (6.2) as

$$h_{\mathrm{dmin}}(\mathbf{x}) = \operatorname*{argmin}_k \{ d_{\mathrm{dmin}}(\mathbf{x}, k) \}. \tag{6.3}$$

This formulation based on minimal distances will be used in the following sections. In pyxvis Library, this classifier is implemented using function clf_model with parameter 'dmin'.   Python Example 6.2 In this example, we show how to train and test a classifier based on Euclidean minimal distance. We use data that was simulated using a mixture of Gaussian distributions. The data consists of 800 samples for training and 400 samples for testing purposes. Each sample has two features x_1 and x_2 and it belongs to class ω_1 or ω_0. Figure 6.2 shows the feature spaces for training and testing.

Listing 6.2 : Classification using Euclidean minimal distance

```
from pyxvis.io.data import load_features
from pyxvis.io.plots import show_clf_results
from pyxvis.learning.classifiers import clf_model, define_classifier
from pyxvis.learning.classifiers import train_classifier, test_classifier

(X,d,Xt,dt)   = load_features('../data/F2/F2')  # load training and testing data
cl_name       = 'dmin'                          # generic name of the classifier
(name,params) = clf_model(cl_name)              # function name and parameters
clf           = define_classifier([name,params]) # classifier definition
clf           = train_classifier(clf,X,d)        # classifier training
d0            = test_classifier(clf,X)           # clasification of training
ds            = test_classifier(clf,Xt)          # clasification of testing
show_clf_results(clf,X,d,Xt,dt,d0,ds,cl_name)   # show performance and feature space
```

The output of this code is shown in Fig. 6.5. In this case, the accuracy, defined as the ratio of samples correctly classified, is 85.50% in the testing dataset. The low performance of this classifier is because the decision line is a straight line. The reader can imagine that the decision line can be computed in three steps: (i) Compute the centers of mass of each class distribution in the training set as $\bar{\mathbf{x}}_1$ and $\bar{\mathbf{x}}_0$ according to (6.1). (ii) Compute ℓ_C the straight line that contains both centers of mass. (iii) Compute the decision line ℓ as the line that is perpendicular to ℓ_C and equidistant to $\bar{\mathbf{x}}_1$ and $\bar{\mathbf{x}}_0$. The decision line is shown in Fig. 6.5. Obviously, the straight line is not able to separate these curved distributions.

The syntax of the use of the classification functions in pyxvis Library is explained in Listing 6.1.   □

6.2.2   Mahalanobis Distance

The Mahalanobis classifier employs the same concept as minimal distance (see Sect. 6.2.1), however, it uses a distance metric based on the 'Mahalanobis distance', in which, by means of the covariance matrix, the features to be evaluated are weighted according to their variances. A test sample $\mathbf{x}$ is assigned to class ω_k if the Mahalanobis distance of $\mathbf{x}$ to class ω_k, denoted as $d_{\mathrm{maha}}(\mathbf{x}, k)$, is minimal. The Mahalanobis

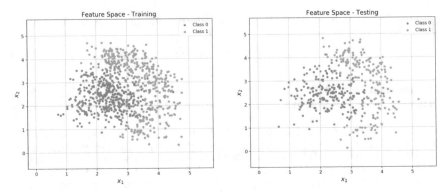

Fig. 6.2 Simulated data that is used in Sect. 6.2. [→ Example 6.2 🐍]

distance is defined as

$$d_{\text{maha}}(\mathbf{x}, k) = (\mathbf{x} - \bar{\mathbf{x}}_k)^{\mathsf{T}} \mathbf{C}_k^{-1}, (\mathbf{x} - \bar{\mathbf{x}}_k), \tag{6.4}$$

where $\mathbf{C}_k$ is the covariance matrix of the kth class. It can be estimated as

$$\mathbf{C}_k = \frac{1}{N_k - 1} \sum_{j=1}^{N_k} (\mathbf{x}_{kj} - \bar{\mathbf{x}}_k)(\mathbf{x}_{kj} - \bar{\mathbf{x}}_k)^{\mathsf{T}}, \tag{6.5}$$

where $\mathbf{x}_{jk}$ is the jth sample of class ω_k of the training data, and N_k is the number of samples of the kth class. Some examples are illustrated in Fig. 6.3. Formerly,

$$h_{\text{maha}}(\mathbf{x}) = \underset{k}{\operatorname{argmin}} \{d_{\text{maha}}(\mathbf{x}, k)\}, \tag{6.6}$$

where distance d_{maha} is defined in (6.4). In pyxvis Library, this classifier is implemented using function clf_model with parameter 'maha'. An example of this classifier is presented in Example 6.4.

6.2.3 Bayes

In Bayes classifier the idea is to assign the test sample $\mathbf{x}$ to the most *probable* class. For this purpose, we use the conditional probability $p(\omega_k|\mathbf{x})$, that gives the probability of class ω_k occurs given sample $\mathbf{x}$. Thus, if $p(\omega_k|\mathbf{x})$ is maximal the $\mathbf{x}$ is assigned to class ω_k:

$$h_{\text{Bayes}}(\mathbf{x}) = \underset{k}{\operatorname{argmax}} \{p(\omega_k|\mathbf{x})\}. \tag{6.7}$$

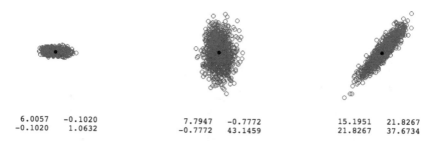

| 6.0057 | -0.1020 | | 7.7947 | -0.7772 | | 15.1951 | 21.8267 |
| -0.1020 | 1.0632 | | -0.7772 | 43.1459 | | 21.8267 | 37.6734 |

Fig. 6.3 Examples of three different Gaussian distributions $p(\mathbf{x}|\omega_k)$ in 2D. The black point represents the mean μ_k and the 2×2 matrices the covariances Σ_k

Using Bayes theorem we can write the conditional probability as

$$p(\omega_k|\mathbf{x}) = p(\omega_k)\frac{p(\mathbf{x}|\omega_k)}{p(\mathbf{x})}, \tag{6.8}$$

where $p(\omega_k|\mathbf{x})$ is known as 'posterior', $p(\omega_k)$ as 'prior', $p(\mathbf{x}|\omega_k)$ as 'likelihood' and $p(\mathbf{x})$ as 'evidence'. Since $p(\mathbf{x})$ is the same by evaluating $p(\omega_k|\mathbf{x})$ for all k we can re-write (6.7) as follows:

$$h_{\text{Bayes}}(\mathbf{x}) = \underset{k}{\text{argmax}} \left\{ p(\mathbf{x}|\omega_k)p(\omega_k) \right\}. \tag{6.9}$$

In order to evaluate (6.9) properly, we need good estimations for $p(\mathbf{x}|\omega_k)$ and $p(\omega_k)$. There are several known approaches to estimate these, some of which will be covered in the following sections under the assumption of Gaussian distributions of the classes (see Sects. 6.2.4 and 6.2.5).

In Naïve Bayes approach, each feature x_i is assumed to make an independent and equal contribution to our output. Obviously, this assumption is not correct in real world, however, in many practical cases it works well enough. Using this assumption, Eq. (6.8) can be formulated as

$$p(\omega_k|\mathbf{x}) = p(\omega_k)\frac{p(x_1|\omega_k)p(x_2|\omega_k)\cdots p(x_n|\omega_k)}{p(x_1)p(x_2)\cdots p(x_n)}, \tag{6.10}$$

and the classification rule for this case is

$$h_{\text{Naïve-Bayes}}(\mathbf{x}) = \underset{k}{\text{argmax}} \left\{ p(\omega_k)\prod_{i=1}^{n} p(x_i|\omega_k) \right\}. \tag{6.11}$$

The prior $p(\omega_k)$ can be estimated by the number of available samples in the training dataset of each class. Thus, $p(\omega_k) = N_k/N$, where N_k is the number of samples that belong to class ω_k and $N = \sum_k N_k$ the total number of samples. Nevertheless, in many cases of X-ray testing the available samples are not balanced, e.g., in defect detection problems there are a reduced number of flaws in comparison with

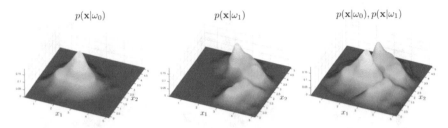

$p(\mathbf{x}|\omega_0)$ $p(\mathbf{x}|\omega_1)$ $p(\mathbf{x}|\omega_0), p(\mathbf{x}|\omega_1)$

Fig. 6.4 Estimation of $p(\mathbf{x}|\omega_k)$ using Kernel Density Estimation (KDE) for distributions of the training set of Fig. 6.2. [→ Example 6.3]

the large number of non-flaws [7]. If we use the estimation $p(\omega_k) = N_k/N$ then the most important class to be detected will have a very low prior, and it will be very difficult to detect. In such cases, the prior must be considerably increased in order to be the more probable.

In order to estimate $p(\mathbf{x}|\omega_k)$, we can use an approach based on Kernel Density Estimation (KDE) [22]:

$$\hat{p}(\mathbf{x}|\omega_k) = \alpha_k \sum_{j=1}^{N_k} K\left(\frac{\mathbf{x} - \mathbf{x}_{jk}}{\Delta}\right), \tag{6.12}$$

where K is a kernel function such as a Gaussian, that has a mean zero and variance of one, Δ is the bandwidth, and α_k is a normalization factor equal to $1/(N_k\Delta)$. Since $K(\mathbf{x}/\Delta)$ integrates to Δ, with this normalization factor we ensure that $\hat{p}(\mathbf{x}|\omega_k)$ integrates to one. Example of KDE can be found in Fig. 5.21 that were estimated using the training data of Fig. 5.23. In pyxvis Library, this classifier is implemented using function clf_model with parameter 'bayes-kde' (for KDE implementation) or 'bayes-naive' (for a naive estimation of the probability density function, where each variable is considered to be statistically independent) (Fig. 6.4).

Python Example 6.3 In this example, we show how to train and test a Bayes classifier using Kernel Density Estimation and Naive Bayes Estimation. We use the same simulated data addressed in Example 6.2 and illustrated in Fig. 6.2.

Listing 6.3 : Classification using Bayes

```python
from pyxvis.io.data import load_features
from pyxvis.io.plots import show_clf_results
from pyxvis.learning.classifiers import clf_model, define_classifier
from pyxvis.learning.classifiers import train_classifier, test_classifier

(X,d,Xt,dt)   = load_features('../data/F2/F2')          # load training and testing data
ss_cl         = ['bayes-naive','bayes-kde']
for cl_name in ss_cl:
    (name,params) = clf_model(cl_name)                  # function name and parameters
    clf           = define_classifier([name,params])    # classifier definition
    clf           = train_classifier(clf,X,d)           # classifier training
```

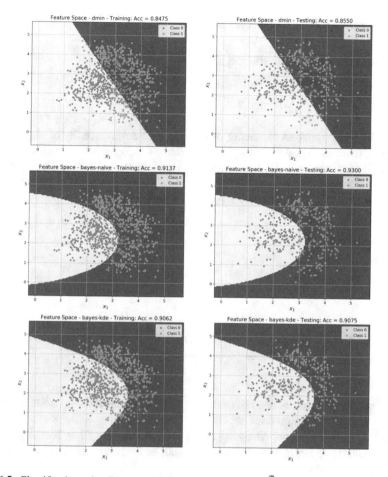

Fig. 6.5 Classification using Bayes and dmin. [→ Example 6.3 🐍]

```
d0              = test_classifier(clf,X)        # clasification of training
ds              = test_classifier(clf,Xt)       # clasification of testing
show_clf_results(clf,X,d,Xt,dt,d0,ds,cl_name)   # display results and decision lines
```

The output of this code is shown in Fig. 6.5. In this case, the accuracy, defined as the ratio of samples correctly classified, is 93.00% and 90.75% for Naive–Bayes and KDE-Bayes respectively. The reader can compare this result with the accuracy obtained by classifier of Example 6.2.   □

6.2.4   Linear Discriminant Analysis

For Gaussian distributions with $\mathbf{x} \in \mathbb{R}^n$:

$$p(\mathbf{x}|\omega_k) = \frac{1}{(2\pi)^{n/2}|\Sigma_k|^{1/2}} \exp\left\{-\frac{1}{2}(\mathbf{x} - \mu_k)^\mathsf{T}\Sigma_k^{-1}(\mathbf{x} - \mu_k)\right\}, \qquad (6.13)$$

where a good estimation for center of mass μ_k and covariance Σ_k of class ω_k can be taken from (6.1) and (6.5) respectively. Since the logarithm is a monotonically increasing function $\mathrm{argmax}_k\{p\} = \mathrm{argmax}_k\{\log(p)\}$. Thus, (6.9) can be written as

$$h(\mathbf{x}) = \underset{k}{\mathrm{argmax}}\left\{\log\left\{p(\mathbf{x}|\omega_k)p(\omega_k)\right\}\right\}. \qquad (6.14)$$

Using some manipulation,

$$\log\left\{p(\mathbf{x}|\omega_k)p(\omega_k)\right\} = \log\left\{p(\mathbf{x}|\omega_k)\right\} + \log\left\{p(\omega_k)\right\} \qquad (6.15)$$

$$= \underbrace{-\frac{1}{2}(\mathbf{x} - \mu_k)^\mathsf{T}\Sigma_k^{-1}(\mathbf{x} - \mu_k)}_{①} \underbrace{-\frac{1}{2}\log(|\Sigma_k|)}_{②} \underbrace{-\frac{n}{2}\log(2\pi)}_{③} + \underbrace{\log(p(\omega_k))}_{④}. \qquad (6.16)$$

It is clear, that we do not need to evaluate ③ because this term is constant and the location of the maximum does not change.

In Linear Discriminant Analysis (LDA) [11], we assume $\Sigma_k = \Sigma$ (constant) for all k, i.e., term ② in (6.16) is constant as well, and it is not necessary to be evaluated. Consequently,

$$\log\left\{p(\mathbf{x}|\omega_k)p(\omega_k)\right\} = \underbrace{-\frac{1}{2}(\mathbf{x} - \mu_k)^\mathsf{T}\Sigma^{-1}(\mathbf{x} - \mu_k) + \log(p(\omega_k))}_{-d_{\mathrm{LDA}}(\mathbf{x},k)} + C, \qquad (6.17)$$

where constant C corresponds to terms ② + ③. Covariance matrix Σ can be computed from training data. A good estimation is the average of the individual covariance matrices $\Sigma = \frac{1}{K}\sum_k \mathbf{C}_k$. Formerly, the LDA classifier is defined as follows:

$$h_{\mathrm{LDA}}(\mathbf{x}) = \underset{k}{\mathrm{argmin}}\left\{d_{\mathrm{LDA}}(\mathbf{x}, k)\right\}, \qquad (6.18)$$

where $d_{\mathrm{LDA}}(\mathbf{x}, k)$ is defined in (6.17). In pyxvis Library, the LDA classifier is implemented using function clf_model with parameter 'LDA'. An example of this classifier is presented in Example 6.4.

A variant of Mahalanobis classifier is obtained by assuming that not only Σ_k is constant, but also $p(\omega_k)$ is constant.[3] Thus, $\Sigma_k = \Sigma$ and $p(\omega_k) = p_c$ for all k. That means that in (6.16) terms ④ is constant as well:

$$\log\{p(\mathbf{x}|\omega_k)p(\omega_k)\} = \underbrace{-\frac{1}{2}(\mathbf{x} - \mu_k)^\mathsf{T}\Sigma^{-1}(\mathbf{x} - \mu_k)}_{-d_{\text{maha}}(\mathbf{x},k)} + C, \tag{6.19}$$

where constant C corresponds to terms ② + ③ + ④. The classification is performed by (6.6) where $d_{\text{maha}}(\mathbf{x}, k)$ is defined in (6.19). The reader can observe that if we assume that $\Sigma = \mathbf{I}$ we obtain the Minimal Distance classifier (6.3).

6.2.5   Quadratic Discriminant Analysis

In Quadratic Discriminant Analysis (QDA) [11], we assume that Σ_k and $p(\omega_k)$ are not constant for all k, i.e., in (6.16) only term ③ is constant:

$$\log\{p(\mathbf{x}|\omega_k)p(\omega_k)\} = \underbrace{-\frac{1}{2}(\mathbf{x} - \mu_k)^\mathsf{T}\Sigma^{-1}(\mathbf{x} - \mu_k) - \frac{1}{2}\log(|\Sigma_k|) + \log(p(\omega_k))}_{-d_{\text{QDA}}(\mathbf{x},k)} + C,$$

$$\tag{6.20}$$

where constant C corresponds to terms ③. Formerly,

$$h_{\text{QDA}}(\mathbf{x}) = \underset{k}{\arg\min}\{d_{\text{QDA}}(\mathbf{x}, k)\}, \tag{6.21}$$

where $d_{\text{QDA}}(\mathbf{x}, k)$ is defined in (6.20). In pyxvis Library, QDA classifier is implemented using function clf_model with parameter 'QDA'.

Python Example 6.4 In this example, we show how to train and test three different classifiers: Mahalanobis (see Sect. 6.2.2), LDA (see Sect. 6.2.4) and QDA (see Sect. 6.2.5). We use the same simulated data addressed in Example 6.2 and illustrated in Fig. 6.2.

> **Listing 6.4 : Classification using Mahalanobis, LDA and QDA**

```
from pyxvis.io.data import load_features
from pyxvis.io.plots import show_clf_results
from pyxvis.learning.classifiers import clf_model, define_classifier
from pyxvis.learning.classifiers import train_classifier, test_classifier

(X,d,Xt,dt)  = load_features('../data/F2/F2')      # load training and testing data
ss_cl        = ['lda','qda','maha-0','maha']
```

---

[3]In pyxvis Library, this classifier is implemented using function clf_model with parameter 'maha-0'.

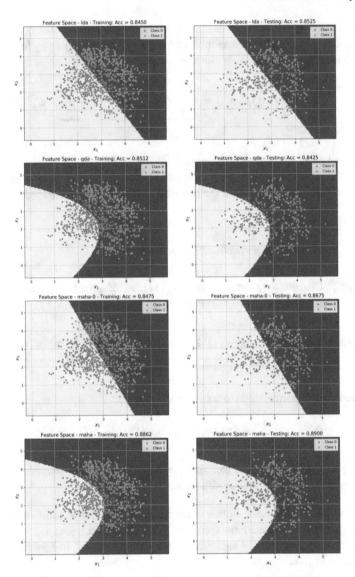

Fig. 6.6 Classification using LDA, QDA and Mahalanobis. [→ Example 6.4 🐍]

```
for cl_name in ss_cl:
    (name,params) = clf_model(cl_name)              # function name and parameters
    clf           = define_classifier([name,params]) # classifier definition
    clf           = train_classifier(clf,X,d)        # classifier training
    d0            = test_classifier(clf,X)           # clasification of training
    ds            = test_classifier(clf,Xt)          # clasification of testing
    show_clf_results(clf,X,d,Xt,dt,d0,ds,cl_name)   # display results and decision lines
```

The output of this code is shown in Fig. 6.6. In these cases on the testing data, we obtain 85.25%, 84.25%, 86.75%, and 89.00% for LDA, QDA, Mahalanobis and Mahalanobis-0 respectively. It is clear that Mahalanobis and QDA achieve a better performance than LDA and Mahalanobis-0 because they can model the curved distributions. □

6.2.6 K-Nearest Neighbors

K-Nearest Neighbors (KNN) is a non-parametric approach, in which the K most similar training samples to a given test feature vector $\mathbf{x}$ are determined [11]. The assigned class is the most frequent class from those K samples [8]. In other words, we find—in the training set—the K nearest neighbors of $\mathbf{x}$ and we evaluate the majority vote of their classes:

$$h_{\text{knn}}(\mathbf{x}) = \text{mode}(y(\mathbf{x}^1), \ldots y(\mathbf{x}^K)),$$ (6.22)

where $\{\mathbf{x}^i\}_{i=1}^K$ are the K nearest neighbors of $\mathbf{x}$, and $y(\mathbf{x}^i)$ the labeled class of $(\mathbf{x}^i)$.

KNN can be implemented (avoiding the exhaustive search of all samples of the training set) by a search using a $k-d$ tree structure [2] to search the nearest neighbors. In pyxvis Library, KNN classifier is implemented with function clf_model with parameter '$\text{knn } K$' where K is the number of neighbors to consider.

Python Example 6.5 In this example, we show how to train and test a Bayes classifier using Kernel Density Estimation. We use the same simulated data addressed in Example 6.2 and illustrated in Fig. 6.2.

```
Listing 6.5 : Classification using KNN

from pyxvis.io.data import load_features
from pyxvis.io.plots import show_clf_results
from pyxvis.learning.classifiers import clf_model, define_classifier
from pyxvis.learning.classifiers import train_classifier, test_classifier

(X,d,Xt,dt)    = load_features('../data/F2/F2')       # load training and testing data
ss_cl          = ['knn1','knn3','knn7','knn15']
for cl_name in ss_cl:
    (name,params) = clf_model(cl_name)                    # function name and parameters
    clf           = define_classifier([name,params])      # classifier definition
    clf           = train_classifier(clf,X,d)             # classifier training
    d0            = test_classifier(clf,X)                # clasification of training
    ds            = test_classifier(clf,Xt)               # clasification of testing
    show_clf_results(clf,X,d,Xt,dt,d0,ds,cl_name)         # display results and decision lines
```

The output of this code is Fig. 6.7 for different number of neighbors. In this case, we obtain 90.75%, 93.50%, 94.25%, and 93.75% for 1, 3, 7, and 15 neighbors respectively. It is clear that KNN classifier can properly model any distribution. The hyper-parameter K, i.e., the number of neighbors is to be estimated for the best performance on the testing dataset. □

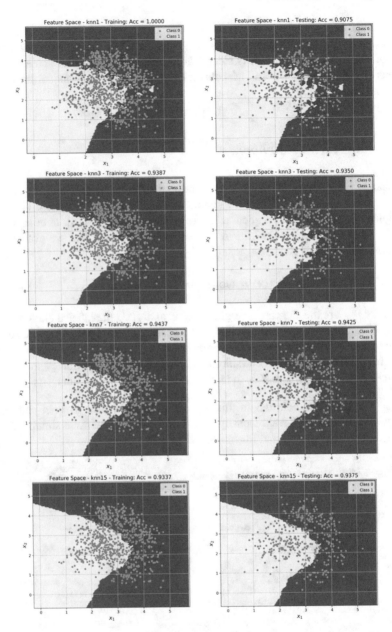

Fig. 6.7 Classification using KNN. [→ Example 6.5 🐍]

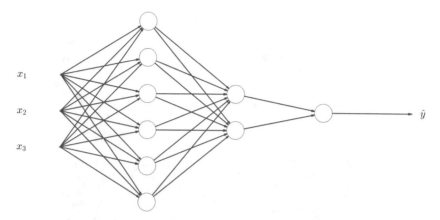

Fig. 6.8 Simple neural network with 3 inputs $\mathbf{x} = (x_1, x_2, x_3)$, one output $\hat{y}$ and two hidden layers (one with 6 nodes and the another with 2). In this example, the input can be classified as class ω_1 if $\hat{y} > 0.5$, and otherwise as class ω_0

6.2.7   Neural Networks

Artificial neuronal networks are mathematical tools derived from what is known about the mechanisms and physical structure of biological learning, based on the function of a neuron. They are parallel structures for the distributed processing of information [3]. A neural networks consists of artificial neurons connected in a network that is able to classify a test feature vector $\mathbf{x}$ evaluating a linear weighted sum of non-linear functions as illustrated in Fig. 6.8. The weights, the functions, and the connections are estimated in a training phase by minimizing the classification error [3, 4]. In this section, we only mention that neural networks have been established as one of the best classification approaches in pattern recognition. The basic structure of the neural networks and the learning strategies developed for training neural networks are the basis of deep learning models. Nowadays, it is well known that deep learning has been successfully used in image and video recognition. For these reasons, we decided to dedicate in this book an entire chapter to deep learning (see Chap. 7), and in Sect. 7.2 of this chapter, we address the theory of neural networks and give some examples.

Python Example 6.6 In this example, we show how to train and test a Neural Network. We use the same simulated data addressed in Example 6.2 and illustrated in Fig. 6.2. In pyxvis Library, neural networks are implemented with function clf_model with parameter '$\text{nn } (n_1, \cdots, n_p)$' where n_i is the number of nodes of hidden layer for an architecture of p hidden layers.

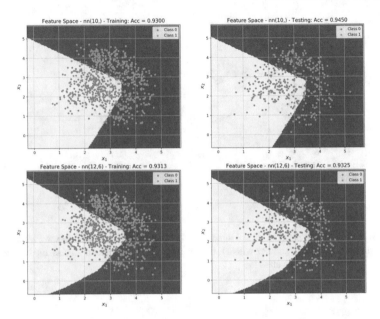

Fig. 6.9 Classification using Neural Networks (NN). [→ Example 6.6]

```
from pyxvis.io.data import load_features
from pyxvis.io.plots import show_clf_results
from pyxvis.learning.classifiers import clf_model, define_classifier
from pyxvis.learning.classifiers import train_classifier, test_classifier

(X,d,Xt,dt)   = load_features('../data/F2/F2')        # load training and testing data
ss_cl         = ['nn(10,)','nn(12,6)']
for cl_name in ss_cl:
    (name,params) = clf_model(cl_name)                # function name and parameters
    clf           = define_classifier([name,params])  # classifier definition
    clf           = train_classifier(clf,X,d)         # classifier training
    d0            = test_classifier(clf,X)            # clasification of training
    ds            = test_classifier(clf,Xt)           # clasification of testing
    show_clf_results(clf,X,d,Xt,dt,d0,ds,cl_name)     # display results and decision lines
```

The output of this code is Fig. 6.9 for different configurations of hidden layers: nn(10,) means one hidden layer with 10 nodes, whereas nn(12,6) means two hidden layers with 12 and 6 nodes respectively.[4] In this case, we obtain 94.50% and 93.25% respectively. The reader can compare this result with the accuracy obtained by classifier of Examples 6.2, 6.3, 6.4, and 6.5. It is clear that classifiers based on neural networks can properly model the curved distributions.   □

---

[4]For the configuration of Fig. 6.8 is nn(6,2).

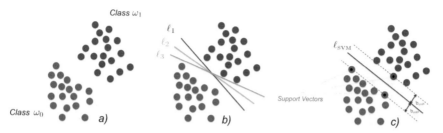

Fig. 6.10 Key idea of support vector machine: **a** Given a two-class problem, find a decision line ℓ. **b** There are many possible decision lines that can separate both classes. **c** In SVM, we search decision line ℓ_{SVM} so that the margin b is maximized. The support vectors are defined as those samples that belong to the margin lines

6.2.8 Support Vector Machines

The original Support Vector Machines (SVM) find a decision line that separate two classes (ω_1 and ω_0) as illustrate in Fig. 6.10a. In this example, we can see that there are many possible decision lines like ℓ_1, ℓ_2, and ℓ_3 among others (see Fig. 6.10b). A relevant question arises: which decision line ℓ can separate both classes at 'best'? In SVM strategy, we define the 'margins' b_1 and b_0 as the minimal distance from the decision line to a sample of class ω_1 and ω_0 respectively. After SVM criterion, the 'best' separation line ℓ_{SVM} is one that (i) it is in the middle, i.e., $b_1 = b_0 = b$, and (ii) its margin is maximal, i.e., $b = b_{max}$. Thus, decision line ℓ_{SVM} is equidistant to the margin lines and the margin is maximal.

In $\mathbb{R}^2$ we have a decision line, however, in general, in $\mathbb{R}^n$, we have a hyperplane that is defined as

$$\ell_{SVM} : \; g(\mathbf{x}) = \mathbf{a}^\mathsf{T}\mathbf{x} + a_0 = 0, \tag{6.23}$$

where $\mathbf{x} = [x_1 \ldots x_n]^\mathsf{T}$ is our feature vector and $\mathbf{a} = [a_1 \ldots a_n]^\mathsf{T}$ and a_0 are the linear parameters to be estimated. The solution for $\{a_j\}_{i=0}^n$ can be found following an optimization approach [21]. In the solution, $\{a_j\}_{i=0}^n$ depends only on the *support vectors*, i.e., the samples of both classes that belong to the margin lines as shown in Fig. 6.10c. The solution of this optimization problem consists of parameter values λ_i corresponding to ith support vector:

$$\mathbf{a} = \sum_{i=1}^m \lambda_i z_i \mathbf{x}_i, \tag{6.24}$$

for m support vectors, where $z_i = \pm 1$ if $\mathbf{x}_i$ belongs to ω_1 and ω_0 respectively. In addition, a_0 can be calculated from any support vector as $a_0 = z_i - \mathbf{a}^\mathsf{T}\mathbf{x}_i$ [11]. In SVM, the classification of a test sample $\mathbf{x}$ can be formulated as follows:

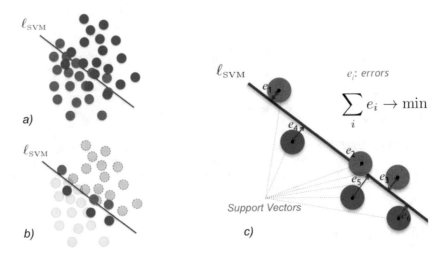

Fig. 6.11 Key idea of support vector machine with overlapping: **a** Given a two-class problem with overlapping, find a decision line ℓ_{SVM}. **b** By choosing a decision line ℓ_{SVM} there will be misclassified samples. **c** The misclassified samples are the support vectors. Each of them has an error e_i defined as the perpendicular distance to the decision line ℓ_{SVM}. In SVM, we search decision line ℓ_{SVM} so that the total error $\sum e_i$ is minimized

$$h_{SVM}(\mathbf{x}) = \begin{cases} 1 \text{ if } & \mathbf{a}^T\mathbf{x} + a_0 > 0 \\ 0 \text{ otherwise} \end{cases}. \qquad (6.25)$$

In practice, however, there is some overlapping between the classes as shown in Fig. 6.11a. If we have a decision line that separates the feature space, we will have misclassified samples. In SVM strategy, we consider only the misclassified samples as illustrated in Fig. 6.11b. They will be the *support vectors*. The ith support vector has a distance e_i to the decision line that corresponds to an error (see Fig. 6.11c). After SVM criterion, the 'best' decision line ℓ_{SVM} is one that minimizes the total error $e = \sum_i e_i$. Again, the solution for $\{a_i\}_{i=0}^n$ depends only on the support vectors, and they can be estimated using an optimization approach [21]. The classification is performed according to (6.25).

The previous approach estimates a straight line decision boundary in feature space. In many cases, however, it is convenient to find a curve that separates the classes as illustrated in Fig. 6.12a. In order to use SVM linear classification, the feature space can be transformed into a new enlarged feature space (Fig. 6.12b) where the classification boundary can be linear. Thus, as shown in Fig. 6.12c, a simple linear classification (6.25) can be designed in the transformed feature space in order to separate both classes [21].

The original feature space is transformed using a function $f(\mathbf{x})$. Thus, according to (6.23) and (6.24) we obtain:

$$\begin{aligned} g(f(\mathbf{x})) &= \mathbf{a}^T f(\mathbf{x}) + a_0 \\ &= \sum_i \lambda_i z_i \langle f(\mathbf{x}_i), f(\mathbf{x}) \rangle + a_0, \end{aligned} \qquad (6.26)$$

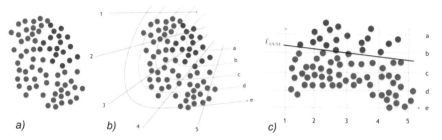

Fig. 6.12 Non-linear decision line. **a** Feature space with two classes that can be separated using a curve. **b** The feature space can be described in a new coordinate system. **c** Transformed coordinate system in which a linear decision line can be used

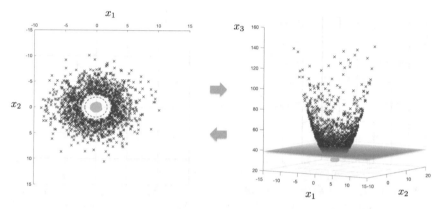

Fig. 6.13 The kernel trick: the original 2D space is transformed into a 3D space where the separation of the classes is linear (this case can be found in Example 6.7 using dataset 'P2')

where $\langle f(\mathbf{x}_i), f(\mathbf{x}) \rangle$ is the inner product $[f(\mathbf{x}_i)]^\mathsf{T} f(\mathbf{x})$. In (6.26), we can observe that for the classification, only the kernel function $\langle f(\mathbf{x}_i), f(\mathbf{x}) \rangle = K(\mathbf{x}_i, \mathbf{x})$ that computes inner products in the transformed space is required. Consequently, using (6.26) we can write (6.25) in general as

$$h_{\text{SVM}}(\mathbf{x}) = \begin{cases} 1 \text{ if } & \sum_i \lambda_i z_i K(\mathbf{x}_i, \mathbf{x}) + a_0 > 0 \\ 0 \text{ otherwise} \end{cases}. \tag{6.27}$$

Table 6.1 shows typical kernel functions that are used by SVM classifiers. They should be a symmetric positive (semi-) definite function [11]. In pyxvis Library, SVM classifier is implemented with function clf_model with parameter 'svm-lin', 'svm-pol', 'svm-rbf', 'svm-sig' for the four kernels of Table 6.1.

Python Example 6.7 In this example, we show how to train and test SVM classifiers. We use the same simulated data addressed in Example 6.2 and illustrated in Fig. 6.2.

Table 6.1 Kernel functions used by SVM

Name	$K(\mathbf{x}_i, \mathbf{x})$
Linear	$\langle \mathbf{x}_i, \mathbf{x} \rangle$
qth degree polynomial	$(1 + \langle \mathbf{x}_i, \mathbf{x} \rangle)^q$
Radial basis (RBF)	$\exp(-\gamma \|\mathbf{x}_i - \mathbf{x}\|^2)$
Sigmoid	$\tanh(\alpha_1 \langle \mathbf{x}_i, \mathbf{x} \rangle + \alpha_2)$

Listing 6.7 : Classification using SVM

```
from pyxvis.io.data import load_features
from pyxvis.io.plots import show_clf_results
from pyxvis.learning.classifiers import clf_model, define_classifier
from pyxvis.learning.classifiers import train_classifier, test_classifier

(X,d,Xt,dt)   = load_features('../data/F2/F2')          # load training and testing data
# (X,d,Xt,dt) = loadFeatures('../data/P2/P2')           # data for the donut example
ss_cl         = ['svm-lin','svm-rbf(0.1,0.05)','svm-rbf(0.03,1)','svm-pol(0.1,0.5,2)']
for cl_name in ss_cl:
    (name,params) = clf_model(cl_name)                  # function name and parameters
    clf           = define_classifier([name,params])    # classifier definition
    clf           = train_classifier(clf,X,d)           # classifier training
    d0            = test_classifier(clf,X)              # clasification of training
    ds            = test_classifier(clf,Xt)             # clasification of testing
    show_clf_results(clf,X,d,Xt,dt,d0,ds,cl_name)       # display results and decision lines
```

The output of this code is Fig. 6.13 (for the donut example) and Fig. 6.14 (for the general example). In this case, we obtain 86.75%, 91.50%, 93.50%, and 91.25% for SVM-LIN, SVM-RBF (gamma=0.1, C=0.05), SVM-RBF (gamma=0.03, C=1), and SVM-POL (gamma=0.1, C=0.5, degree=2).[5] The reader can compare this result with the accuracy obtained by classifier of Examples 6.2, 6.3, 6.4, 6.5, and 6.6. It is clear that (no-linear) SVM classifiers can properly model the curved distributions.   □

Python Example 6.8 In this example, we show how easy is to compare many classifiers in pyxvis Library. The idea of this example is to train and test a list of 30 classifiers given in variable `ss_cl`. We use now a dataset of 3 classes and 2 features as illustrated in Fig. 6.15.

Listing 6.8 : Classification using many classifiers

```
import numpy as np
from sklearn.metrics import accuracy_score
from pyxvis.io.data import load_features
from pyxvis.io.plots import show_clf_results
from pyxvis.learning.classifiers import clf_model, define_classifier
from pyxvis.learning.classifiers import train_classifier, test_classifier
```

---

[5]In sklearn library, 'gamma' defines the influence of the single training examples, 'C' is like a regularization parameter in the optimization, and 'degree' is the the degree of the polynomial for SVM-POL. See https://scikit-learn.org/stable/auto_examples/svm/plot_rbf_parameters.html for further details.

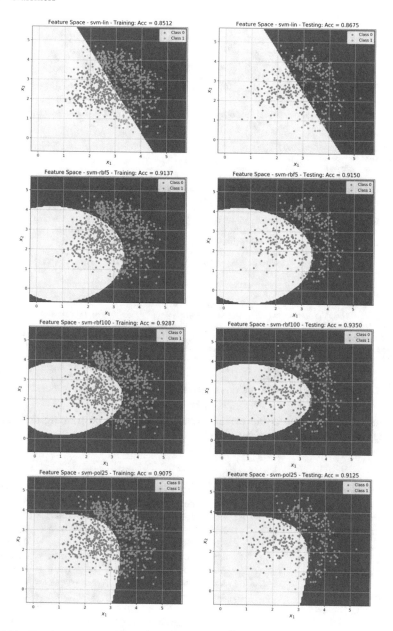

Fig. 6.14 Classification using SVM. [→ Example 6.7 🐍]

```
# List of classifiers
ss_cl      = ['dmin','lda','qda','maha','knn3','knn5','knn7','knn11','knn15',
             'bayes-naive','bayes-kde','adaboost','lr','rf','tree',
             'svm-lin','svm-rbf(0.1,1)','svm-rbf(0.1,0.5)','svm-rbf(0.5,1)',
             'svm-pol(0.05,0.1,2)','svm-pol(0.05,0.5,2)','svm-pol(0.05,0.5,3)',
             'svm-sig(0.1,1)','svm-sig(0.1,0.5)','svm-sig(0.5,1)',
             'nn(10,)','nn(20,)','nn(12,6)','nn(20,10,4)']

(X,d,Xt,dt) = load_features('../data/G3/G3')        # load training and testing data

n = len(ss_cl)
acc_train = np.zeros((n,))
acc_test  = np.zeros((n,))
for k in range(n):
    (name,params) = clf_model(ss_cl[k])             # function name and parameters
    clf           = define_classifier([name,params]) # classifier definition
    clf           = train_classifier(clf,X,d)        # classifier training
    d0            = test_classifier(clf,X)           # clasification of training
    ds            = test_classifier(clf,Xt)          # clasification of testing
    acc_train[k]  = accuracy_score(d,d0)             # accuracy in training
    acc_test[k]   = accuracy_score(dt,ds)            # accuracy in testing
    print(f'{k:3d}'+') '+f'{ss_cl[k]:20s}'+ ': ' +
          f'Acc-Train = {acc_train[k]:.4f}'+ '   ' + f'Acc-Test = {acc_test[k]:.4f}')
ks = np.argmax(acc_test)
print('_____')
print('Best Classifier:')
print(f'{ks:3d}'+') '+f'{ss_cl[ks]:20s}'+ ': ' +
      f'Acc-Train = {acc_train[ks]:.4f}'+ '   ' + f'Acc-Test = {acc_test[ks]:.4f}')
print('_____')
(name,params) = clf_model(ss_cl[ks])                # function name and parameters
clf           = define_classifier([name,params])    # classifier definition
clf           = train_classifier(clf,X,d)           # classifier training
d0            = test_classifier(clf,X)              # clasification of training
ds            = test_classifier(clf,Xt)             # clasification of testing
show_clf_results(clf,X,d,Xt,dt,d0,ds,ss_cl[ks])     # display results and decision lines
```

The output of this code is the evaluation of the accuracy on training and testing subsets of the 30 classifiers as follows:

```
-------------------------------------------------------------------
 0)  dmin              : Acc-Train = 0.8717    Acc-Test = 0.8833
 1)  lda               : Acc-Train = 0.8758    Acc-Test = 0.8883
 2)  qda               : Acc-Train = 0.8808    Acc-Test = 0.8700
 3)  maha              : Acc-Train = 0.9075    Acc-Test = 0.9050
 4)  knn3              : Acc-Train = 0.9467    Acc-Test = 0.9383
 5)  knn5              : Acc-Train = 0.9425    Acc-Test = 0.9417
 6)  knn7              : Acc-Train = 0.9483    Acc-Test = 0.9433
 7)  knn11             : Acc-Train = 0.9442    Acc-Test = 0.9383
 8)  knn15             : Acc-Train = 0.9400    Acc-Test = 0.9383
 9)  bayes-naive       : Acc-Train = 0.9250    Acc-Test = 0.9367
10)  bayes-kde         : Acc-Train = 0.9083    Acc-Test = 0.9133
11)  adaboost          : Acc-Train = 0.7750    Acc-Test = 0.7867
12)  lr                : Acc-Train = 0.8558    Acc-Test = 0.8667
13)  rf                : Acc-Train = 0.9975    Acc-Test = 0.9317
14)  tree              : Acc-Train = 0.9175    Acc-Test = 0.9083
15)  svm-lin           : Acc-Train = 0.8842    Acc-Test = 0.8933
16)  svm-rbf(0.1,1)    : Acc-Train = 0.9342    Acc-Test = 0.9400
17)  svm-rbf(0.1,0.5)  : Acc-Train = 0.9358    Acc-Test = 0.9383
18)  svm-rbf(0.5,1)    : Acc-Train = 0.9367    Acc-Test = 0.9450
```

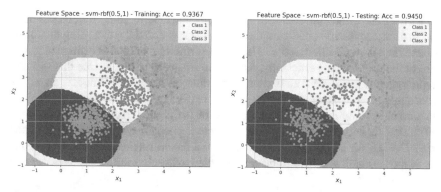

Fig. 6.15 Best classification by evaluating many classifiers 6.8. [→ Example 6.8 🐍]

```
19) svm-pol(0.05,0.1,2)  : Acc-Train = 0.8700    Acc-Test = 0.8600
20) svm-pol(0.05,0.5,2)  : Acc-Train = 0.8933    Acc-Test = 0.9033
21) svm-pol(0.05,0.5,3)  : Acc-Train = 0.8908    Acc-Test = 0.8917
22) svm-sig(0.1,1)       : Acc-Train = 0.2567    Acc-Test = 0.2617
23) svm-sig(0.1,0.5)     : Acc-Train = 0.2692    Acc-Test = 0.2700
24) svm-sig(0.5,1)       : Acc-Train = 0.0058    Acc-Test = 0.0083
25) nn(10,)              : Acc-Train = 0.9358    Acc-Test = 0.9333
26) nn(20,)              : Acc-Train = 0.9342    Acc-Test = 0.9383
27) nn(12,6)             : Acc-Train = 0.9375    Acc-Test = 0.9367
28) nn(20,10,4)          : Acc-Train = 0.9367    Acc-Test = 0.9417
----------------------------------------------------------------

Best Classifier:
18) svm-rbf(0.5,1)       : Acc-Train = 0.9367    Acc-Test = 0.9450
----------------------------------------------------------------
```

In addition, Fig. 6.15 shows the classifier that achieves the best accuracy on testing subset. In this case, the best classifier is #19 – 'svm-rbf(0.5,1)' with an accuracy of 94.50%. □

6.2.9 Classification Using Sparse Representations

In this kind of classifier, the strategy is to use sparse representations of the original data to perform the classification. Thus, the features are first transformed into a sparse representation (see Sect. 5.5) and afterwards, the sparse representation is used by the classifier.

According to Eq. (5.38) it is possible to learn the dictionary $\mathbf{D}$ and estimate the most important constitutive components $\mathbf{Z} = \{\mathbf{z}_i\}_{i=1}^{N}$ of the representative signals $\mathbf{X} = \{\mathbf{x}_i\}_{i=1}^{N}$. In a supervised problem—with labeled data $(\mathbf{x}_i, d_i)$, where d_i is the class of sample $\mathbf{x}_i$—, naturally the classification problem can be stated as follows [1]: given training data $(\mathbf{x}_i, d_i)$, design a classifier h—with parameters θ—which maps the transformed samples $\mathbf{z}_i$ to its classification label d_i, thus, $h(\mathbf{z}_i, \theta)$ should be d_i. In order to classify a new sample data $\mathbf{x}$, it is transformed into $\mathbf{z}$ using dictionary $\mathbf{D}$ and then it is classified as $d = h(\mathbf{z}, \theta)$. Nevertheless, since $\mathbf{Z}$ is estimated to represent the original data efficiently, there is no reason to accept as true that this new representation can ensure an optimal separation of the classes. Another classification strategy uses one dictionary $\mathbf{D}_k$ per class [15], that is learned using the set $\mathbf{X}_k$,[6] that contains only the samples of class ω_k of the training data: $\mathbf{X}_k = \{\mathbf{x}_i | d_i = k\}$. With this strategy, using (5.39) a test sample $\mathbf{x}$ is codified by $\mathbf{z} = \mathbf{z}_k$ with dictionary $\mathbf{D} = \mathbf{D}_k$ for all classes $k = 1 \ldots K$, and a reconstruction error is computed as $e_k = ||\mathbf{x} - \mathbf{D}_k \mathbf{z}_k||$. Finally, sample $\mathbf{x}$ is assigned to the class with the smallest reconstruction error:

$$h_{\text{SPAr}}(\mathbf{x}) = \operatorname*{argmin}_{k} ||\mathbf{x} - \mathbf{D}_k \mathbf{z}_k||. \tag{6.28}$$

This test strategy, however, does not scale well for a large number of classes. For these reasons, new strategies have been developed in order to learn at the same time *reconstructive* and *discriminative* dictionaries (for robustness to noise and for efficient classification respectively) [24]. This can be achieved by adding a new discrimination term in the objective function that includes the representation that is also the most different from the one of signals in other data classes:

$$\operatorname*{argmin}_{\mathbf{D}, \mathbf{Z}, \theta} [||\mathbf{X} - \mathbf{D} \mathbf{Z}||_2^2 + \gamma J(\mathbf{D}, \mathbf{Z}, \mathbf{d}, \theta)] \quad \text{subject to } ||\mathbf{z}||_0 \le T. \tag{6.29}$$

The discrimination term $J(\mathbf{D}, \mathbf{Z}, \mathbf{c}, \theta)$ depends on the dictionary, the coefficient vectors, the labels of the samples $\mathbf{d}$, and the parameters θ of the model used for classification. Parameter γ weights the trade-off between approximation and classification performance. This strategy with a common dictionary has the advantage of sharing some atoms of the dictionary when representing samples of different classes. Equation (6.29) can be solved efficiently by fixed-point continuation methods when the classifier is based on logistic regression methods [16].

Another approach that can be used to classify samples in X-ray testing is based on sparse representations of random patches. This approach, called Adaptive Sparse Representation of Random Patches (ASR+), has been successfully used in other recognition problems [17, 18]. The method consists of two stages (see Fig. 6.16): In the training stage, random patches are extracted from representative images of each class (e.g., in baggage screening we can have handguns, razor blades, etc.) in order to construct representative dictionaries. A stop list is used to remove very common words from the dictionaries [23]. In the testing stage, random test patches of the query image are extracted, and for each non-stopped test patch a dictionary is built concatenating the 'best' representative dictionary of each class. Using this adapted dictionary, each non-stopped test patch is classified following the Sparse Repre-

---

[6]There are some approaches that define the dictionary as the original samples (see Sparse Representation Classification (SRC) [26]), where $\mathbf{D}_k = \mathbf{X}_k$.

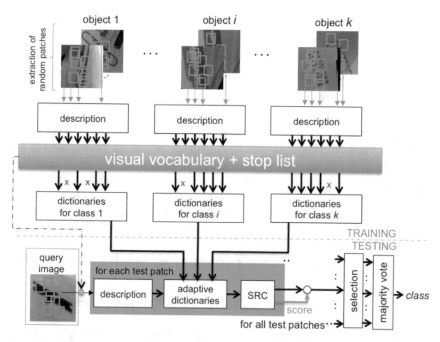

Fig. 6.16 Overview of the proposed method. The figure illustrates the recognition of three different objects. The shown classes are three: clips, razor blades, and springs. There are two stages: Learning and Testing. The stop list is used to filter out patches that are not discriminating for these classes. The stopped patches are not considered in the dictionaries of each class and in the testing stage

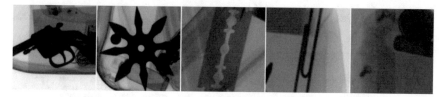

Fig. 6.17 Images used in our experiments. The five classes are: handguns, shuriken, razor blades, clips, and background

sentation Classification (SRC) methodology [26] by minimizing the reconstruction error. Finally, the query image is classified by patch voting. Thus, this approach is able to learn a model for each recognition task dealing with a larger degree of variability in contrast, pose, expression, occlusion, object size, and distance from the X-ray detector.

This method was tested in the recognition of five classes in baggage screening: handguns, shuriken, razor blades, clips, and background (see some samples in Fig. 6.17). In our experiments, there are 100 images per class. All images were resized to

128×128 pixels. The evaluation is performed using leave-one-out (see Sect. 6.3.3). The obtained accuracy was $\eta = 97.17\%$.

6.3   Performance Evaluation

In this section, we will see how to evaluate the performance of a classifier and how to build the datasets 'training data' and 'testing data'. In general, there is a set $\mathbb{D}$ that contains all available data, that is the features of representative samples and their corresponding labels. Sometimes, from set $\mathbb{D}$ a subset $\mathbb{X} \subset \mathbb{D}$ is chosen, however, in most cases $\mathbb{X} = \mathbb{D}$. We call subset $\mathbb{X}$ the 'used data' because it is used to evaluate the performance of a classifier as illustrated in Fig. 6.18. Set $\mathbb{X}$ consists of (i) a matrix $\mathbf{X}$ of size $N \times p$, for N samples and p features; and (ii) a vector $\mathbf{d}$ of N elements with the labels (one label per sample).

In order to estimate the accuracy of a classifier, we can follow this general strategy:

1. From $\mathbb{X}$, select training data $(\mathbf{X}_{\text{train}}, \mathbf{d}_{\text{train}})$ and testing data $(\mathbf{X}_{\text{test}}, \mathbf{d}_{\text{test}})$:

$$(\mathbf{X}_{\text{train}}, \mathbf{d}_{\text{train}}, \mathbf{X}_{\text{test}}, \mathbf{d}_{\text{test}}) = \text{DataSelection}(\mathbb{X}) \qquad (6.30)$$

Typically, a given percentage S of $\mathbb{X}$ is used for training and the rest $(100\text{-}S)$ for testing. That means, we have $N_{\text{train}} = N \times S/100$ samples for training and $N_{\text{test}} = N - N_{\text{train}}$ for training. There are many ways to perform the data selection:

- Random (yes/no): we can choose randomly N_{train} of $\mathbb{X}$ or, for example, the first N_{train} samples of $\mathbb{X}$.

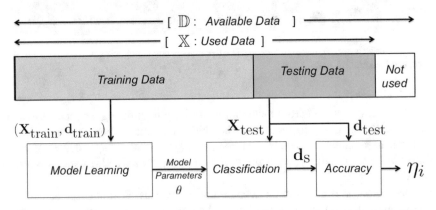

Fig. 6.18 Estimation of the accuracy of a classifier. Figures 6.19, 6.20, and 6.21 show different strategies

- Stratified (yes/no): in stratified case, we select the same S percentage of each class (so the relative number of samples for each class is the same in original dataset and selected dataset), whereas in unstratified cases we select S percentage of $\mathbb{X}$ (so the relative number of samples for each class is not necessarily the same in original dataset and selected dataset).
- Replacement (with/without): Data selection without replacement means that once a sample has been selected, it may not be selected again. In data selection with replacement a sample of $\mathbb{X}$ is allowed to be replicated. It must be ensured that samples in the training data are not in the testing data and viceversa.

2. Using training data $(\mathbf{X}_{\text{train}}, \mathbf{d}_{\text{train}})$ train a classifier:

$$\theta = \text{ClassifierTrain}(\mathbf{X}_{\text{train}}, \mathbf{d}_{\text{train}}), \qquad (6.31)$$

where θ is a vector that contains all parameters of the classifier that was trained. For instance, in a simple classifier like Euclidean minimal distance (see Sect. 6.2.1) we store in θ only the centers of mass of each class in the training set.

3. Using the features of the testing data $\mathbf{X}_{\text{test}}$, the classifier and its parameters θ, we predict the labels of each testing sample and store them in vector $\mathbf{d}_s$ of N_{test} elements:

$$\mathbf{d}_s = \text{Classify}(\mathbf{X}_{\text{test}}, \theta). \qquad (6.32)$$

It is worth mentioning that in this step it is not allowed to use the labels of the testing data $\mathbf{d}_{\text{test}}$.

4. Now, we can compute the accuracy of the testing data defined as

$$\eta_i = \frac{\#\text{ test samples correctly predicted}}{N_{\text{test}}}. \qquad (6.33)$$

5. In (6.33), we use index i because the procedure from steps 1 to 4 can be repeated n times, for $i = 1 \ldots n$. Thus, we can compute the final estimation of the accuracy as

$$\eta = \frac{1}{n} \sum_{i=1}^{n} \eta_i. \qquad (6.34)$$

In the following section, we will explain typical strategies used in the literature.

6.3.1 Hold-Out

In hold-out, we take a percentage S of $\mathbb{X}$ for training and the rest for testing as shown in 6.19. In our general methodology, this strategy corresponds to $n = 1$ in (6.34). This is the simplest way how to evaluate the accuracy. It is recommended just in case

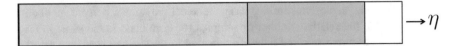

Fig. 6.19 Estimation of the accuracy of a classifier using hold-out. The figure follows the color representation of Fig. 6.18 for training and testing data

the computational time is so enormous that the cost of training a classifier several times is prohibitive. Hold-out can be a good starting point to test if the features and classifier that we are designing are suitable for the recognition task. Nevertheless, the standard deviation of the accuracy estimation can be very high as we will see in next example. An example that evaluates 30 classifiers using hold-out methodology has already been shown in Example 6.8. Additionally, in this section we show a very simple example that evaluates only one classifier.

Python Example 6.9 In this example, we show how to evaluate a classifier using hold-out strategy. We use the same simulated data addressed in Example 6.2 and illustrated in Fig. 6.2.

Listing 6.9 : Hold-out

```
from pyxvis.learning.classifiers import clf_model
from pyxvis.learning.evaluation import hold_out
from pyxvis.io.data import load_features
from pyxvis.io.plots import show_confusion_matrix
from sklearn.model_selection import train_test_split

# load available dataset
(X0,d0)      = load_features('../data/F2/F2',full=1)

# definition of training and testing data
X,Xt,d,dt    = train_test_split(X0,d0,test_size=0.2, stratify=d0)

# definition of the classifier
cl_name      = 'svm-rbf(0.1,1)'    # generic name of the classifier
(name,params) = clf_model(cl_name)  # function name and parameters

# Hold-out (train on (X,d), test on (Xt), compare with dt)
ds,acc,_     = hold_out([name,params],X,d,Xt,dt) # hold out
print(cl_name+ ': ' + f'Accuracy = {acc:.4f}')
# display confusion matrix
show_confusion_matrix(dt,ds,'Testing subset')
```

The output of this code is the value of the estimated accuracy. This number should be around 93%. This method is implemented in function hold_out in pyxvis Library. If we repeat this experiment 1000 times, the mean of the accuracy is 0.9287, the standard deviation is 0.0152, the maximal value is 0.9708 and the minimal value is 0.8792, i.e., the estimation is not very accurate because there is a variation of 9.2% between maximal and minimal value!   □

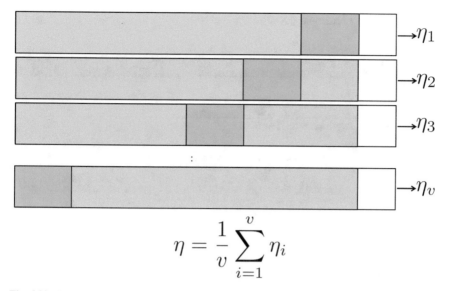

$$\eta = \frac{1}{v} \sum_{i=1}^{v} \eta_i$$

Fig. 6.20 Estimation of the accuracy of a classifier using cross-validation with v folds. The figure follows the color representation of Fig. 6.18 for training and testing data

6.3.2 Cross-Validation

Cross-validation is widely used in machine learning problems [13]. In cross-validation, the data is divided into v folds. A portion $s = (v - 1)/v$ of the whole data is used to train and the rest $(1/v)$ for test. This experiment is repeated v times rotating train and test data to evaluate the stability of the classifier as shown in Fig. 6.20. Then, when training is performed, the samples that were initially removed can be used to test the performance of the classifier on these test data. Thus, one can evaluate the generalization capabilities of the classifier by testing how well the method will classify samples that have not already been examined. The estimated performance, η, is calculated as the mean of the v percentages of the true classifications are tabulated in each case, i.e., $n = v$ (6.34). In our experiments, we use $v = 10$ folds.[7] Confidence intervals, where the classification performance η expects to fall, are obtained from the test sets. These are determined by the cross-validation technique, according to a t—Student test [20]. Thus, the performance and also the confidence can be assessed.

   Python Example 6.10 In this example, we show how to evaluate 30 classifiers using cross-validation strategy with 10 folds. We use the same simulated

---

[7]The number of folds v can be another number, for instance 5-fold or 20-fold cross-validation estimate offers very similar performances. In our experiments, we use 10-fold cross-validation because it has become the standard method in practical terms [25].

data addressed in Example 6.8 with three classes and two features as illustrated in Fig. 6.15.

Listing 6.10 : Cross-validation with many classifiers

```
import numpy as np
from pyxvis.learning.classifiers import clf_model
from pyxvis.learning.evaluation import cross_validation
from pyxvis.io.data import load_features

# List of classifiers
ss_cl        = ['dmin','lda','qda','maha','knn3','knn5','knn7','knn11','knn15',
                'bayes-naive','bayes-kde','adaboost','lr','rf','tree',
                'svm-lin','svm-rbf(0.1,1)','svm-rbf(0.1,0.5)','svm-rbf(0.5,1)',
                'svm-pol(0.05,0.1,2)','svm-pol(0.05,0.5,2)','svm-pol(0.05,0.5,3)',
                'svm-sig(0.1,1)','svm-sig(0.1,0.5)','svm-sig(0.5,1)',
                'nn(10,)','nn(20,)','nn(12,6)','nn(20,10,4)']

(X,d)  = load_features('../data/G3/G3',full=1)      # load training and testing data

n        = len(ss_cl)
folds  = 10
acc    = np.zeros((n,))
for k in range(n):
    (name,params) = clf_model(ss_cl[k])                     # function name and parameters
    acc[k]     = cross_validation([name,params],X,d,folds=folds)
    print(f'{k:3d}'+') '+f'{ss_cl[k]:20s}'+ ': ' + f'CV-Accuracy = {acc[k]:.4f}')
ks = np.argmax(acc)
print('------------------------------------------------')
print('Best Classifier:')
print(f'{ks:3d}'+') '+f'{ss_cl[ks]:20s}'+ ': ' + f'CV-Accuracy = {acc[ks]:.4f}')
print('------------------------------------------------')
```

The output of this code is the estimated accuracy of each classifier. They are presented as follows:

```
-------------------------------------------------
 0)  dmin                 : CV-Accuracy = 0.8800
 1)  lda                  : CV-Accuracy = 0.8828
 2)  qda                  : CV-Accuracy = 0.8811
 3)  maha                 : CV-Accuracy = 0.9067
 4)  knn3                 : CV-Accuracy = 0.9250
 5)  knn5                 : CV-Accuracy = 0.9278
 6)  knn7                 : CV-Accuracy = 0.9356
 7)  knn11                : CV-Accuracy = 0.9344
 8)  knn15                : CV-Accuracy = 0.9378
 9)  bayes-naive          : CV-Accuracy = 0.9228
10)  bayes-kde            : CV-Accuracy = 0.9161
11)  adaboost             : CV-Accuracy = 0.7961
12)  lr                   : CV-Accuracy = 0.8628
13)  rf                   : CV-Accuracy = 0.9328
14)  tree                 : CV-Accuracy = 0.9056
15)  svm-lin              : CV-Accuracy = 0.8833
16)  svm-rbf(0.1,1)       : CV-Accuracy = 0.9339
17)  svm-rbf(0.1,0.5)     : CV-Accuracy = 0.9344
18)  svm-rbf(0.5,1)       : CV-Accuracy = 0.9367
19)  svm-pol(0.05,0.1,2)  : CV-Accuracy = 0.8739
20)  svm-pol(0.05,0.5,2)  : CV-Accuracy = 0.9033
```

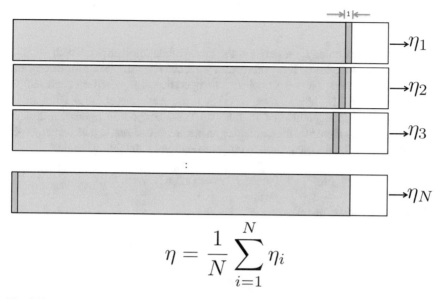

$$\eta = \frac{1}{N} \sum_{i=1}^{N} \eta_i$$

Fig. 6.21 Estimation of the accuracy of a classifier using leave-one-out. The figure follows the color representation of Fig. 6.18 for training and testing data

```
21) svm-pol(0.05,0.5,3)  : CV-Accuracy = 0.9017
22) svm-sig(0.1,1)       : CV-Accuracy = 0.2583
23) svm-sig(0.1,0.5)     : CV-Accuracy = 0.2661
24) svm-sig(0.5,1)       : CV-Accuracy = 0.0089
25) nn(10,)              : CV-Accuracy = 0.9333
26) nn(20,)              : CV-Accuracy = 0.9350
27) nn(12,6)             : CV-Accuracy = 0.9367
28) nn(20,10,4)          : CV-Accuracy = 0.9372
------------------------------------------------
Best Classifier:
 8) knn15                : CV-Accuracy = 0.9378
------------------------------------------------
```

The best result has been achieved by classifier KNN with 15 neighbors. The reader can compare these results with the accuracies presented in Example 6.8. This method is implemented in function cross_validation in pyxvis Library. In order to compare Hold-Out with Cross-Validation variations we can repeat the cross-validation 1000 times for classifier KNN with 15 neighbors. The results are: mean of the accuracy is 93.80%, the standard deviation is 1.65%, the maximal value is 94.28%, and the minimal value is 93.11%, i.e., the estimation is more accurate because there is a variation of 1.2% between maximal and minimal. In hold-out the variation for a similar classifier was 9.2%. □

6.3.3   Leave-One-Out

In leave-one-out strategy, we perform the cross-validation technique with N folds (the number of samples of $\mathbb{X}$). That means, we leave one sample out for testing and we train with the rest ($N - 1$ samples). The operation is repeated for each sample as illustrated in 6.21. The estimated accuracy is the average over the N estimations.

This method is implemented in function leave_one_out in pyxvis Library. In order to illustrate the estimation accuracy using leave-one-out, we can change—in Example 6.10—the line dedicated to cross-validation by the following line:

```
acc[k] = leave_one_out([name,params],X,d)
```

The results are given as follows:

```
---------------------------------------------------
 0) dmin                  : LOO-Accuracy = 0.8800
 1) lda                   : LOO-Accuracy = 0.8828
 2) qda                   : LOO-Accuracy = 0.8811
 3) maha                  : LOO-Accuracy = 0.9067
 4) knn3                  : LOO-Accuracy = 0.9272
 5) knn5                  : LOO-Accuracy = 0.9300
 6) knn7                  : LOO-Accuracy = 0.9367
 7) knn11                 : LOO-Accuracy = 0.9372
 8) knn15                 : LOO-Accuracy = 0.9383
 9) bayes-naive           : LOO-Accuracy = 0.9233
10) bayes-kde             : LOO-Accuracy = 0.9133
11) adaboost              : LOO-Accuracy = 0.8572
12) lr                    : LOO-Accuracy = 0.8661
13) rf                    : LOO-Accuracy = 0.9294
14) tree                  : LOO-Accuracy = 0.9094
15) svm-lin               : LOO-Accuracy = 0.8844
16) svm-rbf(0.1,1)        : LOO-Accuracy = 0.9350
17) svm-rbf(0.1,0.5)      : LOO-Accuracy = 0.9356
18) svm-rbf(0.5,1)        : LOO-Accuracy = 0.9378
19) svm-pol(0.05,0.1,2)   : LOO-Accuracy = 0.8778
20) svm-pol(0.05,0.5,2)   : LOO-Accuracy = 0.9061
21) svm-pol(0.05,0.5,3)   : LOO-Accuracy = 0.9033
22) svm-sig(0.1,1)        : LOO-Accuracy = 0.2589
23) svm-sig(0.1,0.5)      : LOO-Accuracy = 0.2656
24) svm-sig(0.5,1)        : LOO-Accuracy = 0.0067
25) nn(10,)               : LOO-Accuracy = 0.9333
26) nn(20,)               : LOO-Accuracy = 0.9350
27) nn(12,6)              : LOO-Accuracy = 0.9356
28) nn(20,10,4)           : LOO-Accuracy = 0.9400
---------------------------------------------------
Best Classifier:
28) nn(20,10,4)           : LOO-Accuracy = 0.9400
---------------------------------------------------
```

In this example, the best accuracy was achieved by classifier 'nn(20,10,4)' with an accuracy of 94.00%. The reader can compare these results with the accuracies presented in Examples 6.8 and 6.10. It is not necessary to repeat it, because Leave-one-out always obtains the same result. That means, there is no variation of the computed performance, however, leave-one-out is very time-consuming because the number of trainings and testings is very large.

6.3.4 Confusion Matrix

The confusion matrix, **T**, is a $K \times K$ matrix, where K is the number of classes of our data. The element $T(i, j)$ of the confusion matrix is defined as the number of samples that belong to class ω_i and were classified as ω_j. A perfect classification means that $T(i, i)$ is N_i and $T(i, j) = 0$ for $i \neq j$, where N_i is the number of samples of class ω_i.

Python Example 6.11 In this example, we show how to compute the confusion matrix for two classifiers DMIN and SVM-RBF. We use the same simulated data addressed in Example 6.2 and illustrated in Fig. 6.2.

Listing 6.11 : Confusion matrix

```
from pyxvis.learning.classifiers import clf_model,define_classifier
from pyxvis.learning.classifiers import train_classifier,test_classifier
from pyxvis.io.plots import show_confusion_matrix
from pyxvis.io.data import load_features

(X,d,Xt,dt)  = load_features('../data/F2/F2')        # load training and testing data

# Classifier definition
ss_cl = ['dmin','svm-rbf(0.1,1)']
n     = len(ss_cl)
for k in range(n):
    (name,params) = clf_model(ss_cl[k])              # function name and parameters
    clf           = define_classifier([name,params]) # classifier definition
    clf           = train_classifier(clf,X,d)        # classifier training
    ds            = test_classifier(clf,Xt)          # clasification of testing
    show_confusion_matrix(dt,ds,ss_cl[k])            # display confusion matrix
```

The output of this code is two confusion matrices that are illustrated in Fig. 6.22. This method is implemented in function plot_confussion_matrix in pyxvis Library that calls function `confusion_matrix` of sklearn library. □

Typically, in X-ray testing, there are two classes: ω_1 known as the target or object of interest, and ω_0 known as the no-target or background. In this two-class recognition problem (known as 'detection'), we are interested in detecting the target correctly. It is very helpful to build a 2×2 confusion matrix as shown in Table 6.2. We distinguish

- True Positive (TP): number of targets correctly classified.
- True Negative (TN): number of non-targets correctly classified.
- False Positive (FP): number of non-targets classified as targets. The false positives are known as 'false alarms' and 'Type I error'.
- False Positive (FN): number of targets classified as no-targets. The false negatives are known as 'Type II error'.

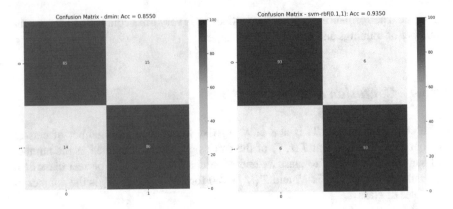

Fig. 6.22 Visualization of confusion matrix of LDA and SVM-RBF. [→ Example 6.11 🐍]

Table 6.2 Confusion matrix for two classes

predicted → actual ↓	ω_1	ω_0
ω_1	TP	FN
ω_0	FP	TN

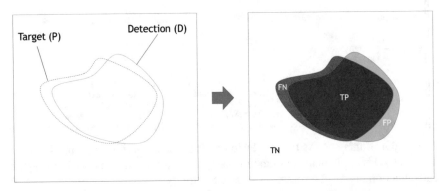

Fig. 6.23 Detection of a target: the ground truth (ideal detection given by an expert) is called in this figure as 'target' (the positive instances). The achieved detection is not a perfect match. For this reason, there are false positives and false negatives

From these statistics, we can obtain following definitions (see Fig. 6.23):
Positive instances:

$$P = TP + FN \tag{6.35}$$

Negative instances:

$$N = TN + FP \tag{6.36}$$

Detections:

$$D = TP + FP \tag{6.37}$$

True positive rate, known as Sensitivity or Recall:

$$TPR = S_n = Re = \frac{TP}{P} = \frac{TP}{TP + FN} \tag{6.38}$$

Precision or Positive Predictive Value:

$$Pr = \frac{TP}{D} = \frac{TP}{TP + FP} \tag{6.39}$$

True negative rate, known as Specificity:

$$TNR = Sp = \frac{TN}{N} = \frac{TN}{TN + FP} \tag{6.40}$$

False positive rate, known as 1-Specificity:

$$FPR = 1 - Sp = \frac{FP}{N} = \frac{FP}{TN + FP} \tag{6.41}$$

False negative rate, known as Miss Rate:

$$FNR = MR = \frac{FN}{P} = \frac{FN}{TP + FN} \tag{6.42}$$

Accuracy:

$$ACC = \frac{TP + TN}{P + N} \tag{6.43}$$

F1-score:

$$F1 = 2\frac{Pr \cdot Re}{Pr + Re} \tag{6.44}$$

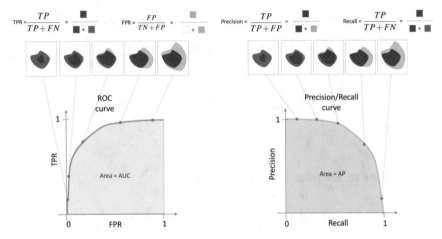

Fig. 6.24 Performance curves for a detection problem (see Fig. 6.23). Left) ROC curve. Right) Precision/Recall curve

Ideally, a perfect detection means all existing targets are correctly detected without any false alarms, i.e., $TP = P$ and $FP = 0$. It is equivalent to: (i) $TPR = 1$ and $FPR = 0$, or (ii) $Pr = 1$ and $Re = 1$, or (ii) $FN = FP = 0$.

6.3.5 ROC and Precision-Recall Curves

It is clear, that the performance of a detector depends on some parameters, e.g., the value of a threshold θ when segmenting a defect in an X-ray image (see Fig. 6.23). An example to see this phenomenon is shown in Fig. 6.24: increasing the sensitivity of the method the target will be 100% detected, however, the false positives will be increased as well. Typically, there is a trade-off between increasing the true positives and decreasing the false positives, because by increasing the first, the second increases as well. In a detector, i.e., a binary classification task, we can analyze the performance of the detector by variating its parameter θ.

As a measure of the performance of a detector, two curves can be plotted:

ROC curve: We can analyze the values TPR and FPR as defined in (6.38) and (6.41) respectively (see Fig. 6.24). In this case, we obtain $TPR(\theta)$ and $FPR(\theta)$ because the values of these variables depend on parameter θ.

The receiver operation characteristic (ROC) curve is a plot of $TPR(\theta)$ versus $FPR(\theta)$. Thus, we choose different values $\{\theta_i\}_{i=1}^n$ and for each value θ_i we plot the corresponding point (x_i, y_i), where $x_i = FPR(\theta_i)$ and $y_i = TPR(\theta_i)$. An example is illustrated in Fig. 6.25. A measure of performance of the detector is the area under the curve (AUC) [6].

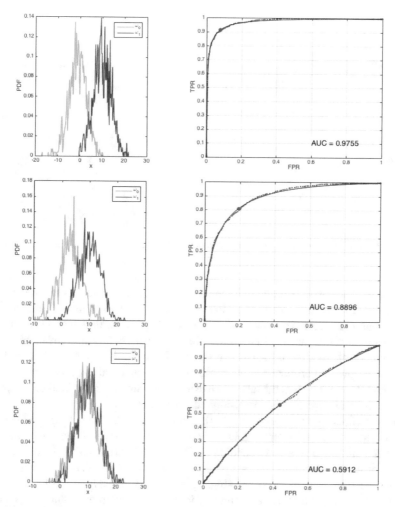

Fig. 6.25 ROC curves (right) for different class distributions (left). The area under the curve (AUC) gives a good measure of the performance of the detection. The obtained points (x_i, y_i) are used to fit the ROC curve to $y = (1 - a^{\gamma x^b})/(1 - a^\gamma)$. In each ROC curve, the 'best operation point' is shown as spscolorred *. This point is defined as the closest point to ideal operation point $(0,1)$

Precision/Recall curve: We can analyze the values Pr and Re as defined in (6.41) and (6.38) respectively (see Fig. 6.24). In this case, we obtain $Pr(\theta)$ and $Re(\theta)$ because the values of these variables depend on parameter θ. As in ROC curve, we choose different values $\{\theta_i\}_{i=1}^n$ and for each value θ_i we plot the corresponding point (x_i, y_i), where $x_i = Re(\theta_i)$ and $y_i = Pr(\theta_i)$. A measure of performance of the detector is the area under the curve, called average precision (PA) [5].

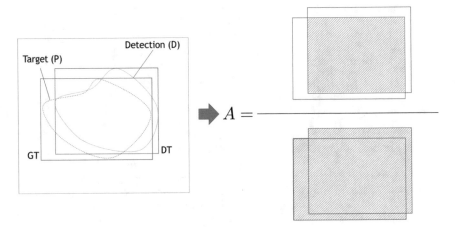

Fig. 6.26 Intersection over Union (IoU). For a perfect detection the normalized area A equals 1

It is worthwhile to mention that the precision and recall values do not depend on the true negatives, like the false positive rate in ROC curve. This is a great advantage when the negative class can be immensely large, e.g., in defect detection, the number of positive instances is limited (there are usually few cases available), and the number of negative instances can be very large. In those cases, FPR will be extremely low, and erroneously we could think that the number of false positives is very low. This is a typical mistake when using ROC curves. In this kind of computer vision problem, typically the precision/recall curve is used.

In object detection, for example, [14], it is very important how to give a measure of the performance of a detector. For this end, there is a set of images with objects to detect, and for each one a bounding box that encloses it has been annotated by a group of human operators. For simplicity, the annotation consists of drawing rectangles (instead of marking every single pixel of the objects). A very established metric in the computer vision community is the 'intersection over union' (IoU) and the PASCAL criterion [9]. For this metric, we need to define two bounding boxes according to Fig. 6.26: GT, the bounding box of the ground truth, i.e., a rectangle that encloses the target region (P), and DT, the bounding box of the detection, i.e., a rectangle that encloses the detection (D). The PASCAL criterion considers a detected object if the normalized area of overlap 'A' between the detected bounding box DT and the ground truth bounding box GT exceeds 0.5, where A is defined as follows:

$$A = \frac{\text{area}(\text{GT} \cap \text{DT})}{\text{area}(\text{GT} \cup \text{DT})}, \tag{6.45}$$

with GT $\cap$ DT the intersection of the detected and ground truth bounding boxes and GT $\cup$ DT their union. An example in the detection of defects in aluminum castings is illustrated in Fig. 6.27.

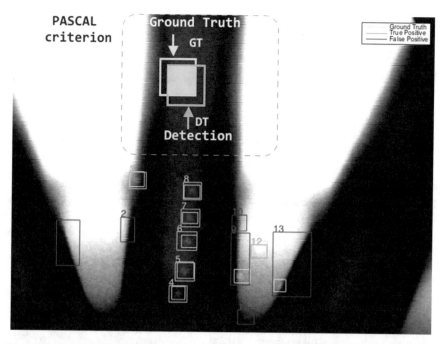

Fig. 6.27 Detection on a single image. A detection is considered as true positive is the normalized area of overlap (6.45) is greater than 50%. In this example, the true positives are shown in green, the false positives in red, and the ground truth in yellow

With PASCAL criterion, the statistics of true positives and negatives, and false positives and negatives are measured, and the precision/recall values are computed in different scenarios. The mean average precision (mPA) is typically used to compare the performance of different object detection algorithms (see details in [14]).

Python Example 6.12 In this example, we show how to compute the ROC curves and Precision/Recall curves for three classifiers based on neural networks in the classification of a two-class problem with two features. We use the same simulated data addressed in Example 6.2 and illustrated in Fig. 6.2.

Listing 6.12 : ROC and Precision/Recall curves

```
from sklearn.metrics import roc_curve, roc_auc_score
from sklearn.metrics import precision_recall_curve, average_precision_score
from pyxvis.learning.classifiers import clf_model,define_classifier,train_classifier
from pyxvis.io.plots import plot_features, plot_ROC, plot_precision_recall
from pyxvis.io.data import load_features

(X,d,Xt,dt)   = load_features('../data/F2/F2')      # load train/test data
plot_features(X,d,'F2 dataset')                      # plot of feature space

ss_cl = ['nn(3,)','nn(4,)','nn(8,)']                 # classifiers to evaluate
```

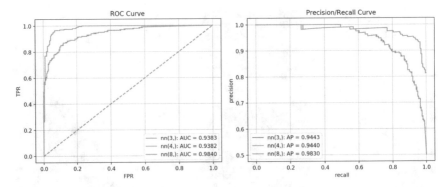

Fig. 6.28 ROC curve and Precision/Recall curve for different neural networks using data distribution of Fig. 6.2. [→ Example 6.12 🐍]

```
curve = 1                                         # 0 = ROC curve,
                                                  # 1 = precision/recall curve

for k in range(len(ss_cl)):
    cl_name      = ss_cl[k]
    (name,params) = clf_model(cl_name)            # function name and parameters
    clf          = define_classifier([name,params]) # classifier definition
    clf          = train_classifier(clf,X,d)      # classifier training
    p            = clf.predict_proba(Xt)[:,1]     # classification probabilities
    if curve == 0: # ROC curve
        auc      = roc_auc_score(dt, p)           # area under curve
        fpr,tpr,_ = roc_curve(dt, p)              # false and true positive rates
        plot_ROC(fpr,tpr,cl_name,auc,[k,n])       # ROC curve
    else:          # precision/recall curve
        ap       = average_precision_score(dt, p) # area under curve
        pr,re,_  = precision_recall_curve(dt, p)  # precision and recall values
        plot_precision_recall(pr,re,cl_name,ap,[k,n]) # precision/recall curve
```

The output of this code are the curves of Fig. 6.28. Variable `curve` must be set to 0 or 1 for ROC curve or Precision/Recall curve respectively. This method is implemented with functions `roc_auc_score`, `roc_curve`, `precision_recall_curve`, and `average_precision_score` of sklearn library and functions plot_ROC and plot_precision_recall of pyxvis Library.   □

6.4   Classifier Selection

In order to select the *best* classifier, we explain in this section a methodology using two examples. Our examples are implemented using powerful functions of pyxvis Library. With these functions, easily, we can (i) extract features, (ii) select features and (iii) select a classifier. Thus, the user can: choose the feature groups that will be extracted, choose the feature selection algorithms to be used, the maximal number of features to be selected, and choose the classifiers that will be evaluated and the number of folds of the cross-validation technique. Using these simple functions, it is possible to design the computer vision system automatically according to the general computer vision framework explained in these three chapters (image processing, image representation and classification, and summarized in Fig. 5.28).

Using this methodology, with a representative set of X-ray images and their labels, we can know which features and which classifier can be used to obtain the *best* performance. The idea is to find a classification strategy (feature extraction, features selection, and classification as shown in Fig. 6.1) that maximizes the accuracy in this dataset. The proposed methodology (based on [19]) evaluates a set of combinations of features (selected by may feature selection algorithms) and trains and tests a set of classifiers to find *best* strategy, i.e., the highest accuracy.

In order to show this methodology, we show two examples, Example 6.13 for the detection of fishbones (that uses intensity features), and Example 6.14 for the

---

Algorithm 1 Feature and Classifier Selection

Input: $(\mathbf{X}, \mathbf{d})$:Training subset; $(\mathbf{X}_t, \mathbf{d}_t)$: Testing subset
Input: $\mathbf{p} = [p_1 \cdots p_n]$: number of features to be selected
Input: $\mathbf{f} = [f_1 \cdots f_m]$: feature selectors algorithms
Input: $\mathbf{h} = [h_1 \cdots h_q]$: classification algorithms
1: $\hat{\eta} = 0$//Initialization of the highest accuracy in training subset
2: $\hat{\eta}_t = 0$//Initialization of the highest accuracy in testing subset
3: **for** $i = 1$ to n **do**
4:   **for** $j = 1$ to m **do**
5:     $s = \text{FeatureSelection}(f_j, p_i, \mathbf{X}, d)$ //Selection of p_i features of $\mathbf{X}$ using f_j
6:     $\mathbf{X}' = \mathbf{X}[:, s]$ //Training subset using selected features
7:     $\mathbf{X}'_t = \mathbf{X}_t[:, s]$ //Testing subset using selected features
8:     **for** $k = 1$ to q **do**
9:       $\eta = \text{CrossValidation}(h_k, \mathbf{X}', d)$ //Accuracy of classifier h_k on data $\mathbf{X}'$
10:       $\eta_t = \text{HoldOut}(h_k, \mathbf{X}', \mathbf{d}, \mathbf{X}'_t, \mathbf{d}_t)$ //Accuracy of classifier h_k on data $\mathbf{X}'_t$
11:       **if** $\eta > \hat{\eta}$ **then**
12:         $\hat{\eta} = \eta$ //Highest performance in training
13:         **if** $\eta_t > \hat{\eta}_t$ **then**
14:           $\hat{\eta}_t = \eta_t$ //Highest performance in testing
15:           $\hat{s} = s$ //Indices of the best selected features
16:           $\hat{p} = p_i$ //Number of selected features
17:           $\hat{j} = j$ //Index of the best feature selector
18:           $\hat{k} = k$ //Index of the best feature classifier
19:         **end if**
20:       **end if**
21:     **end for**
22:   **end for**
23: **end for**
Output: $\hat{\eta}, \hat{\eta}_t, \hat{s}, \hat{p}, \hat{j}, \hat{k}$

classification of three threat objects (that uses geometric features extracted after a segmentation of the threat objects).

In order to find the *best* classification strategy, we use an exhaustive search (Algorithm 1) as follows: we define q classifiers, n feature selection algorithms, and m different numbers of selected features. That means, we valuate the performance of the q classifiers on the $m \times n$ subsets of selected features. For instance, we could have: $q = 3$ classifiers (LDA, KNN with 3 neighbors, and SVM with RBF), $m = 2$ feature selection algorithms (SFS with Fisher criterion and SFS with QDA criterion) with 5, 10, 15, and 20 selected features ($n = 4$). The accuracy is measured on the training dataset using cross-validation, and on the testing dataset using hold-out. According to Algorithm 1, the highest achieved accuracy on training datase (searching in all $q \times m \times n$) is computed as $\hat{\eta}$. In case a maximal value for $\hat{\eta}$, the accuracy on testing dataset is evaluated as $\hat{\eta}_t$. This algorithm is implemented in function best_features_classifier of pyxvis Library.

Python Example 6.13 In this example, we can see the whole process of Algorithm 1: (i) feature extraction, (ii) feature selection, and (ii) classifier selection. pyxvis Library provides a suite of helpful commands that can be used in this process. The idea is to design a classifier that can be used to detect fish bones in X-ray images of salmon filets (see details of the dataset in Example 5.9). In this code, we show how to automatically design a computer vision system for this application. For this example, (i) we extract basic intensity, Gabor, LBP, Haralick with distance of 2 pixels, Fourier and HOG features; (ii) we evaluate four different feature selection algorithms based on Fisher, QDA, SVM-LIN and SVM-RBF with 3, 5, 10, 12, and 15 features to be selected; and (iii) we train and test 8 different classifiers: Mahalanobis, Bayes-KDE, SVM-LIN, SVM-RBF, QDA, LDA, KNN-3, KNN-7, and a Neural Network.

Listing 6.13 : Feature extraction, feature selection and classification selection - 1

```python
import numpy as np
from pyxvis.io.data import load_features,save_features
from pyxvis.learning.evaluation import best_features_classifier
from pyxvis.features.selection import clean_norm,clean_norm_transform
from pyxvis.features.extraction import extract_features_labels

dataname = 'fbdata' # prefix of npy files of training and testing data
fxnew    = 1        # the features are (0) loaded or (1) extracted and saved
if fxnew:
    # features to extract
    fx       = ['basicint','gabor-ri','lbp-ri','haralick-2','fourier','hog']
    # feature extraction in training images
    path     = '../images/fishbones/'
    X,d      = extract_features_labels(fx,path+'train','jpg')
    # feature extraction in testing images
    Xt,dt    = extract_features_labels(fx,path+'test','jpg')
    # backup of extracted features
    save_features(X,d,Xt,dt,dataname)
else:
    X,d,Xt,dt = load_features(dataname)

X,sclean,a,b = clean_norm(X)
Xt           = clean_norm_transform(Xt,sclean,a,b)
```

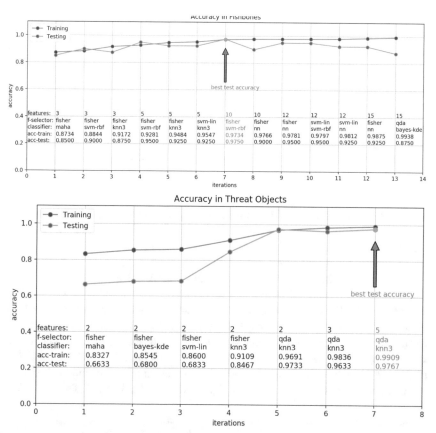

Fig. 6.29 Examples of Algorithm 1 for feature and classification selection. [→ Example 6.13] [→ Example 6.14]

```
# Classifiers to evaluate
ss_cl        = ['maha','bayes-kde','svm-lin','svm-rbf','qda','lda','knn3','knn7','nn']
# Number of features to select
ff           = [3,5,10,12,15]
# Feature selectors to evaluate
ss_fs        = ['fisher','qda','svm-lin','svm-rbf']

clbest,ssbest = best_features_classifier(ss_fs,ff,ss_cl,X,d,Xt,dt,
                                   'Accuracy in Fishbones')
print('    Selected Features: '+str((np.sort(sclean[ssbest]))))
```

The result of this algorithm is illustrated in Fig. 6.29-Top. We can see that the best performance was achieved by classifier SVM-RBF using 10 features that were selected using SFS algorithm with Fisher criterion. The accuracy on testing dataset is in this case 97.50%. The indices of the selected features are shown in following output:

```
--------------------------------------------------------------------------------
       Best iteration: 7 (maximum of testing accuracy)
     Feature Selector: fisher with 10 features
                     : (Fisher, )
           Classifier: svm-rbf
                     : (SVC, kernel = "rbf", gamma=0.1,C=1) CrossVal with 5 folds
         Training-Acc: 0.9734
          Testing-Acc: 0.9750
     Selected Features: [ 3 20 21 39 51 57 63 65 69 71]
--------------------------------------------------------------------------------
```

☐

Python Example 6.14 In this example, we can see the whole process of Algorithm 1 using geometric features: (i) feature extraction, (ii) feature selection, and (ii) classifier selection using pyxvis Library. The idea is to design a classifier that can be used to recognize threat objects in X-ray images (see details of the dataset in Example 5.10). In this code, we show how to automatically design a computer vision system for this application. For this example, (i) we extract basic geometric features, Hu, Flusser and Gupta moments, and Fourier descriptors (the features extracted from the segmented image, for this end we use function seg_bimodal of pyxvis Library as explained in Sect. 4.5.1); (ii) we evaluate four different feature selection algorithms based on Fisher, QDA, SVM-LIN, and SVM-RBF with 2, 3, 5, 10, 15, and 20 features to be selected; and (iii) we train and test 8 different classifiers: Mahalanobis, Bayes-KDE, SVM-LIN, SVM-RBF, QDA, LDA, KNN-3, KNN-7, and a Neural Network.

> **Listing 6.14 : Feature extraction, feature selection and classification selection - 2**

```python
import numpy as np
from pyxvis.io.data import load_features,save_features
from pyxvis.learning.evaluation import best_features_classifier
from pyxvis.features.selection import clean_norm,clean_norm_transform
from pyxvis.features.extraction import extract_features_labels

dataname = 'thdata' # prefix of npy files of training and testing data
fxnew    = 1        # the features are (0) loaded or (1) extracted and saved
if fxnew:
    # features to extract
    fx       = ['flusser','hugeo','basicgeo','fourierdes','gupta']
    # feature extraction in training images
    path     = '../images/threatobjects/'
    X,d      = extract_features_labels(fx,path+'train','jpg',segmentation = 'bimodal')
    # feature extraction in testing images
    Xt,dt    = extract_features_labels(fx,path+'test','jpg',segmentation = 'bimodal')
    # backup of extracted features
    save_features(X,d,Xt,dt,dataname)
else:
    X,d,Xt,dt = load_features(dataname)
Nx            = X.shape[1]
X,sclean,a,b  = clean_norm(X)
Xt            = clean_norm_transform(Xt,sclean,a,b)
# Classifiers to evaluate
ss_cl         = ['maha','bayes-kde','svm-lin','svm-rbf','qda','lda','knn3','knn7','nn']
# Number of features to select
```

```
ff          = [2,3,5,10,15,20]
# Feature selectors to evaluate
ss_fs       = ['fisher','qda','svm-lin','svm-rbf']

clbest,ssbest = best_features_classifier(ss_fs,ff,ss_cl,X,d,Xt,dt,
                                   'Accuracy in Threat Objects')
print('    Extracted Features: '+str(Nx))
print('      Cleaned Features: '+str(len(sclean)))
print('     Selected Features: '+str(len(ssbest))+ ' > '+str((np.sort(sclean[ssbest]))))
```

The result of this algorithm is illustrated in Fig. 6.29-Bottom. We can see that the best performance was achieved by classifier KNN-3 using 5 features selected using QDA criterion. The accuracy on testing dataset is in this case 97.67%. The indices of the selected features are shown in following output:

```
--------------------------------------------------------------------
        Best iteration: 7 (maximum of testing accuracy)
      Feature Selector: qda with 5 features
                     : (QuadraticDiscriminantAnalysis, )
            Classifier: knn3
                     : (KNeighborsClassifier, n_neighbors=3) CrossVal with 5 folds
          Training-Acc: 0.9909
           Testing-Acc: 0.9767
    Extracted Features: 48
      Cleaned Features: 44
     Selected Features: 5 > \cite{
--------------------------------------------------------------------
```

☐

6.5   Summary

In this chapter, we covered the following classifiers:

- Minimal distance (using Euclidean and Mahalanobis distance)
- Bayes
- Linear and quadratic discriminant analysis
- K-nearest neighbors
- Neural networks
- Support vector machines
- Classifiers using sparse representations

In addition, several simple examples were presented using simulated data and real data. The reader can easily modify the proposed implementations in order to test different classification strategies or real data.

Afterwards, we presented how to estimate the accuracy of a classifier using hold-out, cross-validation, and leave-one-out. We covered the well-known confusion matrix and receiver-operation-characteristic curve will be outlined as well.

Finally, we presented an example that involves all steps of a pattern recognition problem, i.e., feature extraction, feature selection, classifier's design, and evaluation.

All steps can be designed automatically using a simple code program of a couple of lines.

References

1. Bar, L., Sapiro, G.: Hierarchical dictionary learning for invariant classification. In: 2010 IEEE International Conference on Acoustics Speech and Signal Processing (ICASSP), pp. 3578–3581 (2010)
2. Bentley, J.: Multidimensional binary search trees used for associative searching. Commun. ACM **18**(9), 509–517 (1975)
3. Bishop, C.: Neural Networks for Pattern Recognition. Oxford University Press, Oxford (2005)
4. Bishop, C.: Pattern Recognition and Machine Learning. Springer, Berlin (2006)
5. Boyd, K., Eng, K.H., Page, C.D.: Area under the precision-recall curve: point estimates and confidence intervals. In: Joint European Conference on Machine Learning and Knowledge Discovery in Databases, pp. 451–466. Springer, Berlin (2013)
6. Bradley, A.P.: The use of the area under the roc curve in the evaluation of machine learning algorithms. Patt. Recogn. **30**(7), 1145–1159 (1997)
7. Carvajal, K., Chacón, M., Mery, D., Acuna, G.: Neural network method for failure detection with skewed class distribution. Insight **46**(7), 399–402 (2004)
8. Duda, R., Hart, P., Stork, D.: Pattern Classification, 2nd edn. Wiley, New York (2001)
9. Everingham, M., Gool, L.V., Williams, C.K.I., Winn, J., Zisserman, A.: The pascal visual object classes (voc) challenge. Int. J. Comput. Vis. **88**(2), 303–338 (2010)
10. Fukunaga, K.: Introduction to Statistical Pattern Recognition, 2nd edn. Academic Press Inc., San Diego (1990)
11. Hastie, T., Tibshirani, R., Friedman, J.: The Elements of Statistical Learning: Data Mining, Inference, and Prediction, 2nd edn. Springer, Berlin (2009)
12. Jain, A., Duin, R., Mao, J.: Statistical pattern recognition: a review. IEEE Trans. Patt. Anal. Mach. Intell. **22**(1), 4–37 (2000)
13. Kohavi, R.: A study of cross-validation and bootstrap for accuracy estimation and model selection. In: International Joint Conference on Artificial Intelligence, vol. 14, pp. 1137–1145. Citeseer (1995)
14. Lin, T.Y., Maire, M., Belongie, S., Hays, J., Perona, P., Ramanan, D., Dollár, P., Zitnick, C.L.: Microsoft COCO: common objects in context. In: European Conference on Computer Vision, pp. 740–755. Springer, Berlin (2014)
15. Mairal, J., Bach, F., Ponce, J., Sapiro, G., Zisserman, A.: Discriminative learned dictionaries for local image analysis. In: IEEE Conference on Computer Vision and Pattern Recognition (CVPR) (2008)
16. Mairal, J., Bach, F., Ponce, J., Sapiro, G., Zisserman, A.: Supervised dictionary learning. Tech. Rep. 6652, INRIA (2008)
17. Mery, D., Bowyer, K.: Face recognition via adaptive sparse representations of random patches. In: IEEE Workshop on Information Forensics and Security (WIFS 2014) (2014)
18. Mery, D., Bowyer, K.: Recognition of facial attributes using adaptive sparse representations of random patches. In: 1st International Workshop on SoftBiometrics, in Conjunction with European Conference on Computer Vision (ECCV 2014) (2014)
19. Mery, D., Pedreschi, F., Soto, A.: Automated design of a computer vision system for visual food quality evaluation. Food Bioprocess Technol. **6**(8), 2093–2108 (2013)
20. Mitchell, T.: Machine Learning. McGraw-Hill, Boston (1997)
21. Shawe-Taylor, J., Cristianini, N.: Kernel Methods for Pattern Analysis. Cambridge University Press, Cambridge (2004)
22. Silverman, B.W.: Density Estimation for Statistics and Data Analysis, vol. 26. CRC Press, Boca Raton (2003)

23. Sivic, J., Zisserman, A.: Video Google: a text retrieval approach to object matching in videos. In: International Conference on Computer Vision (ICCV 2003), pp. 1470–1477 (2003)
24. Tosic, I., Frossard, P.: Dictionary learning. Signal Process. Mag. IEEE **28**(2), 27–38 (2011)
25. Witten, I., Frank, E.: Data Mining: Practical Machine Learning Tools and Techniques, 2nd edn. Morgan Kaufmann, Burlington (2005)
26. Wright, J., Yang, A.Y., Ganesh, A., Sastry, S.S., Ma, Y.: Robust face recognition via sparse representation. IEEE Trans. Patt. Anal. Mach. Intell. **31**(2), 210–227 (2009)

Chapter 7
Deep Learning in X-ray Testing

Abstract Deep learning has been inspired by ideas from neuroscience. The key idea of deep learning is to replace *handcrafted* features (explained in details in Chap. 5) with features that are *learned* efficiently using a hierarchical feature extraction approach. Usually, the learned features are so discriminative that no sophisticated classifiers are required. In last years, deep learning has been successfully used in image and video recognition, and it has been established as the state of the art in many areas such as computer vision, machine translation, and natural language processing. In comparison with other computer vision applications, we have seen that the introduction of techniques based on deep learning in computer vision for X-ray testing has been rather slow. However, there are many methods based on deep learning that have been designed and tested in some X-ray testing applications. In this chapter, we review many relevant concepts of deep learning that can be used in computer vision for X-ray testing. We covered the theory and practice of deep learning techniques in real X-ray testing problems. The chapter explained neural networks, Convolutional Neural Network (CNN) that can be used in classification problems, pre-trained models, transfer learning that are used in sophisticated models, Generative Adversarial Networks (GANs) to generate synthetic images, and modern detection methods that are used to classify and localize objects in an image. In addition, for every method, we give not only the basic concepts but also practical details in real X-ray testing examples that have been implemented in Python.

Cover image: *Synthetic X-ray mages generated by a GAN model that has been trained using X-ray images of backpacks with no threat objects (from series* B0082 *colored with 'jet' colormap).*

© Springer Nature Switzerland AG 2021
D. Mery and C. Pieringer, *Computer Vision for X-Ray Testing*,
https://doi.org/10.1007/978-3-030-56769-9_7

7.1   Introduction

Originally, deep learning is inspired by ideas from neuroscience [19]. In last years, deep learning has been successfully used in image and video recognition (see, for example, [3, 31, 58]), and it has been established as the state of the art in many areas such as computer vision, machine translation, and natural language processing [57].

The key idea of deep learning is to replace *handcrafted* features (explained in details in Chap. 5) with features that are *learned* efficiently using a hierarchical feature extraction approach. Usually, the learned features are so discriminative that no sophisticated classifiers are required. In recent years, we have witnessed tremendous improvements in many fields of computer vision by using complex deep neural network architectures trained with thousands or millions of images (e.g., face recognition [10], object recognition and detection [35, 71], diagnosis of prostate cancer [44], classification of skin cancer [14], among others). Methods based on deep learning have become fundamental in these fields, however, an enormous number of images used for training purposes are required in order to achieve satisfactory results.

In comparison with other computer vision applications, we have seen that the introduction of techniques based on deep learning in computer vision for X-ray testing has been rather slow. In our opinion, this is due to three reasons. The first has to do with the availability of public databases that can be used for these purposes. While in some areas of computer vision (e.g., face recognition), there are hundreds of databases since the 1990s, in X-ray testing, there is only one public database for X-ray testing for general purposes [40] created in the last 5 years with around 20.000 X-ray images, and another one for baggage inspection [42] created recently with around 1 million X-ray images. The rest of the datasets used in the experiments reported by the industry and academia are private. In many cases, the entities (industry, government, or academia) that fund research in X-ray testing do not allow databases to be made public. Sometimes this happens in baggage inspection research (for security reasons) or in other industrial applications (to prevent competitors from having access to data that could improve their processes). The second reason is related to the number of experts working in this field. While in other areas of computer vision almost anyone can be an expert (such as in object recognition), in nondestructive testing the relative number of people working on these subjects is rather low and usually, their work is expensive. In this kind of computer applications, experts are necessary to label the data (make annotations, define bounding boxes, etc.). It is very simple to find people that detect bicycles in photographs, however, it is not so easy to find human operators that can distinguish the anomalies in a welding process by observing an X-ray image. Finally, the last reason is that, in other applications of computer vision, color photos can be acquired with inexpensive equipment (often a cell phone), whereas in X-ray testing, we need expensive equipment.

In this chapter, we review many relevant concepts of deep learning that can be used in computer vision for X-ray testing. This chapter should be considered as an introduction to the subject rather than an in-depth treatise.[1] We will cover many relevant topics, so the reader will be able to understand and apply these techniques in real X-ray testing problems. The chapter begins with the basics, i.e., neural networks (see Sect. 7.2). Afterwards, we will review the Convolutional Neural Network (CNN) (see Sect. 7.3) that can be used in a classification problem. CNNs can be trained from scratch or using pre-trained models (see Sect. 7.4) or transfer learning (see Sect. 7.5). In addition, we cover the Generative Adversarial Networks (GANs) (see Sect. 7.6) that have been proposed to generate synthetic images. Finally, we give an overview of more complex architectures that can be used as detection methods (see Sect. 7.7), i.e., when we want to classify and localize an object in an image. For every section, we will cover the basic concepts, give practical details (e.g., training and testing) and show some examples in X-ray testing using Python.

7.2 Neural Networks

Artificial neuronal networks are mathematical tools derived from what is known about the mechanisms and physical structure of biological learning, based on the function of a neuron. They are parallel structures for the distributed processing of information [4]. A neural network consists of artificial neurons connected in a net-

---

[1]Recommendation for further reading: [1, 18, 31].

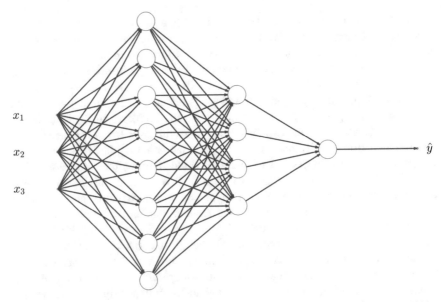

Fig. 7.1 Simple neural network with three inputs $\mathbf{x} = (x_1, x_2, x_3)$, one output $\hat{y}$, and two hidden layers (one with 8 nodes and the another with 4). In this example, the input can be classified as class ω_1 if $\hat{y} > 0.5$, and otherwise as class ω_0

work that is able to classify a test feature vector $\mathbf{x}$ evaluating a linear weighted sum of non-linear functions as illustrated in Fig. 7.1. The weights, the functions, and the connections are estimated in a training phase by minimizing the classification error [4, 5]. Neural networks have been established as one of the best classification approaches in pattern recognition. The basic structure of the neural networks and the learning strategies developed for training neural networks are the basis of deep learning models.

7.2.1   Basics of Neural Networks

The basic processing unit is the neuron, made up of multiple inputs and only one output as shown in Fig. 7.2. This output is determined by an activation function that operates on input values, and a transfer function that operates on the activation value. In other words, if we consider the input vector $\mathbf{x} = [x_1 \ldots x_n]^\mathsf{T}$, the weight vector $\mathbf{w} = [w_1 \ldots w_n]^\mathsf{T}$, the activation value z, and the output value of the neuron a, the values of z and a can be described by a linear projection and an a non-linear function:

$$z = \mathbf{w}^\mathsf{T}\mathbf{x} + b \qquad a = \sigma(z), \tag{7.1}$$

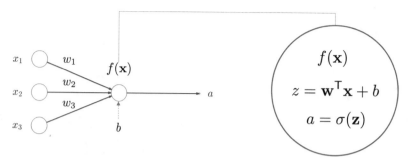

Fig. 7.2 Single neuron with three inputs (x_1, x_2, x_3), three weights, one weight for each input, (w_1, w_2, w_3), one bias value (b), and one output a

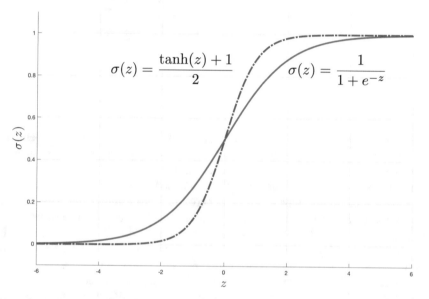

Fig. 7.3 Two typical sigmoids as activation functions

where b is the bias value and $\sigma(z)$ is the so-called transfer function or activation function and is generally a sigmoid such as (see Fig. 7.3)

$$\sigma(z) = \frac{1}{1 + e^{-z}} \quad \text{or} \quad \sigma(z) = \frac{\tanh(z) + 1}{2}. \tag{7.2}$$

A very simple structure, called *logistic regression*, is defined for two classes and no hidden layer, i.e., the output of the model is $y = a$. Thus, class ω_1 is determined when $y > 0.5$, and ω_0 otherwise. This is a linear approach because the separation of both classes corresponds to a hyperplane (or a straight line for a feature space of two dimensions).

Fig. 7.4 Multi-Layer
Perceptron (MLP) with one
input layer with two inputs,
two hidden layers (Layer 1
and Layer 2) with 6 and 12
nodes respectively and one
output layer with four
outputs. [→ Example 7.1
⬛] [→ Example 7.2 ⬛]

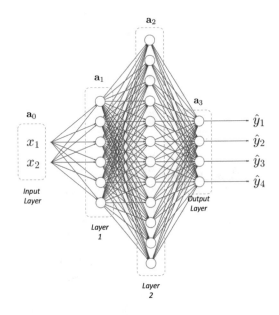

The structure of a neuronal network can have one or more neurons and depending
on the type of problem and the training, these networks receive different names.
They have the capacity to associate and classify patterns, compress data, perform
process control, and approximate non-linear functions [43].

The most often used type of neural network in classification is the Multi-Layer
Perceptron (MLP) which consists of sequential layers of neurons. The structure of
an MLP is shown in Fig. 7.4 where each neuron has Eq. (7.1) associated to it. It
consists of a input layer, hidden layers, and an output layer (in Fig. 7.4, there are two
hidden layers, $\mathbf{a}_1$ and $\mathbf{a}_2$). In a classification problem based on neural networks, the
input $\mathbf{x}$ corresponds to the feature vector, and the output $\hat{\mathbf{y}}$ that is the classification of
$\mathbf{x}$. Output $\hat{\mathbf{y}}$ is defined as a vector of K elements for a classification into K classes.
The value $\hat{y}_i$ can be understood as the probability that sample $\mathbf{x}$ is classifies a class
ω_i. Formerly,

$$h_{\mathrm{nn}}(\mathbf{x}) = \underset{i}{\mathrm{argmax}} \left\{ \hat{y}_1, \ldots, \hat{y}_K \right\}. \tag{7.3}$$

Usually, an index $k = 0, \ldots, m$ is used to denote a layer, where $k = 0$ is the
input layer and $k = m$ is the output layer (in Fig. 7.4, $m = 3$). In addition, index
$i = 1, \ldots, n_k$ is used to denote the node i of layer k. In Fig. 7.4, the two hidden
layers have $n_1 = 6$ and $n_2 = 12$ nodes respectively. We define the output of layer k
as vector $\mathbf{a}_k$, and it is a vector with n_k elements. In this definition, $\mathbf{a}_0$ corresponds to
$\mathbf{x}$, i.e., the input vector of the neural network with n_0 elements (in Fig. 7.4, $n_0 = 2$
for two inputs). Similarly, $\mathbf{a}_m$ corresponds to $\hat{\mathbf{y}}$, i.e., the output vector of the neural
network (in Fig. 7.4, $n_m = 4$ for four outputs). Thus, in general, layer k is defined
by

$$\mathbf{a}_k = \begin{bmatrix} a_k(1) \\ a_k(2) \\ \vdots \\ a_k(n_k) \end{bmatrix}, \tag{7.4}$$

where the input and output layers are respectively:

$$\mathbf{x} = \mathbf{a}_0 = \begin{bmatrix} x_1 \\ x_2 \\ \vdots \\ x_{n_0} \end{bmatrix} = \begin{bmatrix} a_0(1) \\ a_0(2) \\ \vdots \\ a_0(n_0) \end{bmatrix} \tag{7.5}$$

$$\hat{\mathbf{y}} = \mathbf{a}_m = \begin{bmatrix} \hat{y}_1 \\ \hat{y}_2 \\ \vdots \\ \hat{y}_{n_k} \end{bmatrix} = \begin{bmatrix} a_m(1) \\ a_m(2) \\ \vdots \\ a_m(n_m) \end{bmatrix}. \tag{7.6}$$

With these definitions, it is simple to write the equation of each node according to (7.1) for $k = 1, \ldots, m$:

$$\mathbf{z}_k = \mathbf{W}_k \mathbf{a}_{k-1} + \mathbf{b}_k \quad, \quad \mathbf{a}_k = \sigma(\mathbf{z}_k). \tag{7.7}$$

In this equation, $\mathbf{z}_k$, $\mathbf{a}_k$, and $\mathbf{b}_k$ are n_k-element vectors, $\mathbf{a}_{k-1}$ is a n_{k-1}-element vector, and $\mathbf{W}_k$ is a matrix of $n_k \times n_{k-1}$ elements, where $w_k(i, j)$ is the weight of the connection between node i of layer k and node j of layer $k - 1$.

For our example of Fig. 7.4 with $m = 3$, output $\mathbf{y}$ can easily computed using following steps:

$$\begin{bmatrix} \mathbf{z}_1 = \mathbf{W}_1 \mathbf{a}_0 + \mathbf{b}_1 & , & \mathbf{a}_1 = \sigma(\mathbf{z}_1) \\ \mathbf{z}_2 = \mathbf{W}_2 \mathbf{a}_1 + \mathbf{b}_2 & , & \mathbf{a}_2 = \sigma(\mathbf{z}_2) \\ \mathbf{z}_3 = \mathbf{W}_3 \mathbf{a}_2 + \mathbf{b}_3 & , & \mathbf{a}_3 = \sigma(\mathbf{z}_3). \end{bmatrix} \tag{7.8}$$

with $\mathbf{a}_0 = \mathbf{x}$ and $\mathbf{a}_3 = \hat{\mathbf{y}}$.

This procedure is called *forward-propagation*, and it is used to compute output $\hat{\mathbf{y}}$ from input $\mathbf{x}$ and parameters Θ, where the parameters are defined as $\Theta = \{\theta_k\}_{k=1}^{m}$ withe $\theta_k = (\mathbf{W}_k, \mathbf{b}_k)$, i.e., the parameters of each layer. The reader can observe that the computation of $\hat{\mathbf{y}}$ is very fast, because in these equations, there are only multiplications and additions of vectors and matrices.

7.2.2   Training of Neural Networks

In order to train a neural network, parameters Θ are to be estimated. For this end, we have a training dataset of N samples $\{\mathbf{x}_i\}_{i=1}^{N}$ and its corresponding ideal classification $\{\mathbf{y}_i\}_{i=1}^{N}$. We distinguish between the ground truth $\mathbf{y}_i$ (ideal classification for sample $\mathbf{x}_i$) and the output of the neural network $\hat{\mathbf{y}}_i$ (real classification of $\mathbf{x}_i$). The idea of the training is to find parameters Θ so that the difference between $\mathbf{y}_i$ and $\hat{\mathbf{y}}_i$ is minimal for $i = 1, \ldots, N$. In neural networks, a loss function, $f_{\text{loss}}(\mathbf{y}_i, \hat{\mathbf{y}}_i)$ is used too compute the difference between ideal ($\mathbf{y}_i$) and real ($\hat{\mathbf{y}}_i$), so the training process can be stated as an optimization problem in which an objective function is to be minimized:

$$J(\Theta) = \frac{1}{N} \sum_{i=1}^{N} f_{\text{loss}}(\hat{\mathbf{y}}_i, \mathbf{y}_i) \rightarrow \min. \tag{7.9}$$

Intuitively, the loss function (7.9) can be based on the norm:

$$f_{\text{loss}}(\hat{\mathbf{y}}_i, \mathbf{y}_i) = \frac{1}{2}||\hat{\mathbf{y}}_i - \mathbf{y}_i||^2, \tag{7.10}$$

however, more sophisticated loss functions, like *cross-entropy* that minimizes the distance between both probability distributions are typically used [1, 4]:

$$f_{\text{loss}}(\hat{\mathbf{y}}_i, \mathbf{y}_i) = -\mathbf{y}_i \log(\hat{\mathbf{y}}_i) - (1 - \mathbf{y}_i) \log(1 - \hat{\mathbf{y}}_i) \tag{7.11}$$

for a two-class problem.[2] In order to find the parameters, i.e., to minimize objective function J, a method based on gradient descent can be used. We choose start parameters with random values $(\mathbf{W}_k, \mathbf{b}_k)$ for each layer, and they will be updated iteratively by small increments using the opposite direction of the gradient of the objective function J. The iterative method is summarized as follows:

1. Parameters $\boldsymbol{\theta}_k = (\mathbf{W}_k, \mathbf{b}_k)$, for $k = 1, \ldots, m$, are initialized with random values.

$$\mathbf{W}_k := \text{random matrix}(n_k \times n_{k-1}) \quad , \quad \mathbf{b}_k := \text{random vector}(n_k \times 1). \tag{7.12}$$

2. Layer outputs are computed for each training sample i using (7.7):

$$\mathbf{z}_{k,i} = \mathbf{W}_k \mathbf{a}_{k-1,i} + \mathbf{b}_k \quad , \quad \mathbf{a}_{k,i} = \sigma(\mathbf{z}_{k,i}). \tag{7.13}$$

---

[2]For a multi-class problem, we sum the loss for each class.

3. Derivatives of the parameters are computed:

$$\Delta \mathbf{W}_k = \frac{\partial J}{\partial \mathbf{W}_k} \quad , \quad \Delta \mathbf{b}_k = \frac{\partial J}{\partial \mathbf{b}_k}. \tag{7.14}$$

4. Parameters are updated using a learning rate α:

$$\mathbf{W}_k := \mathbf{W}_k - \alpha \, \Delta \mathbf{W}_k \quad , \quad \mathbf{b}_k := \mathbf{b}_k - \alpha \, \Delta \mathbf{b}_k. \tag{7.15}$$

5. The procedure is repeated from step 2 until convergence. For example, when

$$J(\mathbf{W}_1, \ldots, \mathbf{W}_m, \mathbf{b}_1, \ldots, \mathbf{b}_m) < \varepsilon. \tag{7.16}$$

We observe that step 2 corresponds to the forward-propagation of the neural network, that means we have an input $\mathbf{x}_i = \mathbf{a}_{0,i}$, and we evaluate forwards (from left to right) the layers of the network until we have the output $\hat{\mathbf{y}}_i = \mathbf{a}_{m,i}$, and the output $\mathbf{a}_k$ depends on input $\mathbf{a}_{k-1}$ and parameters $\mathbf{W}_k$ and $\mathbf{b}_k$. On the other hand, step 3 computes the increments $\Delta \mathbf{W}_k$ and $\Delta \mathbf{b}_k$ that are required in step 4 to update the parameters. Step 3 is performed using a *backward-propagation* approach, that means, we compute the derivatives at the output and we propagate them backwards (from right to left) through the layers using the chain rule for derivatives. The idea is that in the backward-propagation, we will have in each layer the increments $\Delta \mathbf{W}_k$ and $\Delta \mathbf{b}_k$ that depends on $\partial J / \partial \mathbf{a}_k$ and parameters $\mathbf{W}_k$ and $\mathbf{b}_k$ as illustrated in Fig. 7.5. Formerly,

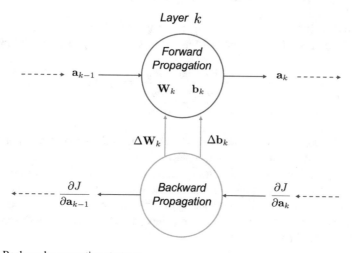

Fig. 7.5 Backward-propagation strategy

$$\Delta \mathbf{W}_k = \frac{\partial J}{\partial \mathbf{W}_k} = \underbrace{\frac{\partial J}{\partial \mathbf{a}_k} \frac{\partial \mathbf{a}_k}{\partial \mathbf{z}_k}}_{\gamma_k} \underbrace{\frac{\partial \mathbf{z}_k}{\partial \mathbf{W}_k}}_{\mathbf{a}_{k-1}} = \gamma_k \, \mathbf{a}_{k-1} \tag{7.17}$$

$$\Delta \mathbf{b}_k = \frac{\partial J}{\partial \mathbf{b}_k} = \underbrace{\frac{\partial J}{\partial \mathbf{a}_k} \frac{\partial \mathbf{a}_k}{\partial \mathbf{z}_k}}_{\gamma_k} \underbrace{\frac{\partial \mathbf{z}_k}{\partial \mathbf{b}_k}}_{1} = \gamma_k \tag{7.18}$$

The last derivatives ($\partial \mathbf{z}_k / \partial \mathbf{W}_k$ and $\partial \mathbf{z}_k / \partial \mathbf{b}_k$) are computed from (7.7), and the term γ_k can be written as:

$$\gamma_k = \underbrace{\frac{\partial J}{\partial \mathbf{a}_k}}_{\text{input}} \underbrace{\frac{\partial \mathbf{a}_k}{\partial \mathbf{z}_k}}_{\sigma_k'} = \frac{\partial J}{\partial \mathbf{a}_k} \mathbf{a}_k (1 - \mathbf{a}_k), \tag{7.19}$$

where the last term is the derivative of the activation function $a = \sigma(z) = 1/(1 + e^{-z})$ and $\sigma'(z) = a(1 - a)$. In this approach, the derivative $\partial J / \partial \mathbf{a}_m$, that is the input of the most right node of the backward-propagation schema, is computed directly from (7.9) and (7.10) with $\hat{\mathbf{y}} = \mathbf{a}_m$:

$$\frac{\partial J}{\partial \mathbf{a}_m} = \frac{\partial}{\partial \mathbf{a}_m} \left\{ \frac{1}{N} \sum_{i=1}^{N} \frac{1}{2} \|\hat{\mathbf{y}}_i - \mathbf{y}_i\|^2 \right\} = \frac{1}{N} \sum_{i=1}^{N} (\mathbf{a}_{m,i} - \mathbf{y}_i). \tag{7.20}$$

Thus, the last four equations can be used to estimate the increments $\Delta \mathbf{W}_k$ and $\Delta \mathbf{b}_k$ of step 3 and the updates of the parameters $\mathbf{W}_k$ and $\mathbf{b}_k$ in step 4, for $k = m, m - 1, \ldots, 1$.

We observe in the backward-propagation approach (Fig. 7.5), that the next layer, at the left, i.e., layer $(k - 1)$, needs $\partial J / \mathbf{a}_{k-1}$, that can be expressed as follows:

$$\frac{\partial J}{\partial \mathbf{a}_{k-1}} = \underbrace{\frac{\partial J}{\partial \mathbf{a}_k} \frac{\partial \mathbf{a}_k}{\partial \mathbf{z}_k}}_{\gamma_k} \underbrace{\frac{\partial \mathbf{z}_k}{\partial \mathbf{a}_{k-1}}}_{\mathbf{W}_k} = \gamma_k \, \mathbf{W}_k. \tag{7.21}$$

Typically, the iteration is stopped when the increments are small enough, that means that no significant update takes place.

Backpropagation is the learning algorithm normally used to train this type of network. Its goal is to minimize the error function constructed from the difference between the desired ($\mathbf{y}$) and modeled ($\hat{\mathbf{y}}$) output. In this section, we explained a simple backpropagation approach in four steps, where the increments $\mathbf{W}_k$ and $\mathbf{b}_k$ can be computed in an easy way.

7.2.3 Examples of Neural Networks

In this section, we give two examples:

1. The first one can be used to understand the basic operation and training of a neural network. The idea of this example is to design a neural network *from scratch*. Here, the reader can find all the details of the implementation of forward- and backward-propagation of a network with no sophisticated neural network library, only linear algebra is required (in our implementation for this end, we use the well-known numpy library[3] [47]). [→ Example 7.1 🐍]
2. The second example is to show, how we can design a neural network using a well-known library dedicated to machine learning, called 'sklearn'.[4] The idea of this example is to give details of the practice in typical applications that can be implemented with a neural network. [→ Example 7.2 🐍]

🐍 Python Example 7.1: In this example, we present a classification problem using simulated with using Gaussian distributions: four classes $(\omega_1, \ldots, \omega_4)$ for two features (x_1, x_2) (see Fig. 7.6). For this classification problem, we design a neural network with two hidden layers as shown in our example of Fig. 7.4. That means, the input layer has two entries, the hidden layers has 6 and 12 nodes respectively and the output layer as four elements $(\hat{y}_1, \ldots, \hat{y}_4)$. This example is provided for those readers that want to learn how a neural network is designed from scratch showing the general five steps explained in Eqs. (7.12)–(7.16): (1) random initialization of the parameters, (2) forward-propagation, (3) backward-propagation, (4) update of the parameters, and (5) repeat from step 2 until convergence. The details of all training steps can be found in `classifiers.py` of pyxvis Library, where the implementation is performed using only 'numpy' library. In this example, we used 80% of the available data for training and 20% for validation purposes.

Listing 7.1 : Neural network from scratch.

```python
import numpy as np
from pyxvis.learning.classifiers import nn_definition
from pyxvis.learning.classifiers import nn_forward_propagation, nn_backward_propagation
from pyxvis.learning.classifiers import nn_parameters_update, nn_loss_function
from pyxvis.io.plots import plot_features_y, show_confusion_matrix, plot_loss
from pyxvis.io.data import load_features

# Load training and testing data
(Xtrain,Ytrain,Xtest,Ytest)   = load_features('../data/G4/G4',categorical=1)
plot_features_y(Xtrain,Ytrain,'Training Subset')

# Definitions
N          = Xtrain.shape[1]   # training samples
n_0        = Xtrain.shape[0]   # number of inputs (X)
n_m        = Ytrain.shape[0]   # number of outputs (Y)
tmax       = 1000              # max number of iterations
alpha      = 10                # learning rate
loss_eps   = 0.01              # stop if loss<loss_eps
```

---

[3] See https://numpy.org.
[4] See https://scikit-learn.org [48].

```
nh        = [6,12]              # nodes of hidden layers
n         = [n_0]+nh+[n_m]      # nodes of each layer
m         = len(n)-1
ltrain    = np.zeros([tmax,1])  # training loss

# Training
t       = -1
train =  1
W,b     = nn_definition(n,N)                          # (step 1)
while train:
    t          = t+1
    a          = nn_forward_propagation(Xtrain,W,b)       # (step 2)
    dW,db      = nn_backward_propagation(Ytrain,a,W,b)    # (step 3)
    W,b        = nn_parameters_update(W,b,dW,db,alpha)    # (step 4)
    ltrain[t]  = nn_loss_function(a,Ytrain)               # (step 5)
    train      = ltrain[t]>=loss_eps and t<tmax-1

# Loss function on training and validation subsets
plot_loss(ltrain)

# Evaluation on training and testing subsets
a = nn_forward_propagation(Xtrain,W,b)      # output layer is a[m]
show_confusion_matrix(a[m],Ytrain,'Training',categorical=1)
a = nn_forward_propagation(Xtest,W,b)       # output layer is a[m]
show_confusion_matrix(a[m],Ytest,'Testing',categorical=1)
```

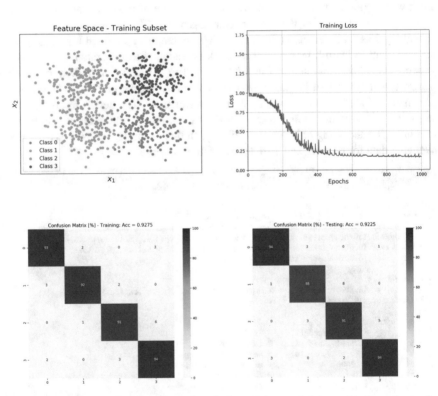

Fig. 7.6 Feature space, loss function, and confusion matrices on training and testing subsets of a four-class problem using a neural network. [→ Example 7.1 🔳]

The output of this code is in Fig. 7.6. We can see how the loss function is minimized and how are the samples of each classified (see confusion matrix). In this example, the accuracy in the testing dataset was 92.25%. It is difficult to obtain better results due to the overlapping of the classes. The reader can evaluate the performance of a new network with only one layer with 12 nodes (the line for nh definition should be nh = [12]).   □

The initially developed backward-propagation algorithm used the steepest descent first-order method as the learning rule. Nonetheless, other more powerful optimization approaches are in common use today. The reader is referred to [1] for more training approaches based on gradient descent strategies, like second-order methods, and stochastic methods like Adam [28], etc.

Python Example 7.2: This example is very similar to the previous one [→ Example 7.1 🐍] with two hidden layers as shown in Fig. 7.4. However, the implementation is given using sklearn library. The reader can study the syntax of Python class MLPClasifier for MLP neural networks. The training stage is performed by function fit and the prediction that evaluates the trained network on input data is performed by function predic. The optimization approach is performed by a *solver* that is in charge of estimate the parameters of the objective function. Similar to previous example, we use two hidden layers (with 6 and 12 nodes) respectively. In MLPClasifier, there are three possible solvers: (1) 'lbfgs' for quasi-Newton methods, (2) 'sgd' for stochastic gradient descent methods, and (3) 'adam' stochastic gradient descent method based on Adam approach [28].

Listing 7.2 : Neural network using sklearn library.

```python
from sklearn.neural_network import MLPClassifier
from pyxvis.io.plots import plot_features2, show_confusion_matrix, plot_loss
from pyxvis.io.data import load_features

# Load training and testing data
(Xtrain, Ytrain, Xtest, Ytest)    = load_features('../data/G4/G4')
plot_features2(Xtrain, Ytrain, 'Training+Testing Subsets')

# Definitions
alpha      = 1e-5      # learning rate
nh         = (6,12)    # nodes of hidden layers
tmax       = 2000      # max number of iterations
solver     = 'adam'    # optimization approach ('lbfgs','sgd', 'adam')

# Training
net = MLPClassifier(solver=solver, alpha=alpha, hidden_layer_sizes=nh,
                    random_state=1, max_iter=tmax)
print(Xtrain.shape)
print(Ytrain.shape)
net.fit(Xtrain, Ytrain)

# Evaluation
Ym  = net.predict(Xtrain)
show_confusion_matrix(Ym, Ytrain, 'Training')

Ys  = net.predict(Xtest)
show_confusion_matrix(Ys, Ytest, 'Testing')
```

The output of this code is the accuracy and confusion matrix evaluated on training and testing data. The results of the confusion matrices are very similar to the results given in the last example (see Fig. 7.6). The reader can evaluate the performance of a new network with only one layer with 12 nodes (the line for nh definition should be nh = (12,)). □

Some examples of neural networks in pyxvis Library are given in previous chapter (see Examples 6.6, 6.12, 6.13, and 6.14). In Example 6.6, the reader can append 'lr' to list ss_cl to evaluate the performance of a logistic regression in the classification of a two-class problem.

7.3 Convolutional Neural Network (CNN)

There are several deep architectures such as deep neural networks, convolutional neural networks, energy-based models, Boltzmann machines, deep belief networks, among others [3]. CNN (CNN), which were inspired by a biological model [30], is a very powerful method for image recognition [29].

In previous chapters, we studied how an X-ray image $\mathbf{X}$ can be classified: in the control quality of salmons, a region of an X-ray image has a fishbone or not, in baggage inspection, a region of interest shows a knife, a razor blade, a shuriken, and so on. The idea is to extract features of $\mathbf{X}$ and to classify them according to a classification strategy (see Fig. 7.7). In a problem of K classes, the output can be a value $y \in \{1 \dots K\}$ that gives the number of the class, or sometimes the output can be a K-element vector $\mathbf{y}$, where element y_k gives the probability that the image belongs to class k. If we use a classical neural network to solve the whole problem (representation and classification), the number of parameters to be learned could be so high, that the training process turns completely impractical (see Fig. 7.8). For this reason, CNNs have been developed, in which a strategy of concatenated layers is used (see Fig. 7.9). Using CNN, the number of parameters decreases considerably, the model is trained faster and the classification is more effective.

In this section, we review the basic concepts of CNN, and how a model is trained and tested. Finally, we give an example that can be used in the automated detection of casting defects (Fig. 7.10).

7.3.1 Basics of CNN

An X-ray testing method based on CNNs can be used to recognize an object of interest in an X-ray image. For example, we can have a region of interest $\mathbf{X}$ of an X-ray image of a casting to determine if this region has a defect or not. In this case, the CNN replaces feature extraction and classification with a single neural network.

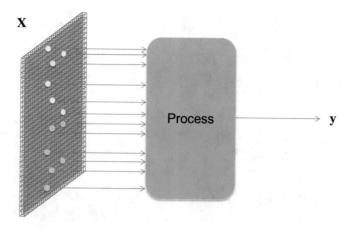

Fig. 7.7 An image **X** classified as vector **y** after a pattern recognition approach where features are extracted and classified using a classification strategy

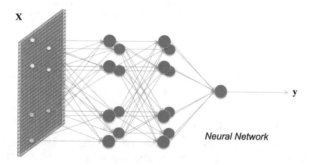

Fig. 7.8 Using a classical neural network approach (as shown in Fig. 7.4), every pixel of an image **X** can be connected to a node of a neural network of some hidden layers, however, the numbers of parameters of this architecture can be prohibited (in the first layer it could be N^4 connections for a $N \times N$-pixel input image) and a layer with $N \times N$ nodes

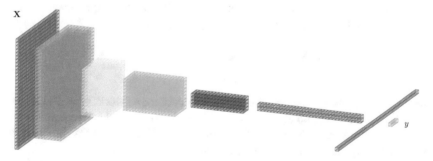

Fig. 7.9 Image classification using a convolutional neural network (CNN): concatenation of layers

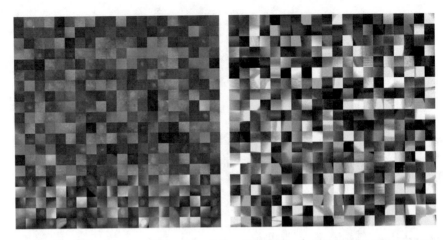

Fig. 7.10 Example of defects (lefts) and no-defects (right). It is clear that there are some patterns that can be easily detected (see, for example, defects that are bright bubbles with dark background and no-defects that are regular structures with edges), however, the recognition of both classes can be very difficult for low-contrast defects because they are very similar to homogenous no-defects

Thus, the CNN maps an input image $\mathbf{X}$ onto an output vector $\mathbf{y}$ of K elements, for K classes:

$$\mathbf{y} = \mathscr{F}_L(\mathbf{X}, \mathbf{w}). \tag{7.22}$$

Typically, element y_k gives the probability that image $\mathbf{X}$ belongs to class k. In our example, $K = 2$ (for two classes: defects and no-defects), and image $\mathbf{X}$ will be classified as defect if $y_1 > y_2$. Function $\mathscr{F}_L$ can be viewed as feed-forward network with L linear and non-linear layers f_l, for $l = 1 \ldots L$. The functions contain parameters

$$\mathbf{w} = (\mathbf{w}_1, \ldots, \mathbf{w}_L) \tag{7.23}$$

that can be discriminatively learned from training data: a set of input images $\mathbf{X}_i$ and their corresponding labels $\mathbf{z}_i$, for $i = 1, \ldots, n$, so that

$$\sum_i f_{\text{loss}}(\mathscr{F}_L(\mathbf{X}_i, \mathbf{w}), \mathbf{z}_i)/n \rightarrow \min. \tag{7.24}$$

Ideally, for an input of training data $(\mathbf{X}_i)$ the output of the network $(\mathbf{y}_i = \mathscr{F}_L(\mathbf{X}_i, \mathbf{w}))$ should be the the the corresponding label $(\mathbf{z}_i)$. Thus, f_{loss} is defined as *loss function* that gives a measurement of the error of the classification. This optimization problem can be solved using the backward-propagation approach [1, 18].

A method based on CNN can be understood as a set of L layers. Layer l (for $l = 1 \ldots L$), is a function $\mathbf{f}_l$ (with parameters $\mathbf{w}_l$) that processes an input image $\mathbf{X}_{l-1}$ in order to obtain an output image $\mathbf{X}_l$ (see Fig. 7.11):

$$\mathbf{X}_l = \mathbf{f}_l(\mathbf{X}_{l-1}, \mathbf{w}_l), \tag{7.25}$$

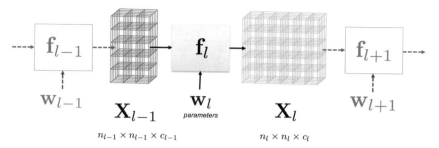

Fig. 7.11 Structure of layer l of a CNN according to (7.25): input image $\mathbf{X}_{l-1}$ is transformed into output image $\mathbf{X}_l$ using function $\mathbf{f}_l$ with parameters $\mathbf{w}_l$

where $\mathbf{X}_0 = \mathbf{X}$ is the input image of the whole CNN that we want to recognize. In our case, $\mathbf{X}_0$ is a grayscale X-ray image,[5] for this reason, the number of channels is one, $c_0 = 1$.

In our example, the input image corresponds to a cropped image of (e.g., 32×32 pixels as illustrated in Fig. 7.10 for two classes). In this CNN, the output of a layer is the input of the next layer. Thus, the output of each layer of the CNN can be defined as follows:

$$\mathbf{X}_l = \mathscr{F}_l(\mathbf{X}, \mathbf{w}) = f_l(f_{l-1}(\ldots f_1(\mathbf{X}, \mathbf{w}_1), \ldots, \mathbf{w}_{l-1}), \mathbf{w}_l)), \qquad (7.26)$$

that is a concatenation of l functions $f_1 \ldots f_l$. Without loss of generality, we will assume that the images are square, where the height and the width are n_l pixels. The images have one or more channels, i.e., image $\mathbf{X}_l$ is a 3D data structure with $n_l \times n_l \times c_l$ pixels, where c_l is the number of channels. Channel k of $\mathbf{X}_l$ is a matrix of $n_l \times n_l$ elements, and it is denoted as $\mathbf{X}_{k,l}$, for $k = 1 \ldots c_l$. The key idea of the CNN is that the output of the last layers correspond to high-level representations of the input image $\mathbf{X}$. These representations can be used in a classification process to recognize automatically the class of $\mathbf{X}$.

There are several types of layers that are normally used in CNN. Typically, the used layers are: convolution layer, pooling layer, rectified linear unit, and fully connected layer. They will be explained in further details.

• **Convolution Layer [conv]:** This layer corresponds to a linear convolution of input image $\mathbf{X}_{l-1}$ with a bank filter $\mathbf{F}_l$ and a bias $\mathbf{b}_l$. The filter bank $\mathbf{F}_l$ consists of a set of m_l 3D filters $\mathbf{F}_{k,l}$ of $p_l \times p_l \times q_l$ elements and a bias $b_{k,l}$ for $k = 1 \ldots m_l$. The parameters $\mathbf{w}_l$ of this layer are the elements of $\mathbf{F}_l$ and $\mathbf{b}_l$. Therefore, the number of parameters of each filter is $p_l \times p_l \times q_l + 1$, that means that the filter bank of layer l has $m_l(p_l^2 q_l + 1)$ parameters. These parameters are to be estimated in a learning process (as shown in Sect. 7.3.2). It is worth noting that the number of channels of the filter bank is the number of channels of the input image ($q_l = c_{l-1}$), and the

---

[5]For an X-ray image with pseudocolors, the number of channels of the input image can be three, $c_0 = 3$.

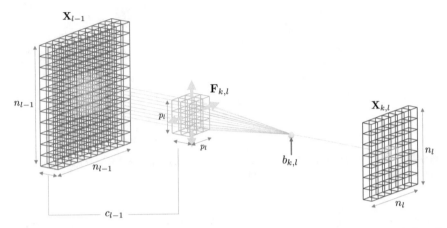

Fig. 7.12 Linear convolution: channel k of the output image $\mathbf{X}_l$ is computed as the convolution of the input image $\mathbf{X}_{l-1}$ with a filter $\mathbf{F}_{k,l}$ adding a bias $b_{k,l}$. In this example, $c_{l-1} = 2$ and $p_l = 3$

number of channels of the output image is the number of filters of the filter bank ($c_l = m_l$). Thus, the output for filter k is channel k of image $\mathbf{X}_l$:

$$\mathbf{X}_{k,l} = \mathbf{X}_{l-1} * \mathbf{F}_{k,l} + b_{k,l} \quad \text{for } k = 1 \dots m_l, \tag{7.27}$$

where '$*$' denotes the convolution operator. In other words, pixel (i, j) of channel k of the output image $\mathbf{X}_l$ is

$$X_{k,l}(i, j) = b_{k,l} + \sum_{u=1}^{p_l} \sum_{v=1}^{p_l} \sum_{w=1}^{q_l} X_{w,l-1}(i + u, j + v) F_{k,l}(u, v, w) \tag{7.28}$$

That means, that the size of the output image $\mathbf{X}_l$ will be a 3D matrix of $n_l \times n_l \times c_l$, with $n_l = n_{l-1} - p_l + 1$. The filtering process (for one channel of the output image) is illustrated in Fig. 7.12.

• **Pooling Layer [pool]:** This process is independently performed for each channel of input image $\mathbf{X}_{l-1}$. Therefore, the number of channels of input and output images are the same ($c_l = c_{l-1}$). In this case, the size of the image is reduced by representing a region of a channel with a scalar value. The output for each channel is defined as

$$X_{k,l}(i, j) = f_{\text{pool}}\{X_{k,l-1}(u, v) : (u, v) \in \Omega(i, j)\}. \tag{7.29}$$

Typically, the set of pixels $\Omega(i, j)$ is a sub-window of $\mathbf{X}_{k,l-1}$ of size $p_l \times p_l$ pixels which first pixel corresponds to the pixel (i, j) as illustrated in Fig. 7.13. The f_{pool} function can be the maximum, the mean, the ℓ^2 norm, etc. In our approach, we use the maximum operator, known as max-pooling', with no overlap, that means, each

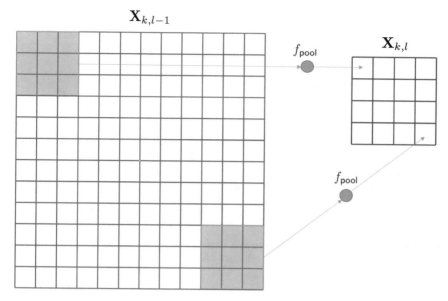

Fig. 7.13 Pooling. In this example, the dimension $n_{l-1} \times n_{l-1}$ of channel k of input image $\mathbf{X}_{l-1}$ is 12×12, and the dimension $n \times n$ of the neighborhood Ω is 3×3. Hence, the size of channel k of the output image $\mathbf{X}_l$ is 4×4, i.e., $n_l = 4$. Function f_{pool} could be in this example the maximum

channel is down-sampled non-linearly. Therefore, the use of this layer can efficiently reduce the computational time for upper layers.

• **Rectified Linear Unit [relu]:** Similar to pooling layer, this process is independently performed for each channel of input image $\mathbf{X}_{l-1}$ ($c_l = c_{l-1}$). In this case, the information of $\mathbf{X}_{l-1}$ is rectified by setting to zero all negative values. The key idea of a ReLU layer is to produce more discriminative representations avoiding negative scores [45]. Thus,

$$X_{k,l}(i, j) = \max\{0, X_{k,l-1}(i, j)\}. \tag{7.30}$$

The ReLU process is illustrated in Fig. 7.14 for channel k.

• **Fully Connected Layer [fc]:** This layer corresponds to a classic layer in a neuronal network (multi-layer perceptron), in which each output of previous layer is connected to new layer as explained in Sect. 7.2 and shown in Fig. 7.4: that means, each input node of a fully connected layer is the weighted sum of all outputs of previous layer plus a bias, and the output is this result after an activation function (see Fig. 7.15). The output is considered as a vector of n_l elements. Thus, if input layer has $n_{l-1} \times n_{l-1} \times c_{l-1}$, then there are $n_l \times n_{l-1}^2 \times c_{l-1}$ weights and n_l bias parameters that must be learned.

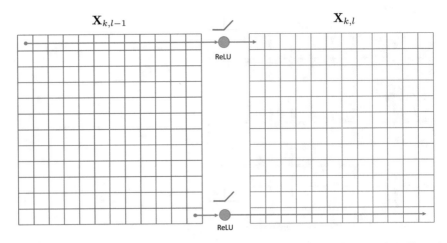

Fig. 7.14 Rectified linear unit. In this example, channel k of input image $\mathbf{X}_l$ is rectified. Channel k of the output image $\mathbf{X}_{l+1}$ has the same dimension: 6×6

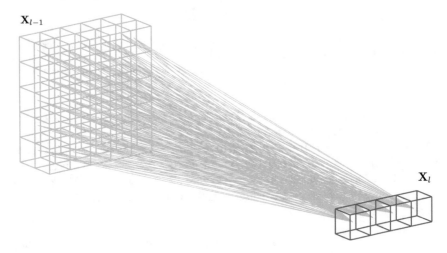

Fig. 7.15 Fully connected layer: all outputs of previous layer are connected one to one to the next layer. In this example, the input (green) and output (red) layers have $5 \times 5 \times 2$ and 4 pixels respectively that means, there are $5 \times 5 \times 2 \times 4 = 200$ connections (see gray lines)

In [39], a CNN model called Xnet is proposed to detect defects in aluminum castings (in Sect. 7.3.4, a similar example is given in Python).[6] The whole CNN is shown in Fig. 7.16. It includes a dropout block (**dropout**) that randomly turns off connections of the neural network during training. It has been shown that this technique reduces significantly the overfitting [59]. Typically, in a CNN model, layer $L - 1$ corresponds to a vector $\mathbf{s}$ with K elements, $[s_1 \ldots s_K]$:

---

[6] Another use of CNN in defects detection in castings can be found in [62].

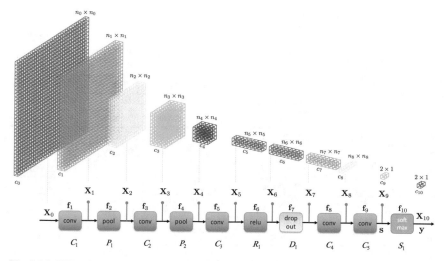

Fig. 7.16 XNet: CNN architecture proposed for automated detection of defects in castings [39]

$$\mathbf{s} = \mathbf{X}_{L-1} \qquad (7.31)$$

In the detection of defects, $K = 2$ because there are only two classes: defects and no-defects. In this approach, the output layer (layer L) is a 'softmax' block that is used to convert the scores of $\mathbf{s}$ into probabilities. Thus, $\mathbf{X}_L = \mathbf{y} = [y_1 \dots y_K]^\mathsf{T}$, where

$$y_k = f_{\text{softmax}}(s_k) = \frac{e^{s_k}}{\sum_{j=1}^{K} e^{s_i}} \quad \text{for } k = 1 \dots K. \qquad (7.32)$$

Using (7.26), it is clear that

$$\mathbf{y} = \mathscr{F}_L(\mathbf{X}, \mathbf{w}) \quad \text{and} \quad \mathbf{s} = \mathscr{F}_{L-1}(\mathbf{X}, \mathbf{w}). \qquad (7.33)$$

Table 7.1 summarizes Xnet [39], where the input image $\mathbf{X}_0 = \mathbf{X}$ is an image of $32 \times 32 \times 1$ pixels. The CNN consists of ten layers with five linear convolutional layers ($C_1, \dots, C_5$), two pooling layers with maximum operator (P_1, P_2), one ReLU layer (R_1), one dropout layer (D_1), and one softmax layer (S_1). As we can see in Table 7.1, our CNN has 5.7×10^5 parameters that must be estimated in a learning stage.

7.3.2 Learning in CNN

As we mentioned in previous section, CNN maps an input image $\mathbf{X}$ on an output vector $\mathbf{y} = \mathscr{F}_L(\mathbf{X}, \mathbf{w})$, where function $\mathscr{F}_L$ can be viewed as a sequence of linear

Table 7.1 Convolutional neural network Xnet [39]

Layer		Function $\mathbf{f}_l$			Output $\mathbf{X}_l$
l	Name	Type	$m_l(p_l \times p_l \times q_l)$	Parameters	$n_l \times n_l \times c_l$
0	Input	–	–	–	$32 \times 32 \times 1$
1	C_1	conv	$64(7 \times 7 \times 1)$	3.200	$26 \times 26 \times 64$
2	P_1	pool-max	2×2	0	$13 \times 13 \times 64$
3	C_2	conv	$128(5 \times 5 \times 64)$	204.928	$9 \times 9 \times 128$
4	P_2	pool-max	2×2	0	$4 \times 4 \times 128$
5	C_3	conv	$256(3 \times 3 \times 128)$	295.168	$2 \times 2 \times 256$
6	R_1	relu	–	0	$2 \times 2 \times 256$
7	D_1	dropout	–	0	$2 \times 2 \times 256$
8	C_4	conv	$64(2 \times 2 \times 256)$	65.600	$1 \times 1 \times 64$
9	C_5	conv	$2(1 \times 1 \times 64)$	130	$1 \times 2 \times 1$
10	S_1	softmax	–	0	$1 \times 2 \times 1$
Total				569.026	

and non-linear functions $f_1, \ldots, f_L$, that depend on parameters $\mathbf{w} = (\mathbf{w}_1, \ldots, \mathbf{w}_L)$ as defined in (7.26) and (7.33).

Learning consists of estimating parameters $\mathbf{w}$ from 'learning data'. The output of this process is the set of parameters $\mathbf{w}$. On the other hand, testing is used to evaluate the performance of the trained model on 'testing data', i.e., the learned model (with fixed parameters $\mathbf{w}$) is used to classify new data. The output of this process is the classification of each testing sample. For this end, a set of annotated X-ray images is available. Thus, for each image, $\mathbf{X}$, vector $\mathbf{z}$—the ground truth of the classification— is given by an expert. Similar to vector $\mathbf{y}$, the output of the CNN, vector $\mathbf{z}$ has K elements. The value z_k is '1' if image $\mathbf{X}$ belongs to class k, otherwise z_k is '0'.

In order to reduce the computation time of learning process, typically, a hold-out protocol is used. The standard hold-out evaluation protocol is based on disjoint learning and testing data, i.e., images that are present in the learning set are not allowed to be in the testing set. We denote the X-ray images and their labels $(\mathbf{X}, \mathbf{z})$ as

- Learning: $\{\mathbf{X}_{\text{learn}}^{(i)}, \mathbf{z}_{\text{learn}}^{(i)}\}_{i=1}^{n_{\text{learn}}}$, with n_{learn} learning samples.
- Testing: $\{\mathbf{X}_{\text{test}}^{(i)}, \mathbf{z}_{\text{test}}^{(i)}\}_{i=1}^{n_{\text{test}}}$, with n_{test} testing samples.

The learning set is subdivided into two disjoint subsets: training set $(\mathbf{X}_{\text{train}}^{(i)}, \mathbf{z}_{\text{train}}^{(i)})$, for $i = 1, \ldots, n_{\text{train}}$, and validation set $(\mathbf{X}_{\text{val}}^{(i)}, \mathbf{z}_{\text{val}}^{(i)})$, for $i = 1, \ldots, n_{\text{val}}$, with $n_{\text{learn}} = n_{\text{train}} + n_{\text{val}}$. Typically, 75–80% of the learning data for training and 25–20% for validation.

The training data is used to estimate the parameters $\mathbf{w}$ of our model as follows. The output of the CNN is $\mathbf{y}_{\text{train}}^{(i)} = \mathscr{F}_L(\mathbf{X}_{\text{train}}^{(i)}, \mathbf{w})$. It is clear, that ideally $\mathbf{y}_{\text{train}}^{(i)}$ should be $\mathbf{z}_{\text{train}}^{(i)}$. Thus, parameters $\mathbf{w}$ can be estimated by minimizing the objective function:

$$e_{\text{train}} = \frac{1}{n_{\text{train}}} \sum_{i=1}^{n_{\text{train}}} f_{\text{loss}}(\mathbf{y}_{\text{train}}^{(i)}, \mathbf{z}_{\text{train}}^{(i)}), \tag{7.34}$$

where f_{loss} is a loss function. This optimization problem can be iteratively solved using the backward-propagation approach [1, 18] as explained for a classic neural network in Sect. 7.2: We start with initial random values $\mathbf{w}_{(0)}$ for the first iteration, and the parameters in epoch j are estimated according to the parameters in previous epoch $j - 1$ and an incremental update:

$$\mathbf{w}_{(j)} = \mathbf{w}_{(j-1)} + \Delta\mathbf{w}_{(j)}. \tag{7.35}$$

For each epoch of the training process, a new version of the parameters $\mathbf{w}_{(j)}$ is estimated. For validation purposes, the subset $(\mathbf{X}_{\text{val}}^{(i)}, \mathbf{z}_{\text{val}}^{(i)})$ for $i = 1, \ldots, n_{\text{val}}$ is used. The error

$$e_{\text{val}} = \frac{1}{n_{\text{val}}} \sum_{i=1}^{n_{\text{val}}} f_{\text{loss}}(\mathbf{y}_{\text{val}}^{(i)}, \mathbf{z}_{\text{val}}^{(i)}), \tag{7.36}$$

is computed, with $\mathbf{y}_{\text{val}}^{(i)} = \mathscr{F}_L(\mathbf{X}_{\text{val}}^{(i)}, \mathbf{w})$, where $\mathscr{F}_L$ is evaluated using $\mathbf{w} = \mathbf{w}_{(j)}$. At the beginning of the training, both errors e_{train} and e_{val} usually decrease. Nevertheless, when the learning process starts overfitting, the error e_{val} starts to increase. This epoch will be denoted by j^*. Thus, the training process is stopped when e_{val} is minimum, and our parameter vector will be $\mathbf{w} = \mathbf{w}_{(j^*)}$.

7.3.3   Testing in CNN

After learning stage, we can test the CNN using the testing dataset: $(\mathbf{X}_{\text{test}}^{(i)}, \mathbf{z}_{\text{test}}^{(i)})$ for $i = 1, \ldots, n_{\text{test}}$. The images of this dataset were not used in the learning stage. There are several approaches that can be used to classify $\mathbf{X}_{\text{test}}^{(i)}$. Obviously, one of them is to use the representation of the last layer:

$$\text{class}(\mathbf{X}_{\text{test}}^{(i)}) = \underset{k}{\text{argmax}} \left\{ \mathbf{y}_{\text{test}}^{(i)}(k) \right\} \tag{7.37}$$

where $\mathbf{y}_{\text{test}}^{(i)}(k)$ is the kth element of $\mathbf{y}_{\text{test}}^{(i)} = \mathscr{F}_L(\mathbf{X}_{\text{test}}^{(i)}, \mathbf{w})$.

In addition, the high-level representations that are present in layers $l < L$ can be used in a classification approach as well. For this purpose, a descriptor $\mathbf{d}$ can be defined as a vector of $n_l^2 c_l \times 1$ elements that contains all elements of $\mathbf{X}_l$ by stacking its columns:

$$\mathbf{d} = s(\mathbf{X}_l, \mathbf{w}) = s(\mathscr{F}_l(\mathbf{X}, \mathbf{w})), \tag{7.38}$$

where $s(\cdot)$ is the stack function. Descriptor $\mathbf{d}$ can be used to train another kind of classifier (a KNN, for example). In [39], the best results were obtained using $\mathbf{d} = s(\mathbf{X}_7)$, i.e., $l = L - 3$.

A classifier h can be designed using the descriptors and the labels of the learning set. Thus, the classifier can be learned using $(\mathbf{d}_{\text{learn}}^{(i)}, \mathbf{z}_{\text{learn}}^{(i)})$, where $\mathbf{d}_{\text{learn}}^{(i)} = s(\mathbf{X}_{\text{learn}}^{(i)})$ for $i = 1 \ldots n_{\text{learn}}$ according to (7.38). After training, $h(\mathbf{d}_{\text{learn}}^{(i)})$ should ideally be $\mathbf{z}_{\text{learn}}^{(i)}$.

7.3.4 Example of CNN

Python Example 7.3: In this example, we test a very simple CNN architecture for the detection of casting defects. The dataset used in this example, called `C1`, is a subset of the dataset used in [39].[7] It contains the *easiest* patches of the original dataset, i.e., those patches that are easy to classify. In this example, there are 8200 patches of 32×32 pixels, 80% for testing and 20% for testing (both subsets with 50% defects and 50% no-defects). The idea of this example is to train an easy dataset with a simple CNN architecture. Thus, the reader in less than 2 min (with around 20–30 epochs) can have a trained model with excellent performance. After this training, it is possible to train more complex architectures with more challenging datasets. In this example, we define the CNN architecture using CNN of pyxvis Library. In this definition, there are n typical 2D convolutional blocks of layers with Keras[8] functions `Conv2D` with a ReLU activation, `BatchNormalization`, `MaxPooling2D`, and `Dropout` with a rate of 25%. The 2D convolution of block i, for $i = 1, \ldots, n$, is with d_i kernels of $p_i \times p_i$ pixels. In our example, we define the three blocks using variables $\mathbf{p} = [7, 5, 3]$ and $\mathbf{d} = [4, 12, 8]$. After the n blocks of convolutional layers, we add m fully connected layers, each layer has f_j elements. In our case, $m = 1$ and $\mathbf{f} = [12]$. If we want to have two fully connected layers, one with 12 elements and another with 4, we define $\mathbf{f} = [12, 4]$. Finally, CNN of pyxvis Library includes a fully connected layer of the number of classes to be recognized (in our case is 2), and a 'softmax' block. In Fig. 7.17, we can see the architecture.

---

[7] The original dataset has 47.520 patches, and it can be downloaded from https://domingomery.ing. puc.cl/material/.

[8] Keras is a library built on top of TensorFlow. It consists of a set of API functions written in Python for building deep learning models. See https://keras.io.

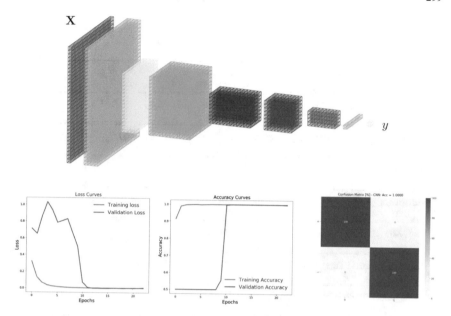

Fig. 7.17 CNN architecture, learning curves, and confusion matrix. In this example, all defects and no-defects from testing subset are correctly classified. [→ Example 7.3 🐍]

Listing 7.3 : Convolutional Neural Network.

```
from   pyxvis.learning.cnn import CNN

# execution type
type_exec    = 0 # training & testing

# patches' file for training and testing
patches_file = '../data/C1/C1'

# architecture
p = [7,5,3]    # Conv2D mask size
d = [4,12,8]   # Conv2D channels
f = [12]       # fully connected

# training and testing
CNN(patches_file,type_exec,p,d,f)
```

The first output of this code is the definition of the architecture (that corresponds to the diagram of Fig. 7.17):

Layer (type)	Output Shape	Param #
conv2d_1 (Conv2D)	(None, 4, 32, 32)	200
batch_normalization_1 (Batch	(None, 4, 32, 32)	128

max_pooling2d_1 (MaxPooling2	(None, 4, 16, 16)	0
dropout_1 (Dropout)	(None, 4, 16, 16)	0
conv2d_2 (Conv2D)	(None, 12, 16, 16)	1212
batch_normalization_2 (Batch	(None, 12, 16, 16)	64
max_pooling2d_2 (MaxPooling2	(None, 12, 8, 8)	0
dropout_2 (Dropout)	(None, 12, 8, 8)	0
conv2d_3 (Conv2D)	(None, 8, 8, 8)	872
batch_normalization_3 (Batch	(None, 8, 8, 8)	32
max_pooling2d_3 (MaxPooling2	(None, 8, 4, 4)	0
dropout_3 (Dropout)	(None, 8, 4, 4)	0
flatten_1 (Flatten)	(None, 128)	0
dense_1 (Dense)	(None, 12)	1536
batch_normalization_4 (Batch	(None, 12)	48
activation_1 (Activation)	(None, 12)	0
dropout_4 (Dropout)	(None, 12)	0
dense_2 (Dense)	(None, 2)	26
activation_2 (Activation)	(None, 2)	0

```
==================================================================
Total params: 4,118.0
Trainable params: 3,982.0
Non-trainable params: 136.0
```

The architecture of our model has almost 4000 trainable parameters. The learning curves and the confusion matrix are shown in Fig. 7.17. We can see that in this very simple example, the accuracy is 100% (a perfect detection). The reader that wants to try a more difficult example can download the dataset of [39] (see footnote 7) and implement a network similar to Xnet (see Table 7.1). In the original code of Example 7.3, the lines of the new code should be:

```
# execution type
type_exec    =  0 # training & testing

# patches' file for training and testing
patches_file = 'wacv_castings.mat'

# architecture
p = [7,5,3]          # Conv2D mask size
d = [64,128,256]     # Conv2D channels
f = [64,32]          # fully connected

# training and testing
CNN(patches_file,type_exec,p,d,f)
```

For this dataset, the model is trained after 1 h (and 30 epochs)> The achieved accuracy is 87.78% very similar to the reported accuracy in [39].   □

7.4 Pre-trained Models

Pre-trained models are deep learning models that have been already trained on large datasets of one domain and can be used as-is on other domains with no additional training. In this section, we explain how to use pre-trained models in X-ray testing.

7.4.1 Basics of Pre-trained Models

In X-ray testing, it is possible to use sophisticated models that have been already trained on other domains (e.g., recognition of common objects in color images). The idea is to use part of the trained model on new domains, such as X-ray images. One of the most popular datasets of color images of common objects is ImageNet [55]. ImageNet consists of an annotated collection of color images of very common objects (like cars, bicycles, trucks, cat, dogs, etc.). ImageNet has 1000 classes of objects with approx. 1000 images per class for training purposes. The dataset has been widely used in competitions of object recognition algorithms. The trained models are typically available as open-source models. The architecture of these models has many layers, and the last one corresponds to a structure of 1000 elements that are used to distinguish the 1000 classes. In the testing stage, if an input image contains

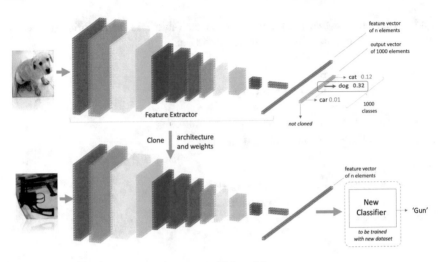

Fig. 7.18 Classification strategy using a pre-trained model

a dog, the output, i.e., a vector with 1000 element, should have the element corresponding to the class 'dog' the maximal value as illustrated in Fig. 7.18-Top. This pre-trained model can be used not only to recognize images that content objects that belong to these 1000 classes but also to recognize other objects. The key idea of this approach is as follows: The last layer of the pre-trained models is not used because it has been trained to recognize objects that do not belong to the new domain. The last layer is replaced by a new fully connected layer or a simple classifier, such as KNN or a SVM, that is designed to classify the classes of the new domain.

The strategy behind this idea is that the pre-trained model should extract in the first layers relevant visual information of the input image and could give us a good representation of the images of the new domain.

To illustrate this idea, we present now a well-known approach that shows what happens inside the layers of a CNN [13, 67]. This approach can visualize how the images are represented in the network and give insight into the layers. It consists of the estimation of a synthetic input image that maximizes the activation of a certain element (pixel) of a layer.

Thus, we can visualize what kind of images activates each element of our network. The estimation is an optimization problem, that starts with a random input image, and after some iterations, the solution converges using a gradient descent algorithm.[9] In this process, the weights of the CNN are fixed, i.e., we do not train the CNN, we only find an input image that maximizes a certain pixel of a layer of the CNN. To illustrate this visualization, we use a pre-trained CNN called VGG16 [58]. The architecture of VGG16 is the following:

---

[9]An implementation of this idea can be found in https://keras.io/examples/ conv_filter_visualization/. The Figs. 7.19 and 7.23 were done using this implementation.

```
Layer (type)                    Output Shape                    Param #
=======================================================================
input_1 (InputLayer)            (None, None, None, 3)     0

block1_conv1* (Conv2D)          (None, None, None, 64)    1792

block1_conv2 (Conv2D)           (None, None, None, 64)    36928

block1_pool (MaxPooling2D)      (None, None, None, 64)    0

block2_conv1 (Conv2D)           (None, None, None, 128)   73856

block2_conv2 (Conv2D)           (None, None, None, 128)   147584

block2_pool (MaxPooling2D)      (None, None, None, 128)   0

block3_conv1* (Conv2D)          (None, None, None, 256)   295168

block3_conv2 (Conv2D)           (None, None, None, 256)   590080

block3_conv3 (Conv2D)           (None, None, None, 256)   590080

block3_pool (MaxPooling2D)      (None, None, None, 256)   0

block4_conv1* (Conv2D)          (None, None, None, 512)   1180160

block4_conv2 (Conv2D)           (None, None, None, 512)   2359808

block4_conv3 (Conv2D)           (None, None, None, 512)   2359808

block4_pool (MaxPooling2D)      (None, None, None, 512)   0

block5_conv1 (Conv2D)           (None, None, None, 512)   2359808

block5_conv2* (Conv2D)          (None, None, None, 512)   2359808

block5_conv3 (Conv2D)           (None, None, None, 512)   2359808

block5_pool (MaxPooling2D)      (None, None, None, 512)   0
=======================================================================
```

Table 7.2 Images of GDXray+ used in our experiments

Set		Gun	Shuriken	Blade	Others
Training	Series	B0049	B0050	B0051	B0078
	Images	1–200	1–100	1–100	1–500
Validation	Series	B0079	B0080	B0081	B0082
	Images	1–50	1–50	1–50	1–200
Testing	Series	B0079	B0080	B0081	B0082
	Images	51–150	51–150	51–150	201–600

VGG16 was developed in the year 2014 and it is one of the most powerful CNN architecture for vision problems.[10] It consists of five blocks with conv and pool-max layers (the last fully connected layers are not given in the previous description). In Fig. 7.19, we show some images that activate certain elements of the layers that have a '*' in this description, i.e., block1_conv1, block3_conv1, block4_conv1, and block5_conv2. In this case, VGG16 was trained for ImageNet [55]. We can observe in this figure, the complexity of the generated patterns: the more complex is the image, the deeper is the layer. Moreover, for the last layers, we can recognize some patterns like birds and feathers!

The idea of using pre-trained models is simple and powerful as we will show in our examples. Simple because the weights of the pre-trained models are available in public repositories or deep learning libraries (like Keras (see footnote 8)) and powerful because good results can be achieved with no implementation difficulty.

7.4.2 Example of Pre-trained Models

In this section, we show how to use pre-trained models in the recognition of threat objects (in baggage inspection). In our experiments, there are three objects: handguns, shuriken (ninja stars), and razor blades. Each category of objects defines a class (Gun, Shuriken, and Blade). Furthermore, there is a fourth class called Other for other objects and background. All X-ray images used in our experiments belong to the GDXray+. As shown in Table 7.2, there are three different sets of images: training, testing, and validation sets. For training, X-ray images of GDXray+ series B0049, B0050, B0051, and B0078 must be used for classes Gun, Shuriken, Blade, and Others respectively. For validation, in case that a method has some parameters to be tuned, it is allowed to use the first 50 images of GDXray+ series B0079, B0080, and B0081 for Gun, Shuriken, and Blade respectively and the first 200 images of folder B0082 for Others. For testing, the last 100 images of GDXray+ series B0079, B0080 and B0081 for Gun, Shuriken, and Blade respectively and the last 400 images of folder B0082 for Others have to be used.

---

[10]See an application in the automated weld defect recognition based on VGG16 in [34].

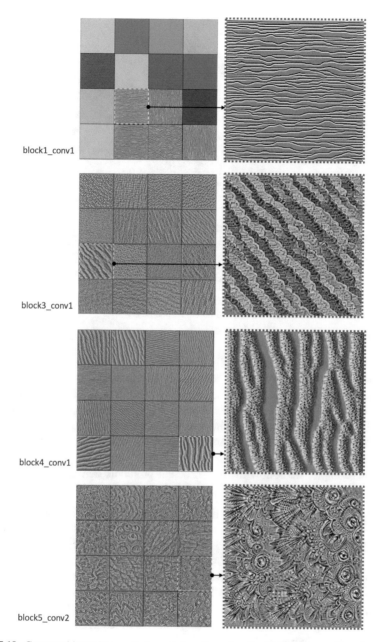

block1_conv1

block3_conv1

block4_conv1

block5_conv2

Fig. 7.19 Generated input images that maximizes the activation of 16 elements (pixels) of some layers in VGG16. A zoom of image with a blue square is presented in Fig. 7.19

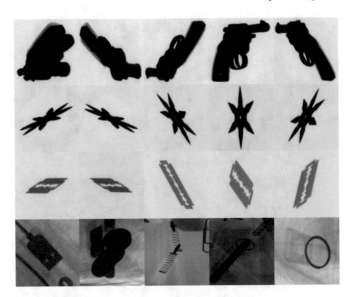

Fig. 7.20 Some training X-ray images used in our experiments. Each row represents a labeled class (handguns, shuriken, razor blades, and others respectively)

The GDXray+ dataset is especially challenging due to the high intra-class variability between training and testing images of positive classes (see some examples for guns, shuriken, and razor blades in Figs. 7.20 and 7.21 for training and testing respectively). Indeed, training images of positive classes contain just the object with a clean background. In contrast, testing images corresponding to these classes show a noisy background that may allow any discriminative model to classify them as the class Others.

In our example, we follow the experimental protocols defined in [41] for two recognition tasks:

- **Four-class Classification:** In the first task, we have to design a classifier that is able to recognize the four mentioned classes: (1) Gun, (2) Shuriken, (3) Blade, and (4) Others. We define $K = 4$ as the number of classes. The classifier has to be trained using the trained data. The parameters of the classifier (if any) can be tuned using the validation only. The performance of the method is reported using the testing data as follows: The elements of the $m \times m$ confusion matrix are defined as $C(i, j)$ for $i = 1 \ldots K$ and $j = 1 \ldots K$, where $C(i, j)$ means the number of images of class i (in the testing data) classified as class j. The accuracy of each class is defined as

$$\eta_i = \frac{C(i, i)}{\sum_{j=1}^{4} C(i, j)}. \tag{7.39}$$

The total accuracy is the average:

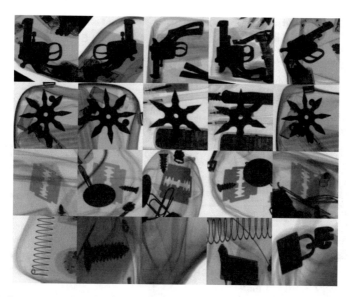

Fig. 7.21 Some testing X-ray images used in our experiments. Each row represents a labeled class (handguns, shuriken, razor blades, and others respectively)

$$\eta = \frac{1}{4} \sum_{i=1}^{4} \eta_i. \tag{7.40}$$

- **Detection of three threat objects:** In the second task, we have to design three different detectors (binary classifiers) : (1) one for Gun, (2) one for Shuriken, and (3) one for Blade. For each detector, there is a target (e.g., Shuriken for second detector). Each detector can be understood as a two-class problem: one class (called the positive class) is the target, and the another class (called the negative class) is the rest. Similar to previous problem, training data must be used to train the detectors, validation data can be used to tune the detectors' parameters (if any), and testing data have to be used to measure the final performance of the detectors. For the second detector (i.e., Shuriken), for example, in our database according to Table 7.2, there are 100 images for the positive class and $200 + 100 + 500 = 800$ images for the negative class that can be used for training purposes. In this example, the validation can be performed using 50 images for the positive class and $50 + 50 + 200 = 300$ images for the negative class. Finally, for the testing of the second detector, there 100 images for the positive class and $100 + 100 + 400$ for the negative class. The performance must be given in terms of precision–recall (Pr, Re) considering all images of the testing set. The variables precision and recall are defined in Eqs. (6.41) and (6.38) respectively. Ideally, a perfect detection means all existing targets are correctly detected without any false alarms, i.e., $Pr = 1$ and $Re = 1$. The values (Pr, Re) that maximizes the score $Q = \sqrt{Pr \times Re}$ are reported. As average performance, we define

Table 7.3 Precision and recall for each detector using pre-trained models [→ Example 7.4 🐍]

Method*	Classifier	Features	Gun		Shuriken		Blade		Q	All
			Pr	Re	Pr	Re	Pr	Re	η_Q	η
AlexNet[2] [29]	svm-rbf	4096	1.00	1.00	1.00	1.00	1.00	1.00	1.00	1.00
DenseNet121[2] [24]	svm-rbf	1024	1.00	1.00	1.00	1.00	1.00	1.00	1.00	1.00
GoogleNet[2] [60]	svm-lin	1024	1.00	1.00	1.00	1.00	0.86	1.00	0.98	0.99
InceptionV3[0] [61]	svm-rbf	2048	1.00	1.00	1.00	1.00	1.00	1.00	1.00	1.00
MobileNet[2] [23]	svm-rbf	1280	1.00	1.00	1.00	1.00	1.00	1.00	1.00	1.00
RCNN_ILSVRC$_{13}^2$ [16]	knn1	4096	1.00	1.00	1.00	1.00	0.62	0.70	0.89	0.95
ResNet50[2] [20]	svm-rbf	2048	1.00	1.00	1.00	1.00	1.00	1.00	1.00	1.00
ShuffleNet[2] [69]	svm-rbf	544	1.00	1.00	1.00	1.00	1.00	1.00	1.00	1.00
SqueezeNet[2] [25]	svm-rbf	1000	1.00	1.00	1.00	1.00	1.00	1.00	1.00	1.00
VGG16[1] [58]	svm-rbf	1000	1.00	1.00	1.00	1.00	1.00	1.00	1.00	1.00
VGG19[1] [58]	knn1	1000	0.97	1.00	0.99	1.00	0.93	1.00	0.98	1.00
Xception[0] [8]	knn1	2048	0.71	0.50	0.89	0.85	0.57	0.40	0.65	0.80
ZfNet512[2] [67]	nn	1024	1.00	1.00	1.00	1.00	0.91	1.00	0.98	1.00
AISM [41, 54]**	–	–	0.97	0.97	0.95	0.96	0.99	0.99	0.94	0.96

(*)Output layer 0: Keras, layer before softmax, 1: ONNX, layer after softmax, 2: ONNX, layer before softmax

(**)Best non-deep-learning method

$$\eta_Q = \frac{1}{3} \sum_{i=1}^{3} Q_i, \qquad (7.41)$$

where $i = 1 \ldots 3$ means the classes Gun, Shuriken and Blade respectively.

🐍 Python Example 7.4: In this example, we follow the experimental protocol defined above according to Table 7.2 in two recognition tasks: four-class classification and detection three threat objects. For these tasks, we evaluate pre-trained model MobilNet [23] using five different classifiers (knn1, knn3, svm-lin, svm-rbf, and nn). The idea is to use the pre-trained model to extract features of each image, and classify the images according to the extracted features as illustrated in Fig. 7.18. The features are extracted by function extract_prt_features[11] of pyxvis Library. Other pre-trained models (such as AlexNet [29], GoogleNet [60], VGG16 and VGG19 [58] among others) are implemented in pyxvis Library as well. In the following code, the reader can see how easy is to define the training, validation, and testing datasets (see Table 7.2) using init_data and append_data of pyxvis Library. In this implementation, the validation set is used to evaluate the performance of each classifier (defined as the average of η in (7.40) and η_Q in (7.41)). Thus, the classifier that achieves the best performance on validation subset is used to report the performance on testing dataset.

---

[11]This function is used to extract the features of all images that are in a folder. For a single image, function extract_prt_features_img of pyxvis Library can be used.

Listing 7.4 : Pre-trained models.

```python
import numpy as np
from sklearn.metrics import accuracy_score
from pyxvis.learning.pretrained import prt_model, extract_prt_features
from pyxvis.io.gdxraydb import DatasetBase
from pyxvis.learning.classifiers import clf_model, define_classifier
from pyxvis.learning.classifiers import train_classifier, test_classifier
from pyxvis.learning.evaluation import precision_recall
from pyxvis.io.data import init_data, append_data
from pyxvis.io.plots import print_confusion

gdxray      = DatasetBase()
path        = gdxray.dataset_path + '/Baggages/'
model_id    = 6 # 0 ResNet50, 1 VGG16, 2 VGG19, ... 6 MobileNet, ... 13 RCNN_ILSVRC13
output_layer = 2 # 0 Keras-Last, 1 ONNX-Last, 2 ONNX-Previous

# Classifiers to evaluate
ss_cl       = ['knn1','knn3','svm-lin','svm-rbf','nn']
(model,size,model_name) = prt_model(model_id,output_layer)
X49 = extract_prt_features(model_id,output_layer,model,size,model_name,path+'B0049/')
X50 = extract_prt_features(model_id,output_layer,model,size,model_name,path+'B0050/')
X51 = extract_prt_features(model_id,output_layer,model,size,model_name,path+'B0051/')
X78 = extract_prt_features(model_id,output_layer,model,size,model_name,path+'B0078/')
X79 = extract_prt_features(model_id,output_layer,model,size,model_name,path+'B0079/')
X80 = extract_prt_features(model_id,output_layer,model,size,model_name,path+'B0080/')
X81 = extract_prt_features(model_id,output_layer,model,size,model_name,path+'B0081/')
X82 = extract_prt_features(model_id,output_layer,model,size,model_name,path+'B0082/')

best_performance = 0 # initial value
for i in range(len(ss_cl)):
    cl_name = ss_cl[i]
    print('\nEvaluation of '+cl_name+' using '+model_name+'...')
    (Q_v,Q_t) = (0,0) # initial score Q values for validation and testing
    for j in range(4):
        if j==0:
            (c0,c1,c2,c3) = (1,0,0,0)
            st = 'Gun'
        elif j==1:
            (c0,c1,c2,c3) = (0,1,0,0)
            st = 'Shuriken'
        elif j==2:
            (c0,c1,c2,c3) = (0,0,1,0)
            st = 'Blade'
        elif j==3:
            (c0,c1,c2,c3) = (0,1,2,3)
            st = 'All'

        print('building dataset for '+st+' using ' + model_name +' ...')
        # Training data
        (X,d)   = init_data(X49[0:200],c0)            # Gun
        (X,d)   = append_data(X,d,X50[0:100,:],c1)    # Shuriken
        (X,d)   = append_data(X,d,X51[0:100,:],c2)    # Blade
        (X,d)   = append_data(X,d,X78[0:500,:],c3)    # Other
        # Validation data
        (Xv,dv) = init_data(X79[0:50,:],c0)           # Gun
        (Xv,dv) = append_data(X,d,X80[0:50,:],c1)     # Shuriken
        (Xv,dv) = append_data(X,d,X81[0:50,:],c2)     # Blade
        (Xv,dv) = append_data(X,d,X82[0:200,:],c3)    # Other
        # Testing data
        (Xt,dt) = init_data(X79[50:150],c0)           # Gun
        (Xt,dt) = append_data(X,d,X80[50:150,:],c1)   # Shuriken
        (Xt,dt) = append_data(X,d,X81[50:150,:],c2)   # Blade
        (Xt,dt) = append_data(X,d,X82[200:600,:],c3)  # Other

        print('training '+cl_name+' for '+st+' using ' + model_name +' ...')
```

```
    (name,params) = clf_model(cl_name)              # function name and parameters
    clf           = define_classifier([name,params]) # classifier definition
    clf           = train_classifier(clf,X,d)       # classifier training
    ds_v          = test_classifier(clf,Xv)         # clasification of validation
    ds_t          = test_classifier(clf,Xt)         # clasification of testing
    print('Results - ' + st + ' ('+cl_name+') for the detectors:')
    if j<3: # detection of three treat objects
        # performance on validation subset
        (pr_v,re_v) = precision_recall(dv,ds_v)
        Q_v         = Q_v + np.sqrt(pr_v*re_v)
        print(f'Pr_val   = {pr_v:.4f}')
        print(f'Re_val   = {re_v:.4f}')
        # performance on testing subset
        (pr_t,re_t) = precision_recall(dt,ds_t)
        Q_t         = Q_t + np.sqrt(pr_t*re_t)
        print(f'Pr_test  = {pr_t:.4f}')
        print(f'Re_test  = {re_t:.4f}')
    else:
        # summary of three detections
        Q_v = Q_v/3            # score Q on validation
        print(f'Q_val    = {Q_v:.4f} of all detectors')
        Q_t = Q_t/3            # score Q on testing
        print(f'Q_test   = {Q_v:.4f} of all detectors')
        # four-class classification
        print('Results - ' + st + ' ('+cl_name+') for the 4-class classifier:')
        acc_v = accuracy_score(dt,ds_t)
        print(f'Acc_val  = {acc_v:.4f}')
        acc_t = accuracy_score(dv,ds_v)
        print(f'Acc_test = {acc_t:.4f}')
        print(f'Acc_t = {acc_t:.4f}')
        print_confusion(dt,ds_t)
    performance = (acc_v+Q_v)/2
    if performance>best_performance:
        print(f'performance = {performance:.4f} *** new max ***')
        best_performance = performance
        best_Q           = Q_t
        best_acc         = acc_t
        best_clf         = cl_name
print('Best result: classifier = '+best_clf)
print(f'                  Q_test = {best_Q:.4f}')
print(f'                acc_test = {best_acc:.4f}')
```

The output of this pre-trained model and other ones implemented in pyxvis Library is shown in Table 7.3. We can see that many of the pre-trained models achieve a perfect performance of 100%. This result is very relevant because the implementation of this solution can be performed in a couple of hours (the pre-trained models are already trained, we only need to extract the features and train a classifier like SVM). It is worthwhile to mention that the best non-deep learning method based on handcrafted features (see AISM [54] in Table 7.3), developed after several months of work for this task, achieves 4–6% less of performance.   □

7.5 Transfer Learning

The use of transfer learning in X-ray testing is similar to the use of pre-trained models (explained in Sect. 7.4). Here, however, the pre-trained model is *re-trained* in a smart way using a low number of X-ray images [66].

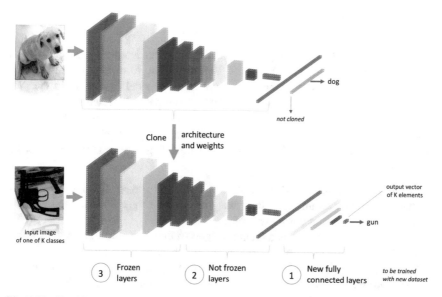

Fig. 7.22 Classification strategy using a transfer learning (see explanation in Sect. 7.5.2)

7.5.1 Basics of Transfer Learning

The idea is to use, in X-ray images, models that have been trained on other domains (e.g., ImageNet [55]). The main difference with pre-trained models (see Sect. 7.4) is that we can re-train the models using a *fine-tuning* approach. In fine-tuning, we re-train a model for the new domain using as initial weights the pre-trained weights of the model (instead of random initial values). Thus, we can take advantage of the pre-trained weights that have been obtained after a sophisticated training process with millions of images. In fine-tuning, the initial (pre-trained) weights are updated using a training approach with images of the new domain (X-ray images). A good example is given in [2], where transfer learning has been used in baggage inspection by fine-tuning AlexNet and GoogleNet.

Usually, sophisticated deep learning models can be trained successfully thanks to the great power of today's computers and also because there are a huge number of annotated images available. However, sometimes it is very difficult to have both of them. For example, in X-ray testing, it is very common to have datasets with hundreds or thousands (and not millions) of X-ray images. In addition, there are many students or universities that do not have access to such a powerful computer. For these reasons, transfer learning is a very attractive alternative: we can re-train a sophisticated model using a low number of X-ray images on a regular computer.

7.5.2   Training in Transfer Learning

In order to train a deep learning model using a transfer learning strategy, we can use the approach illustrated in Fig. 7.22. Before starting to re-train the model it is necessary to clone the first layers of the pre-trained model (as we do in the pre-trained model strategy outlined in Sect. 7.4). Following Fig. 7.22, we add new layers (typically fully connected layers) to the cloned model (see layers ①). Now, we divided the cloned model into two parts: the not frozen layers (see layers ②) and the frozen layers (see layers ③). Usually, in the training stage, we can use the following three steps:

1. Layers ① are trained and the rest of the weights (layers ② and ③) are not changed during training, i.e., their weights are the original pre-trained weights (they remain constant during training).
2. Layers ① + ② are fine-tuned (using weights of first step as initial weight values) and the rest of the weights (layers ③) are not changed during training.
3. (Optional Step) Layers ① + ② + ③ are fine-tuned (using weights of second step as initial weight values). This step can be performed in case we have enough images to train the whole model.

Using the same approach addressed in Sect. 7.4.1 to visualize the activation of the elements of the CNN layers, in Fig. 7.23, we show the synthetic input image generated for a specific element of one of the last layers for the original VGG16 model (trained for ImageNet) and for the fine-tuned model (trained with threat objects). The reader can observe that the patterns are very similar, however, the second one seems to be adapted to the new domain.

7.5.3   Example of Transfer Learning

Python Example 7.5: In this example, we follow the strategy outlined in Fig. 7.22 for transfer learning in a problem of recognition of threat objects. For this end, we use a set of images of threat objects that has four classes (Guns, Shuriken, Blades, and Others) divided into training (with 600 images per class) and testing subsets (with 100, 100, 100, and 400 images for each corresponding class). In our example, we use MobileNet [23] as base model (that has 87 layers) and four extra layers: the first one is `GlobalAveragePooling2D` (that joins the base model with the extra fully connected layers), two fully connected layers defined by variable `fc_layers` with 32 and 16 nodes each, and a final fully connected layer with softmax as output with 4 nodes (because, in this example, there are four classes). Thus, the new model has totally 91 layers. In this example, we have ① the new layers, ② the not frozen layers, and ③ the frozen layers. The training strategy follows the method mentioned above in three steps: in ①, we train 4 layers, in ① + ②, we train 9 layers, and in ① + ② + ③, we train 91 layers. The number of epochs in each step

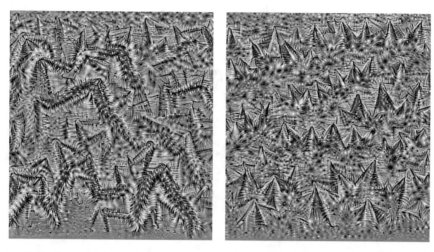

Fig. 7.23 Generated synthetic input images for VGG16 that maximize the activation of layer block5_conv2 filter '0': (left) using base model VGG16 trained with ImageNet, (right) using VGG16 fine-tuned with threat objects, the shapes of the shuriken are remarkable. Left image is illustrated in Fig. 7.19 in a blue square

is defined by variable `nb_epochs`, in our case is $[40,40,40]$. We could decide that the last step is not necessary by defining `nb_epochs` $=[40,40,0]$.

Listing 7.5 : Transfer learning.

```
from pyxvis.learning.transfer import generate_training_set, tfl_train
from pyxvis.learning.transfer import tfl_model, tfl_define_model, tfl_testing_accuracy
from pyxvis.io.plots import plot_confusion

# Definitions
path_dataset = '../images/objects'
nb_classes    = 4              # number of classes of the recognition problem
batch_size    = 10             # batch size in training
nb_epochs     = [40,40,40]     # epochs for Training-1, Training-2, Training-3
                               # 1st value: epochs for new layers only,
                               # 2nd value: epochs for new and top layers of base model,
                               # 3rd value: epochs for all layers
                               # (eg [50,0,0], [40,50,0], etc.)
train_steps  = 10
val_steps    = 5
fc_layers    = [32, 16]        # fully connected layers after froozen layers
img_size     = [224,224]       # size of the used images
val_split    = 0.2             # portion of training set dedicated to validation,
                               # 0 means path_dataset/val is used for validation
opti_method  = 1               # optimzer > 1: Adam, 3: SGD
base_model   = 1               # 1: MobileNet, 2: InceptionV3, 3: VGG16, 4: VGG19,
                               # 5: ResNet50, 6: Xception, 7: MobileNetV2,
                               # 8: DenseNet121, 9: NASNetMobile, 10: NASNetLarge
nb_layers    = -5              # layers 0... nb_layers-1 will be frozen, negative
                               # number means the number of top layers to be unfrozen
augmentation = 0.05            # 0 : no data augmentation, otherwise it is range for
                               # augmentation (see details in generate_training_set)

# Base model (last layer is not included removed)
```

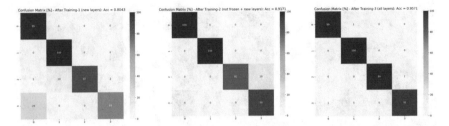

Fig. 7.24 Confusion matrix and accuracy on testing subset using transfer learning: after training-1 (layers ①), the accuracy is 80.43%, training-2 (layers ① + ②), the accuracy is 91.71%, and training-3 (layers ① + ② + ③), the accuracy is 95.71% according to diagram of Fig. 7.22. [→ Example 7.5 📎]

```
bmodel      = tfl_model(base_model)

# New model with dense fully connected layers
model       = tfl_define_model(bmodel,fc_layers,nb_classes)

# Training and validation sets
(train_set,
 val_set)   = generate_training_set(val_split, augmentation, batch_size,
                                     path_dataset, img_size)

# Training: Transfer learning
(model,
 confusion_mtx,
 acc)       = tfl_train(bmodel,model,opti_method, nb_layers,
                        train_set,train_steps,val_set,val_steps,nb_epochs,
                        path_dataset,nb_classes,img_size)

# Accuracy in testing set using best trained model
plot_confusion(confusion_mtx,acc,'Top Model: Testing in Threat Objects',0,nb_classes)
```

The output of this code is in Fig. 7.24 in which we show the confusion matrices and accuracy on testing dataset after each training step. We observe how the accuracy is incremented after each step.   □

7.6 Generative Adversarial Networks (GANs)

Generative Adversarial Networks (GANs) have been used successfully in the last years to generate realistic synthetic data [7, 9, 27]. In X-ray testing, we use GAN to simulate X-ray images, for example, as data augmentation in training data to increase the number of samples of some underrepresented class, or as new data in a training course for human inspectors. Some applications of simulated X-ray images using GAN can be found in [38] for the simulation of casting defects[12] and in [56, 65, 70, 72] for the simulation of threat objects. The simulated X-ray images using GAN are very realistic as we can see in Figs. 7.25 and 7.26 for defects and shuriken respectively.

---

[12]GAN solutions have been used in other kinds of defects, see for example, [46].

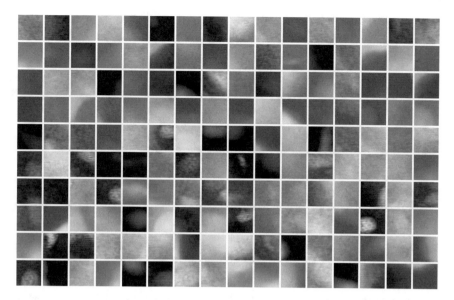

Fig. 7.25 Simulated defects in aluminium castings using GAN. [→ Example 7.6 🐍]

7.6.1 Basics of GAN

The key idea of GAN is simple as we can see in Fig. 7.27: it consists of a *generator* and a *discriminator* working together. The generator will be used to produce a synthetic X-ray image from a noise source, whereas the discriminator will be in charge to determine if an input image is real or fake. Thus, the discriminator should differentiate the real (training) images from the synthetic ones generated by the generator.

Both generator and discriminator are trained following a zero-sum game schema [17]. In the zero-sum game schema, the generator and the adversary (discriminator) compete against other. For this end, we define a noise source as an image $\mathbf{Z}$ of $p \times p$ pixels. The generator ($\mathscr{G}$) is a neural network function based on auto-encoders [1] that takes noise image $\mathbf{Z}$ and transforms it into a fake image $\mathbf{X_F}$ of $n \times n$ pixels:

$$\mathbf{X_F} = \mathscr{G}(\mathbf{Z}). \tag{7.42}$$

On the other hand, the discriminator ($\mathscr{D}$) is a function based on a neural network that takes an input image $\mathbf{X}$ of $n \times n$ pixels and gives as output a value y that corresponds to the probability that $\mathbf{X}$ is a real image; 1 means that $\mathbf{X}$ is real, 0 is fake:

$$y = \mathscr{D}(\mathbf{X}). \tag{7.43}$$

Fig. 7.26 Simulated shuriken using DCGAN from 0 to 15000 iterations. [→ Example 7.6 🐍]

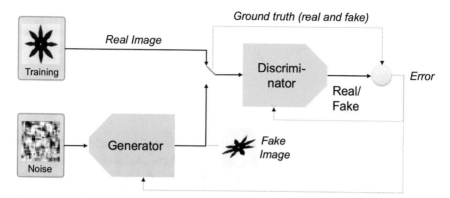

Fig. 7.27 Diagram of a GAN model. The generator produces a fake image from a noise source, whereas the discriminator distinguishes real from fake images. In the training stage, the error is used to increase the performance of both generator and discriminator (see dashed lines). Once the model is learned, the generator alone can be used to generate realistic synthetic images

7.6.2   Training of GAN

In training stage, two sets are available: *(i)* a set of m real images, $\mathbf{X}_R(1)\ldots\mathbf{X}_R(m)$, and *(ii)* a set of m noisy images $\mathbf{Z}(1)\ldots\mathbf{Z}(m)$, from them a set of m fake images are computed $\mathbf{X}_F(1)\ldots\mathbf{X}_F(m)$ using (7.42). Thus, the discriminator is learned by maximizing:

$$J_D = \sum_{i=1}^{m} \log\left[\mathscr{D}(\mathbf{X}_R(i)\right] + \sum_{i=1}^{m} \log\left[1 - \mathscr{D}(\mathbf{X}_F(i)\right] \rightarrow \max \qquad (7.44)$$

The first sum is maximal when the real images are classified as 'real', whereas the second sum is maximal when the synthetic images are classified as 'fake'. Thus, the idea of (7.44) is to classify as 1 the real images and as 0 the synthetic ones.

The aim of the generator is to model the distribution of the training dataset. Since the goal of the generator is to generate fake images that fool the discriminator, the generator will do a good job if the generated fake images are classified by the discriminator as 'real'. Thus, the generator is trained by minimizing the following objective function:

$$J_G = \sum_{i=1}^{m} \log\left[1 - \mathscr{D}(\mathbf{X}_F(i)\right] \rightarrow \min. \qquad (7.45)$$

In this case, the sum is minimal when the synthetic images are classified as 'real'. That means that the generated fake images will be so realistic that the discriminator will classify them as 'real'. In the training stage, objective functions J_D and J_G are playing a *min-max* game, the reader can see that the second sum of (7.44) is equal to the sum of (7.45), however, in the first case, we are trying to maximize it (the discriminator should recognize that the synthetic images are fake), whereas in the second one, the aim is to minimize it (the generator wants to fool the discriminator).

7.6.3 Implementation of GAN

GAN models can be easily implemented using Deep Convolutional Generative Adversarial Networks (DCGAN) [49], where both generator and discriminator are sequential models [49]. In DCGAN, the architectures of discriminator $\mathscr{D}$ and generator $\mathscr{G}$ are CNNs as illustrated in Figs. 7.9 and 7.28 respectively. In each step of the generator, the $\mathbf{Z}$ is upsampled and convoluted. The upsampling process can be achieved using the 2D Transposed Convolution [68]:

- **2D Transposed Convolution [trans_conv]:** This layer corresponds to a convolution that increases the dimension of the input image as illustrated in Fig. 7.29. In general, for an input image $\mathbf{X}$ of $n \times n$ pixels and a convolutional kernel $\mathbf{K}$ of $m \times m$ pixels, the output $\mathbf{Y} = \mathbf{X} \star \mathbf{K}$ is defined as follows:

$$\mathbf{Y}(i_1 : i_2, j_1 : j_2) = \mathbf{Y}(i_1 : i_2, j_1 : j_2) + X(i, j)\mathbf{K}, \qquad (7.46)$$

where

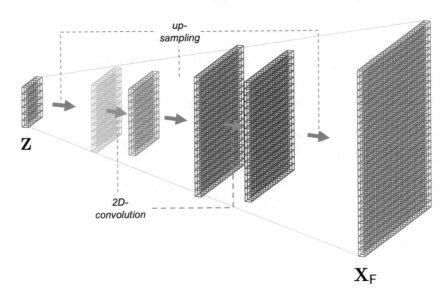

Fig. 7.28 Architecture of a generator based on deep convolutional neural networks in a GAN model. Input $\mathbf{Z}$ is a small noise image, and output $\mathbf{X_F}$ is a (larger) synthetic image

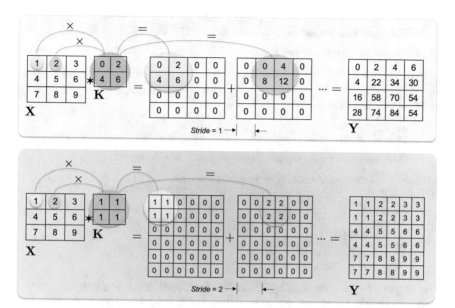

Fig. 7.29 Example of 2D transposed convolution: (Top) with stride $r = 1$, (Bottom) with stride $r = 2$

$$
\begin{aligned}
i_1 &= (i-1)r + 1 \\
i_2 &= (i-1)r + m \\
j_1 &= (j-1)r + 1 \\
j_2 &= (j-1)r + m
\end{aligned}
\tag{7.47}
$$

for $i = 1 \ldots n$ and $j = 1 \ldots n$. The stride, i.e., the number of pixels that the kernel moves to right and down in each step, is given by variable r. As shown in Fig. 7.29, by setting r to m (the size of the kernel), and defining all elements of $\mathbf{K}$ as 1, we can repeat the rows and columns of $\mathbf{X}$ by size $m \times m$.

7.6.4 Example of GAN

Python Example 7.6: In this example, we simulate X-ray images of shuriken using class DCGAN[13] of pyxvis Library. As training data, we use a dataset of 10.640 real images of shuriken of 32×32 pixels (stored in file shuriken_32x32.npy). The real images were extracted from GDXray+ and augmented using rotation and reflection.

---

[13]Based on the implementation of https://github.com/eriklindernoren/Keras-GAN/blob/master/dcgan/dcgan.py.

Listing 7.6 : Generative Adversarial Network.

```
from pyxvis.learning.gan import DCGAN

gan_proc = 0 # 0 training , # 1 testing

# Training
if gan_proc == 1: # Training
    path_file = '../data/shuriken_32x32.npy' # file of real patches
    epochs    = 15000                 # number of epochs
    interval  = 250                   # saving intervals
    dcgan     = DCGAN(path_file)
    dcgan.train(epochs=epochs, batch_size=32, save_interval=interval)

else: # Testing (one generation of simulated images)
    size = 32               # size of the image, eg. 32 for 32x32 pixels
    # trained model h5 file
    gan_weights_file = '../output/GAN/models/gan_model_015000.h5'
    N = 200                 # number of synthetic images to be generated
    dcgan = DCGAN(size)
    dcgan.load_gan_model(gan_weights_file)
    dcgan.save_gan_examples()
    dcgan.save_synthetic_images('output',N)
```

The output of this code is in Fig. 7.26. We can see that the similarity of the synthetic generated X-ray images with trained GAN model is very high after 5000 iterations. In the generation, only the generator is used (and not the discriminator). If we want to generate N new images with the trained generator, in our code, variable `gan_proc` must be set to 1 and variable N must be set to the number N (e.g., $N = 200$). Figure 7.25 shows a GAN simulation of casting defects with this code, in which the dataset `casting_defects_28x28.npy`[14] was use.   □

7.7   Detection Methods

In this section, we address relevant methods of *object detection* that have been published in the last years. The idea of the detection methods is to locate and recognize object instances in real images.

7.7.1   Basics of Object Detection

In computer vision, we distinguish between *image classification* and *image detection* as shown in Fig. 7.30. In X-ray testing, both concepts can be explained as follows:

- **Image classification:** The purpose of *image classification* in X-ray testing is to assign an X-ray image to one class. For example, in image classification, an X-ray

---

[14]The file can be downloaded from https://domingomery.ing.puc.cl/material/.

image can be classified as a 'handgun', that means, in the X-ray image the classifier has found a handgun (here, the classes could be 'handgun', 'knife', and other threat objects), or, in another example, a small sub-image of an X-ray image of an aluminum castings is classified as 'defect' (here, the classes could be 'defect' and 'no-defect'). Image classification is typically used when there is only one object per image to be recognized. An example is illustrated in Fig. 7.30a. The location of the recognized object is not given in image classification, it is well assumed that the object is in the center of the image, but obviously, this is not always true. In image classification using deep learning, as explained in Sect. 7.3, the input image is fed into a CNN that gives a feature vector (of dimension 4096, for example). The vector is the input of a classifier, e.g., a fully connected layer, with K outputs, where K is the number of classes to be recognized. Thus, the input is an image and the output is a category label. In case that both classification and localization of the object in the input image are required, there are some approaches with one fully connected layer for the classification and another fully connected layer for the localization that gives the coordinates and the dimensions of a rectangle that contains the recognized object [26], where the second fully connected layer is treated as a regression problem, where the output is continuous values instead of a class.

- **Image Detection:** On the other hand, in *image detection*, more than one object can be recognized in an X-ray image and the location of each recognized object is given by a *bounding box*, i.e., a rectangle that encloses the detected object defined by the coordinates of the center of the rectangle (x, y) and its dimen-

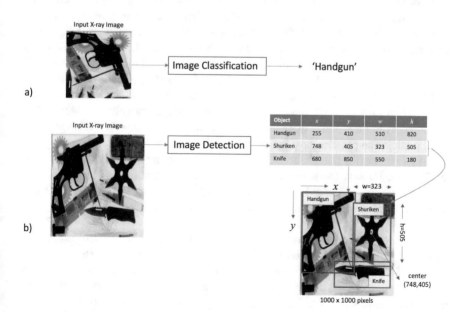

Fig. 7.30 Image detection and image classification

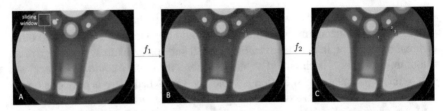

Fig. 7.31 Object detection using CNN and sliding-windows: The original X-ray image 'A' with three defects (see red rectangles) is processed by f_1 using a sliding-window approach. For each position of the detection windows (orange square), a patch is extracted and classified as 'defect' or 'no-defect' by a trained CNN. The dots of image 'B' show in red and green the center of the patches that were classified as 'defect' and 'no-defect' respectively. Using image processing approaches the dots of image 'B' are processed by f_2 to detect the 'defect' regions. In this example, all defects were correctly detected with no false alarm

sions, width and height, (w, h), where all four variables are given in pixels (see for example, the red rectangle of the shuriken in Fig. 7.30b, where the center is in $(x = 748, y = 405)$ and the dimensions are $w = 323$ and $h = 505$ pixels). In example of Fig. 7.30b, image detection is able to recognize a set of objects (see table and red bounding boxes). Typically, a probability of detection is computed for each recognized object (that can be understood as a new column in the table), so that the final output corresponds to those objects that have a probability greater than a threshold.

A simple strategy based on sliding-window methodology has been proposed some years ago for image detection based on image classification. An example is illustrated in Fig. 7.31 for defect detection in aluminum castings [38]. In this approach, a detection window (see the orange square in Fig. 7.31-A) is sledded over an input image in both horizontal and vertical directions, and for each localization of the detection window, a classifier decides to which class belongs the corresponding portion of the image according to its representation. Here, the classifier is a CNN, as explained in Sect. 7.3.4, that is used to identify one of two classes: 'defects' or 'no-defects'. For this end, a huge number of patches of each class is used to train the CNN model. The patches have the same size of the detection window, and they can contain a defect (for the 'defect' class) or not (for the 'no-defect' class) as shown in Fig. 7.10. Finally, the locations of the X-ray image that have been detected as 'defects' (see green dots in Fig. 7.31-B) are analyzed using image processing. Thus, we can determine which regions of the image are defects or not (see detected regions in Fig. 7.31-C).

It is worthwhile to mention, that this approach requires the classification of a huge number of patches. In addition, if the size of the objects to be detected varies, the sliding-windows approach must be performed for different patch-sizes. In this case, the computational time could be prohibited. For these reasons, new approaches that overcome this problem have been developed in the last years. In this section, we will cover them. They can be subdivided into two groups [26]:

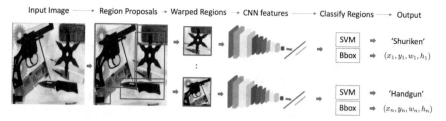

Fig. 7.32 R-CNN strategy

(*i*) **Detection in two stages:** In the first step of these approaches, called *region proposal*, a method is used to determine regions of the input image in which an object can be present. In sliding-windows, for example (explained above), this step corresponds to an exhaustive search, however, there are other methods, e.g., R-CNN [16] that propose some regions instead of analyzing all possible patches of the input image. In the second step of these approaches, called *final classification*, a CNN is used to classify the regions that have been proposed by the first step. In Sect. 7.7.2 of this chapter, we address these region-based methods like R-CNN [16], Fast R-CNN [15], and Faster R-CNN [53] that uses this two-stage strategy.

(*ii*) **Detection in one stage:** In these approaches, there is a single CNN that is trained to both location and classification, i.e., prediction of bounding boxes and estimation of the class probabilities of the detected bounding boxes. This group of approaches corresponds to the state of the art in detection methods because they are very effective and very fast. They are the best-performing and most representative deep learning-based object detection models, as stated in [71]. In this chapter, we address most representative methods, namely, YOLO [52] in Sect. 7.7.3 (versions YOLOv2 [50], YOLOv3 [51] and YOLOv4 [6]), SSD [37] in Sect. 7.7.4 and RetinaNet [32] in Sect. 7.7.5. We give a brief description of these detection models and their principal differences.

7.7.2   Region Based Methods

In this section, we address those methods from the first group that perform object detection in two stages. These methods are region-based methods because the first step is the region proposal, and the second is the final classification. To this group belong R-CNN [16], Fast R-CNN [15] and Faster R-CNN [53]. We include in this section an additional method called Mask R-CNN [22] that is an instance segmentation approach. They will be described in further details.

• **R-CNN:** In R-CNN (Regions with CNN features), there is a step that proposes potential regions and another step that classify them into the classes to be recognized [16], as shown in Fig. 7.32. By selecting regions in the first step, we avoid to classify of a huge number of patches as mentioned in Sect. 7.7.1 for sliding-windows approach.

The first step of R-CNN, called *Selective Search*, is based on a method proposed in [63] that generates candidates of bounding boxes for use in object recognition, i.e., they are regions that have high probability of being an object. They are called regions of interest or RoIs. The method uses complementary image regions that consider many image conditions. Selective search is based on image processing and it consists of three stages: *(i)* Capture all scales: Many potential regions are generated in all possible scales. *(ii)* Diversification: a diverse set of strategies is used to merge similar regions together. *(iii)* Fast to compute: Final regions are proposed in a hierarchical order. In the proposed approach [16], 2000 RoIs are selected, many of them are noisy, but the recall is high, that means that most of the true objects are selected. One of the problems of this method is that the selective search is not learned, it is fixed, and it could be useful to learn which regions are relevant for a given application. In addition, the approach can be very slow because each of the 2000 RoIs must be analyzed independently. In order to speed up this step, sharing computing with Spatial Pyramid Pooling networks (SPP) can be used as proposed in [21]. In the second step of R-CNN, a trained CNN model based on AlexNet [29] is used to extract from each RoI a feature vector of 4096 elements as explained in Sect. 7.4. All RoIs are warped to 227×227 pixels because the CNN requires a fixed square size for the input images. The 4096-element feature vector extracted of a RoI is used by a SVM classifier that is trained to determine the class of the region. In addition, CNN predicts a correction of the bounding boxes because originally they are not correctly located by the selective search approach. Thus, SVM classifier is in charge of class determination, whereas the location is given by the corrected location of the original RoI that has been detected by the SVM.

R-CNN is much faster than a sliding-window approach, however, to analyze 2000 RoIs is still very computationally expensive and cannot be implemented in real time. It has been reported that for the testing stage, it requires around 50 s per image [16].

• **Fast R-CNN:** In order to avoid the mentioned problems, the same author proposed a faster approach called Fast R-CNN [15], as shown in Fig. 7.33. In this approach, two improvements are presented: *(i)* The selective selection of RoIs is performed by using a CNN that gives a feature map of the same size of the input image. The RoIs are partitions of this feature map that are warped into fixed-length vectors using a 'RoI pooling layer', i.e., a max-pooling layer with a pool size that does not depend on the input size. *(ii)* Instead of a SVM that classifies the extracted feature vector for every single RoI, fully connected layers are used for each RoI to determine both the category and the location of the object. Thus, the objects in the image are detected by using *two sibling output layers*, one for establishing the category of the detected bounding box, and another to correct the location of the bounding box. In comparison with R-CNN, the computational time of Fast R-CNN is significantly decreased (to 2.3 s per image) mainly because the CNN is executed just once for the input image and not for every RoI. Moreover, the accuracy of the detection is increased and the training time is around ten times faster.

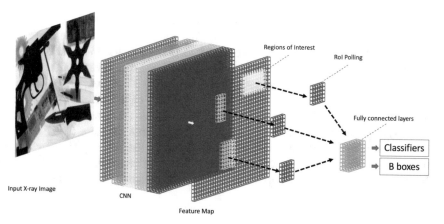

Fig. 7.33 Fast R-CNN strategy

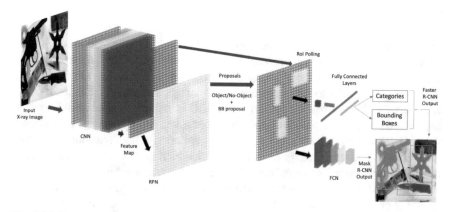

Fig. 7.34 Faster R-CNN and mask R-CNN strategies

• **Faster R-CNN:** The main drawback of Fast R-CNN is the computational time of
the first step dedicated to region proposal, it is around 85% of the total detection
time. In order to speed up the first step, Faster R-CNN (see Fig. 7.34) was proposed
in [53]. Faster R-CNN includes an attention mechanism called Region Proposal
Network (RPN), that is used to predict the RoIs from CNN features. That means the
input image is fed into a CNN to obtain a feature map that is fed into the RPN that
is trained to infer regions proposal. RPN outputs are two for each RoI: *(i)* a probability that the proposal is an object (it is a score that is used to determine whether
the detection is an object or not) and *(ii)* a preliminary bounding box. Afterwards,
a RoI pooling layer makes the final classification of the object in one of the categories and gives a correction of the preliminary bounding box. In Faster R-CNN,
the detection time is decreased to 0.2 s per image. An example of Faster R-CNN in
defect detection in aluminum casting can be found in [11, 12] very good results.

- **Mask R-CNN:** Another approach that is related to R-CNN is the well-known Mask R-CNN [22]. Mask R-CNN is a method that belongs to the category of 'Instance Segmentation'. Whereas in object detection the goal is to detect bounding boxes, in instance segmentation, the goal is to perform a segmentation of an object at a pixel level. That means, the output is not a bounding box, it is the boundaries of the detected object. Mask R-CNN is a combination of R-CNN and Fully Convolutional Network (FCN). It consists of a faster attention mechanism (like Faster R-CNN) to generate RoIs with a FCN that runs on each of the RoIs. The FCN has convolutional layers that are used to predict the mask on the RoI, i.e., a binary image of the same size of the RoI where a pixel equals 1 (or 0) means that the pixel of the RoI belongs (or does not belong) to the detected object. An example is illustrated in Fig. 7.34.

7.7.3   YOLO

In region-based approaches, as explained in Sect. 7.7.2, object detection is performed in two stages: region proposal and final classification. That means the classification is not performed by *looking* at the complete image but at selected regions of the image. In order to overcome this disadvantage, a new method called YOLO, *You-Only-Look-Once* was proposed [52]. YOLO is a single (and powerful) convolutional neural network that *looks* the image once, i.e., the input image is fed into a single CNN which output is the simultaneous prediction of both the bounding boxes (localization) and the category probabilities (classification) of the detected objects. It is very fast because the input image is processed in a single pass by the CNN.

The key idea of YOLO is very simple: The input image is divided into a grid of $S \times S$ cells, and for each cell, YOLO can detect B objects. For each detected bounding box, YOLO computes:

- (x, y, w, h): variables that define the detected bounding box, i.e., location (x, y) and dimension (width, height),
- p: confidence score that gives the probability that the bounding box encloses an object (Pr(Object)), and
- p_i : for $i = 1 \ldots K$: probability distribution over all K possible classes, i.e., p_i is a conditional class probability (Pr(Class$_i$|Object)).

That means, for each bounding box, YOLO provides an array of $R = 4 + 1 + K$ elements: $(x, y, w, h, p, p_1, p_2, \ldots, p_K)$, as illustrated in Fig. 7.35. In testing stage, an object of class i is detected if Pr(Object) $\times$ Pr(Class$_i$|Object) is greater than a threshold.

Since in a grid cell, B bounding boxes can be detected, for each cell, an array of $Q = B \times R$ elements is computed.

The simplicity of YOLO (see Fig. 7.35) is due to *(i)* the architecture has only standard convolution layers with 3×3 kernes and max-pooling layers with 2×2 kernels, and *(ii)* the output of the CNN is a tensor of $S \times S \times Q$, that means, for

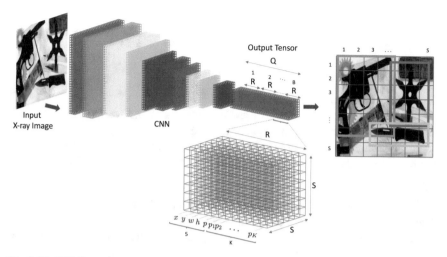

Fig. 7.35   YOLO strategy

each grid cell we have $5 + K$ elements per bounding box that give us information about the localization of the bounding box and the category probability.

In the last years, many versions of YOLO have been developed. In this section, we address the most relevant of them: YOLOv2 [50], YOLOv3 [51], and YOLOv4 [6].

• **YOLOv2:** The improvements proposed in YOLOv2 [50] focus on expanding the subdivision of the image, and the use of anchor boxes of different dimensions in each subdivision of the image (as proposed in the Faster R-CNN [53] model). These anchor boxes are pre-configured using the 'k-means' algorithm with Euclidean distance in the training set. Then, for each cell of the feature map extracted using the DarkNet-19 model, its anchor boxes are created with predictions for the objects inside [50].

Python Example 7.7: In this example, we show how to use YOLOv2 in the detection of threat objects. For this task, we use the implementation of [56] (see footnote 17). In this implementation, there are four options: *(i)* object detection using a model that has been already trained for this task, *(ii)* training a new model using a set of training images, *(iii)* testing the model trained in the previous step, and *(iv)* evaluation of a model on a set of images:

Listing 7.7 :   Threat object detection in baggage inspection using YOLOv2.

```
# Pre-trained model
python3 predict_yolo2.py -c config_full_yolo2_infer.json -i input_path/folder -o save/
    folder/detection

# Training
python3 train_yolo2.py -c config_full_yolo2.json
```

```
# Testing
python3 predict_yolo2.py —c config_full_yolo2.json —i input_path/folder —o save/folder/
    detection

# Evaluation
python3 evaluate_yolo2.py —c config_full_yolo2.json
```

The output of this code is in Figs. 7.36 and 7.37.   □

• **YOLOv3:** In comparison with previous versions, YOLOv3 [51] includes two main updates: *(i)* the use of different scales (three scales) using a pyramidal architecture that aims to solve the problem of detection of small objects, and *(ii)* the use of a new feature extractor architecture called DarkNet-53 that improves upon DarkNet-19.

Python Example 7.8: In this example, we show how to use YOLOv3 in the detection of threat objects. For this task we use the implementation of [56] (see footnote 17). In this implementation, there are four options: *(i)* object detection using a model that has been already trained for this task, *(ii)* training a new model using a set of training images, *(iii)* testing the model trained in the previous step, and *(iv)* evaluation of a model on a set of images:

Listing 7.8 : Threat object detection in baggage inspection using YOLOv3.

```
# Pre—trained model
python3 predict_yolo3.py —c config_full_yolo3_infer.json —i input_path/folder —o save/
    folder/detection

# Training
python3 train_yolo3.py —c config_full_yolo3.json

# Testing
python3 predict_yolo3.py —c config_full_yolo3.json —i input_path/folder —o save/folder/
    detection

# Evaluation
python3 evaluate_yolo3.py —c config_full_yolo3.json
```

The output of this code is in Figs. 7.36 and 7.37.   □

• **YOLOv4:** YOLOv4 has recently proposed in [6]. In this version, the feature map is extracted using a new architecture called Cross Stage Partial Network [64] that decreases the computation by reducing redundant gradient information. In YOLOv4, this network is called CSPDarknet-53. An additional increment of the performance is obtained by using Spatial Pyramid Pooling networks (SPP) [21] for sharing computing for pyramid features and a Path Aggregation Network (PAN) [36] for parameter aggregation from different levels of the CSPDarknet-53. Finally, the final prediction is performed as in YOLOv3 [51]. With these improvements, in comparison to YOLOv3, the accuracy is increased by 10% and the computation time is reduced by 11%.[15]

---

[15] In the last week (June 2020), YOLOv5 was released. See https://github.com/ultralytics/yolov5.

7.7.4 SSD

Another architecture contemporary to Faster R-CNN [53] and YOLO [52] is the SSD (Single-Shot Multi-Box Detector) [37]. Using direct image transformations, like YOLO, it predicts the location of the desired objects. The major difference is the use of map features in different depths, in order to obtain the analysis at different scales of the image. SSD combines the use of anchor boxes, like Faster R-CNN [53] and YOLOv2 [50], to predict the desired frames and uses a loss function for multi-tasking, as in the aforementioned detectors.

Python Example 7.9: In this example, we show how to use SSD7 in the detection of threat objects. For this task, we use the implementation of [56] (see footnote 17). In this implementation, there are four options: *(i)* object detection using a model that has been already trained for this task, *(ii)* training a new model using a set of training images, *(iii)* testing the model trained in the previous step, and *(iv)* evaluation of a model on a set of images:

Listing 7.9 : Threat object detection in baggage inspection using SSD7.

```
# Pre-trained model
python3 predict_ssd.py -c config_7_infer.json -i input_path/folder -o save/folder/
    detection

# Training
python3 train_ssd.py -c config_7.json

# Testing
python3 predict_ssd.py -c config_7.json -i input_path/folder -o save/folder/detection

# Evaluation
python3 evaluate_ssd.py -c config_7_infer.json
```

The output of this code is in Figs. 7.36 and 7.37.   □

Python Example 7.10: In this example, we show how to use SSD300 in the detection of threat objects. For this task we use the implementation of [56] (see footnote 17). In this implementation, there are four options: *(i)* object detection using a model that has been already trained for this task, *(ii)* training a new model using a set of training images, *(iii)* testing the model trained in the previous step, and *(iv)* evaluation of a model on a set of images:

Listing 7.10 : Threat object detection in baggage inspection using SSD300.

```
# Pre-trained model
python3 predict_ssd.py -c config_300_infer.json -i input_path/folder -o save/folder/
    detection

# Training
python3 train_ssd.py -c config_300.json

# Testing
python3 predict_ssd.py -c config_300.json -i input_path/folder -o save/folder/detection
```

```
# Evaluation
python3 evaluate_ssd.py —c config_300_infer.json
```

The output of this code is in Figs. 7.36 and 7.37.   □

7.7.5   RetinaNet

Together with YOLOv3 [51] and YOLOv4 [6], the RetinaNet architecture [32] is
one of the most recent object detection models and combines the pyramidal feature extraction structure [33] with a residual architecture (ResNet) [20] that has
obtained promising results in image classification. The pyramidal approach consists
of decreasing the size of the image several times and making predictions for each of
those sizes. Another novelty of this architecture is the shift from the typical cross-
entropy to a 'focal loss'-based objective that reduces the penalty for well classified
classes while punishing misclassifications more aggressively for the rest.[16]

 Python Example 7.11: In this example, we show how to use RetinaNet in
the detection of threat objects. For this task, we use the implementation of [56]
(see footnote 17). In this implementation, there are four options: *(i)* object detection
using a model thàt has been already trained for this task, *(ii)* training a new model
using a set of training images, *(iii)* testing the model trained in the previous step,
and *(iv)* evaluation of a model on a set of images:

> **Listing 7.11 : Threat object detection in baggage inspection using RetinaNet.**

```
# Pre—trained model
python3 predict_retinanet.py —c config_resnet50_infer.json —i input_path/folder —o save/
        folder/detection

# Training
python3 train_retinanet.py —c config_resnet50.json

# Testing
python3 predict_retinanet.py —c config_resnet50.json —i input_path/folder —o save/folder/
        detection

# Evaluation
python3 evaluate_retinanet.py —c config_resnet50.json
```

The output of this code is in Figs. 7.36 and 7.37.   □

---

[16]An implementation of RetinaNet for casting defect detection in GDXray+, can be found on
https://github.com/aurotripathy/GDXray-retinanet by Auro Tripathy.

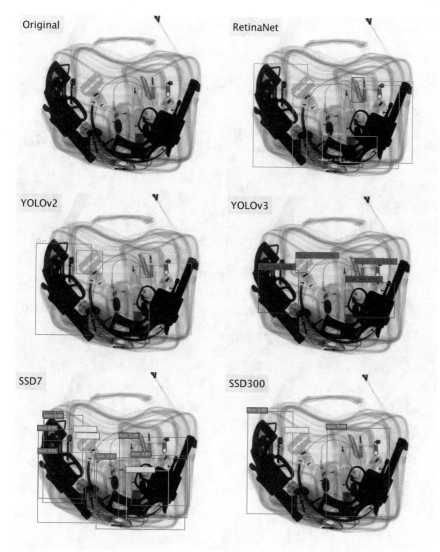

Fig. 7.36   Object detection on $\mathbb{GDX}$ray+ image B0046_0151

Fig. 7.37 Object detection on GDXray+ image B0046_0184

7.7.6 Examples of Object Detection

In this section, we show detection of threat objects in GDXray+ using the implementation of [56].[17] This implementation contains the following four detectors for use in the detection of threat objects in baggage inspection:

---

[17] See https://github.com/dlsaavedra/Detector_GDXray. In addition, all examples are implemented in Google Colab on https://github.com/computervision-xray-testing/pyxvis.

- Yolov2 (see Sect. 7.7.3 and Example 7.7)
- Yolov3 (see Sect. 7.7.3 and Example 7.8)
- SSD7 (see Sect. 7.7.4 and Example 7.9)
- SSD300 (see Sect. 7.7.4 and Example 7.10)
- RetinaNet (see Sect. 7.7.5 and Example 7.11)

In order to compare the implemented methods, in Figs. 7.36 and 7.37 we can observe the performance of each method visually. The reader is referred to [56] for more details of the training and the evaluation protocol.

7.8 Summary

In this chapter, we review many relevant concepts of deep learning that can be used in computer vision for X-ray testing. We covered the theory and practice of deep learning techniques in real X-ray testing problems. The chapter explained

- Neural Networks,
- Convolutional Neural Network (CNN) that can be used in classification problems,
- Pre-trained Models,
- Transfer Learning that is used in sophisticated models,
- Generative Adversarial Networks (GANs) to generate synthetic images, and
- modern detection methods that are used to classify and localize objects in an image.

In addition, for every method, we gave not only basic concepts but also practical details in real X-ray testing examples implemented in Python.

References

1. Aggarwal, C.C.: Neural Networks and Deep Learning. Springer International Publishing, Cham (2018)
2. Akçay, S., Kundegorski, M.E., Devereux, M., Breckon, T.P.: Transfer learning using convolutional neural networks for object classification within X-ray baggage security imagery. In: 2016 IEEE International Conference on Image Processing (ICIP), pp. 1057–1061. IEEE (2016)
3. Bengio, Y., Courville, A., Vincent, P.: Representation learning: a review and new perspectives. IEEE Trans. Pattern Anal. Mach. Intell. **35**(8), 1798–1828 (2013)
4. Bishop, C.: Neural Networks for Pattern Recognition. Oxford University Press, Oxford (2005)
5. Bishop, C.: Pattern Recognition and Machine Learning. Springer, Berlin (2006)
6. Bochkovskiy, A., Wang, C.Y., Liao, H.Y.M.: YOLOv4: optimal speed and accuracy of object detection (2020)
7. Brock, A., Donahue, J., Simonyan, K.: Large scale GAN training for high fidelity natural image synthesis. In: The International Conference on Learning Representations (ICLR 2019), pp. 1–35 (2019)

8. Chollet, F.: Xception: deep learning with depthwise separable convolutions. In: Proceedings of the IEEE Conference on Computer Vision and Pattern Recognition, pp. 1251–1258 (2017)
9. Creswell, A., White, T., Dumoulin, V., Arulkumaran, K., Sengupta, B., Bharath, A.A.: Generative adversarial networks: an overview. IEEE Signal Process. Mag. **35**(1), 53–65 (2018)
10. Deng, J., Guo, J., Xue, N., Zafeiriou, S.: ArcFace: additive angular margin loss for deep face recognition. In: Proceedings of the IEEE Conference on Computer Vision and Pattern Recognition, pp. 4690–4699 (2019)
11. Du, W., Shen, H., Fu, J., Zhang, G., He, Q.: Approaches for improvement of the X-ray image defect detection of automobile casting aluminum parts based on deep learning. NDT & E Int. **107**, 102,144 (2019)
12. Du, W., Shen, H., Fu, J., Zhang, G., Shi, X., He, Q.: Automated detection of defects with low semantic information in X-ray images based on deep learning. J. Intell. Manuf. 1–16 (2020)
13. Erhan, D., Bengio, Y., Courville, A., Vincent, P.: Visualizing higher-layer features of a deep network. Technical report, Univeriste de Montreal (2009)
14. Esteva, A., Kuprel, B., Novoa, R.A., Ko, J., Swetter, S.M., Blau, H.M., Thrun, S.: Dermatologist-level classification of skin cancer with deep neural networks. Nature **542**(7639), 115–118 (2017)
15. Girshick, R.: Fast R-CNN. In: Proceedings of the IEEE International Conference on Computer Vision, pp. 1440–1448 (2015)
16. Girshick, R., Donahue, J., Darrell, T., Malik, J.: Rich feature hierarchies for accurate object detection and semantic segmentation. In: Proceedings of the IEEE Conference on Computer Vision and Pattern Recognition, pp. 580–587 (2014)
17. Goodfellow, I., Pouget-Abadie, J., Mirza, M., Xu, B., Warde-Farley, D., Ozair, S., Courville, A., Bengio, Y.: Generative adversarial nets. In: Advances in Neural Information Processing Systems, pp. 2672–2680 (2014)
18. Goodfellow, I., Bengio, Y., Courville, A.: Deep Learning. MIT Press, Cambridge (2016)
19. Hassabis, D., Kumaran, D., Summerfield, C., Botvinick, M.: Neuroscience-inspired artificial intelligence. Neuron **95**(2), 245–258 (2017)
20. He, K., Zhang, X., Ren, S., Sun, J.: Deep residual learning for image recognition. CoRR (2015). arXiv:1512.03385
21. He, K., Zhang, X., Ren, S., Sun, J.: Spatial pyramid pooling in deep convolutional networks for visual recognition. IEEE Trans. Pattern Anal. Mach. Intell. **37**(9), 1904–1916 (2015)
22. He, K., Gkioxari, G., Dollár, P., Girshick, R.: Mask R-CNN. In: Proceedings of the IEEE International Conference on Computer Vision, pp. 2961–2969 (2017)
23. Howard, A.G., Zhu, M., Chen, B., Kalenichenko, D., Wang, W., Weyand, T., Andreetto, M., Adam, H.: MobileNets: efficient convolutional neural networks for mobile vision applications (2017). arXiv:1704.04861
24. Huang, G., Liu, Z., Weinberger, K.Q.: Densely connected convolutional networks. CoRR (2016). arXiv:1608.06993
25. Iandola, F.N., Moskewicz, M.W., Ashraf, K., Han, S., Dally, W.J., Keutzer, K.: SqueezeNet: AlexNet-level accuracy with 50x fewer parameters and <1mb model size. CoRR (2016). arXiv:1602.07360
26. Jiang, X., Hou, Y., Zhang, D., Feng, X.: Deep learning in face recognition across variations in pose and illumination. Deep Learning in Object Detection and Recognition, pp. 59–90. Springer, Berlin (2019)
27. Karras, T., Laine, S., Aila, T.: A style-based generator architecture for generative adversarial networks. In: Proceedings of the IEEE Conference on Computer Vision and Pattern Recognition, pp. 4401–4410 (2019)
28. Kingma, D.P., Ba, J.: Adam: a method for stochastic optimization (2014). arXiv:1412.6980
29. Krizhevsky, A., Sutskever, I., Hinton, G.E.: ImageNet classification with deep convolutional neural networks. In: NIPS, pp. 1106–1114 (2012)
30. LeCun, Y., Bottou, L., Bengio, Y.: Gradient-based learning applied to document recognition. In: Proceedings of the Third International Conference on Research in Air Transportation (1998)

31. LeCun, Y., Bengio, Y., Hinton, G.: Deep learning. Nature **521**(7553), 436–444 (2015)
32. Lin, T., Goyal, P., Girshick, R.B., He, K., Dollár, P.: Focal loss for dense object detection. CoRR (2017). arXiv:1708.02002
33. Lin, T.Y., Dollár, P., Girshick, R., He, K., Hariharan, B., Belongie, S.: Feature pyramid networks for object detection. In: Proceedings of the IEEE Conference on Computer Vision and Pattern Recognition, pp. 2117–2125 (2017)
34. Liu, B., Zhang, X., Gao, Z., Chen, L.: Weld defect images classification with VGG16-based neural network. In: International Forum on Digital TV and Wireless Multimedia Communications, pp. 215–223. Springer (2017)
35. Liu, L., Ouyang, W., Wang, X., Fieguth, P., Chen, J., Liu, X., Pietikäinen, M.: Deep learning for generic object detection: a survey. Int. J. Comput. Vis. **128**(2), 261–318 (2020)
36. Liu, S., Qi, L., Qin, H., Shi, J., Jia, J.: Path aggregation network for instance segmentation. In: Proceedings of the IEEE Conference on Computer Vision and Pattern Recognition, pp. 8759–8768 (2018)
37. Liu, W., Anguelov, D., Erhan, D., Szegedy, C., Reed, S.E., Fu, C., Berg, A.C.: SSD: single shot multibox detector. CoRR (2015). arXiv:1512.02325
38. Mery, D.: Aluminum casting inspection using deep learning: a method based on convolutional neural networks. J. Nondestruct. Eval. **39**(1), 12 (2020)
39. Mery, D., Arteta, C.: Automatic defect recognition in X-ray testing using computer vision. In: 2017 IEEE Winter Conference on Applications of Computer Vision (WACV), pp. 1026–1035. IEEE (2017)
40. Mery, D., Riffo, V., Zscherpel, U., Mondragón, G., Lillo, I., Zuccar, I., Lobel, H., Carrasco, M.: GDXray: the database of X-ray images for nondestructive testing. J. Nondestruct. Eval. **34**(4), 1–12 (2015)
41. Mery, D., Svec, E., Arias, M., Riffo, V., Saavedra, J.M., Banerjee, S.: Modern computer vision techniques for X-ray testing in baggage inspection. IEEE Trans. Syst. Man Cybern.: Syst. **47**(4), 682–692 (2016)
42. Miao, C., Xie, L., Wan, F., Su, C., Liu, H., Jiao, J., Ye, Q.: SIXray: a large-scale security inspection X-ray benchmark for prohibited item discovery in overlapping images. In: Proceedings of the IEEE Conference on Computer Vision and Pattern Recognition, pp. 2119–2128 (2019)
43. Mitchell, T.: Machine Learning. McGraw-Hill, Boston (1997)
44. Nagpal, K., Foote, D., Liu, Y., Chen, P.H.C., Wulczyn, E., Tan, F., Olson, N., Smith, J.L., Mohtashamian, A., Wren, J.H., et al.: Development and validation of a deep learning algorithm for improving Gleason scoring of prostate cancer. NPJ Digit. Med. **2**(1), 1–10 (2019)
45. Nair, V., Hinton, G.E.: Rectified linear units improve restricted Boltzmann machines. In: Proceedings of the 27th International Conference on Machine Learning (ICML-10), pp. 807–814 (2010)
46. Niu, S., Li, B., Wang, X., Lin, H.: Defect image sample generation with GAN for improving defect recognition. IEEE Trans. Autom. Sci. Eng. (2020)
47. Oliphant, T.E.: A Guide to NumPy, vol. 1. Trelgol Publishing, New York (2006)
48. Pedregosa, F., Varoquaux, G., Gramfort, A., Michel, V., Thirion, B., Grisel, O., Blondel, M., Prettenhofer, P., Weiss, R., Dubourg, V., Vanderplas, J., Passos, A., Cournapeau, D., Brucher, M., Perrot, M., Duchesnay, E.: Scikit-learn: machine learning in Python. J. Mach. Learn. Res. **12**, 2825–2830 (2011)
49. Radford, A., Metz, L., Chintala, S.: Unsupervised representation learning with deep convolutional generative adversarial networks. CoRR (2015). arXiv:1511.06434
50. Redmon, J., Farhadi, A.: YOLO9000: better, faster, stronger. CoRR (2016). arXiv:1612.08242
51. Redmon, J., Farhadi, A.: Yolov3: an incremental improvement. CoRR (2018). arXiv:1804.02767
52. Redmon, J., Divvala, S.K., Girshick, R.B., Farhadi, A.: You only look once: unified, real-time object detection. CoRR (2015). arXiv:1506.02640
53. Ren, S., He, K., Girshick, R., Sun, J.: Faster R-CNN: towards real-time object detection with region proposal networks. In: Advances in Neural Information Processing Systems, pp. 91–99 (2015)

54. Riffo, V., Mery, D.: Automated detection of threat objects using adapted implicit shape model. IEEE Trans. Syst. Man Cybern.: Syst. **46**(4), 472–482 (2016)
55. Russakovsky, O., Deng, J., Su, H., Krause, J., Satheesh, S., Ma, S., Huang, Z., Karpathy, A., Khosla, A., Bernstein, M., Berg, A., Fei-Fei, L.: ImageNet large scale visual recognition challenge. Int. J. Comput. Vis. **115**(3), 211–252 (2015). https://doi.org/10.1007/s11263-015-0816-y
56. Saavedra, D., Banerjee, S., Mery, D.: Detection of threat objects in baggage inspection with X-ray images using deep learning. Neural Comput. Appl. pp. 1–17. Springer (2020)
57. Shrestha, A., Mahmood, A.: Review of deep learning algorithms and architectures. IEEE Access **7**, 53040–53065 (2019)
58. Simonyan, K., Zisserman, A.: Very deep convolutional networks for large-scale image recognition. CoRR (2014). arXiv:1409.1556
59. Srivastava, N., Hinton, G., Krizhevsky, A., Sutskever, I., Salakhutdinov, R.: Dropout: a simple way to prevent neural networks from overfitting. J. Mach. Learn. Res. **15**, 1929–1958 (2014)
60. Szegedy, C., Liu, W., Jia, Y., Sermanet, P., Reed, S., Anguelov, D., Erhan, D., Vanhoucke, V., Rabinovich, A.: Going deeper with convolutions. In: CVPR 2015 (2015)
61. Szegedy, C., Vanhoucke, V., Ioffe, S., Shlens, J., Wojna, Z.: Rethinking the inception architecture for computer vision. CoRR (2015). arXiv:1512.00567
62. Tang, Z., Tian, E., Wang, Y., Wang, L., Yang, T.: Non-destructive defect detection in castings by using spatial attention bilinear convolutional neural network. IEEE Trans. Ind. Inform. 1–1 (2020)
63. Uijlings, J.R., Van De Sande, K.E., Gevers, T., Smeulders, A.W.: Selective search for object recognition. Int. J. Comput. Vis. **104**(2), 154–171 (2013)
64. Wang, C.Y., Liao, H.Y.M., Yeh, I.H., Wu, Y.H., Chen, P.Y., Hsieh, J.W.: CSPNet: A new backbone that can enhance learning capability of CNN (2019). arXiv:1911.11929
65. Yang, J., Zhao, Z., Zhang, H., Shi, Y.: Data augmentation for X-ray prohibited item images using generative adversarial networks. IEEE Access **7**, 28894–28902 (2019)
66. Yosinski, J., Clune, J., Bengio, Y., Lipson, H.: How transferable are features in deep neural networks? In: Advances in Neural Information Processing Systems, pp. 3320–3328 (2014)
67. Zeiler, M.D., Fergus, R.: Visualizing and understanding convolutional networks. In: European Conference on Computer Vision, pp. 818–833. Springer (2014)
68. Zhang, A., Lipton, Z.C., Li, M., Smola, A.J.: Dive into deep learning. Unpublished draft. Retrieved **3**, 319 (2019)
69. Zhang, X., Zhou, X., Lin, M., Sun, J.: ShuffleNet: an extremely efficient convolutional neural network for mobile devices. CoRR (2017). arXiv:1707.01083
70. Zhao, Z., Zhang, H., Yang, J.: A GAN-based image generation method for X-ray security prohibited items. In: Chinese Conference on Pattern Recognition and Computer Vision (PRCV), pp. 420–430. Springer (2018)
71. Zhao, Z., Zheng, P., Xu, S., Wu, X.: Object detection with deep learning: a review. IEEE Trans. Neural Netw. Learn. Syst. **30**(11), 3212–3232 (2019). https://doi.org/10.1109/TNNLS.2018.2876865
72. Zhu, Y., Zhang, Y., Zhang, H., Yang, J., Zhao, Z.: Data augmentation of X-ray images in baggage inspection based on generative adversarial networks. IEEE Access **8**, 86536–86544 (2020)

Chapter 8
Simulation in X-ray Testing

Abstract In order to evaluate the performance of computer vision techniques, computer simulation can be a useful tool. In this chapter, we review some basic concepts of the simulation of X-ray images, and present simple geometric and imaging models that can be used in the simulation. We explain the basic simulation principles and we address some techniques of simulated defects (that can be used to assess the performance of a computer vision method for automated defect recognition) and simulation of threat objects (that can be used to assess the performance of computer vision methods, to enhance the training dataset, or to improve a training program for human inspectors). Afterwards, the chapter gives an overview of the use of Generative Adversarial Networks (GANs) in the simulation of realistic X-ray images. Finally, we present 'aRTist', a simulation software that can be used to generate very realistic X-ray images. The chapter also has some Python examples that the reader can run and follow easily.

Cover image: *X-ray image of a wood located in 1, 4, 6, 36, 72 and 180 positions (image* `N0010_0051` *colored with 'hot' colormap).*

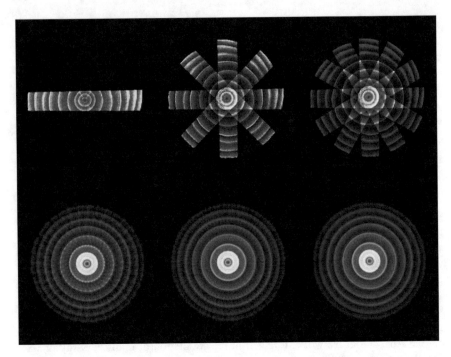

8.1   Introduction

In order to evaluate the performance of computer vision techniques, e.g., an automated defect recognition system, computer simulation can be a useful tool [8, 13].

The simulated X-ray images, however, should be as similar as possible to real X-ray images. For this purpose, the simulation should model the physics of the X-ray formation (generation, interaction, and detection) and handle complex 3D objects efficiently [29]. State of the art of computer modeling of X-ray testing methods is able to simulate different X-ray spectrum and X-ray source size, varied photon–matter interactions, and several X-ray detector responses. Special attention has been given to general purpose Monte Carlo methods that are able to calculate higher order scattering events [24, 27, 32]. A computer simulator for X-ray testing should include the following modules [28]:

- Source model: generates the spectra of X-ray tubes and isotopic sources.
- Ray-tracing engine: determines ray paths in complex geometries of test objects.
- Material database: contains cross-sectional data.
- Straight line attenuation model: determines the contribution of direct radiation, a scatter model, and a post-processor, combining both contributions.
- Detector model: converts radiation to an optical density and a digital X-ray image.

In this chapter, we review some basic concepts of simulation of X-ray images that can be used to understand other complex and more realistic approaches such as [24, 27]. In Sect. 8.2, we give the simple (geometric and imaging) models that can be used in the simulation. In Sect. 8.3, we explain the basic simulation principles providing some Python examples that the reader can run and follow. In Sect. 8.4, we address some techniques of simulated defects. In Sect. 8.5, we show a simulation method that can be used to generate realistic X-ray images with several threat objects. The simulated X-ray images can be used to assess the performance of a computer vision method for automated defect recognition. Examples of simulated defects in castings and welds are also given. In Sect. 8.6, we show how GAN models (as explained in Sect. 7.6, can be used to generate some synthetic X-ray images. Finally, in Sect. 8.7, we show a simulation software that can generate very realistic X-ray images by modeling the whole process using CAD models.

8.2 Modeling

In this section, we will explain the geometric model and the imaging model that we will use in the simulation.

8.2.1 Geometric Model

The model is based on a theoretical approach of Chap. 3 and follows the diagram of Fig. 8.1. The reader will be referred to the corresponding Sects. 3.2.4 and 3.3 to see the details.

As explained in Fig. 3.6, a 3D point M can be represented in world coordinate system $(\bar{X}, \bar{Y}, \bar{Z})$ as $\mathbf{M} = [\bar{X} \ \bar{Y} \ \bar{Z} \ 1]^T$ or in object coordinate system (X, Y, Z) as $\mathbf{M} = [X \ Y \ Z \ 1]^T$ in homogenous coordinates. There is an Euclidean 3D $\rightarrow$ 3D transformation defined by (3.10)

$$\bar{\mathbf{M}} = \mathbf{HM}, \tag{8.1}$$

where $\mathbf{H}$ is a 4×4 matrix that includes the rotation $\mathbf{R}$ and translation $\mathbf{t}$ between both coordinate systems (3.11):

$$\mathbf{H} = \begin{bmatrix} \mathbf{R} \ \mathbf{t} \\ \mathbf{0} \ 1 \end{bmatrix}. \tag{8.2}$$

Point M is projected into projection plane Π as point m using a perspective transformation. Applying intercept theorem (3.14), the coordinates of m in this 2D system are $(\bar{x}, \bar{y})$, with

$$\bar{x} = f\bar{X}/\bar{Z} \quad \text{and} \quad \bar{y} = f\bar{Y}/\bar{Z}. \tag{8.3}$$

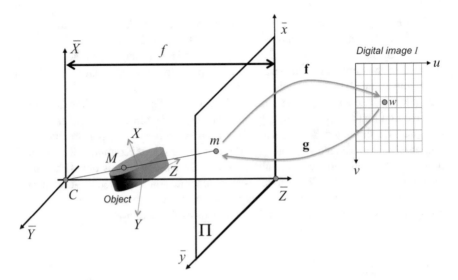

Fig. 8.1 Simplified geometric model taken from Fig. 3.6

This equation can be rewritten as (3.17): $\lambda \mathbf{m} = \mathbf{PM}$, where λ is a scale factor $\lambda \neq 0$. Again m is given in homogeneous coordinates $\bar{\mathbf{m}} = [\bar{x}\ \bar{y}\ 1]^T$. Perspective matrix $\mathbf{P}$ depends on the focal length f. Thus, a point M given in (X, Y, Z) is projected as point m in $(\bar{x}, \bar{y})$ as in (3.18):

$$\lambda \underbrace{\begin{bmatrix} \bar{x} \\ \bar{y} \\ 1 \end{bmatrix}}_{\bar{\mathbf{m}}} = \underbrace{\begin{bmatrix} f & 0 & 0 & 0 \\ 0 & f & 0 & 0 \\ 0 & 0 & 1 & 0 \end{bmatrix}}_{\mathbf{P}} \underbrace{\begin{bmatrix} \mathbf{R} & \mathbf{t} \\ \mathbf{0} & 1 \end{bmatrix}}_{\mathbf{H}} \underbrace{\begin{bmatrix} X \\ Y \\ Z \\ 1 \end{bmatrix}}_{\mathbf{M}}, \tag{8.4}$$

where $\mathbf{0} = [0\ 0\ 0]$. In image coordinate system (u, v), point m is viewed as point w that can be represented in homogenous coordinates as $\mathbf{w} = [u\ v\ 1]^T$. The transformation $\bar{\mathbf{m}} \rightarrow \mathbf{w}$ is defined by function $\mathbf{f}$, and the back transformation $\mathbf{w} \rightarrow \bar{\mathbf{m}}$ by function $\mathbf{g}$. In linear case, where no distortion takes place, transformation $\mathbf{f}$ can be defined by (3.24)

$$\mathbf{f}: \underbrace{\begin{bmatrix} u \\ v \\ 1 \end{bmatrix}}_{\mathbf{w}} = \underbrace{\begin{bmatrix} k_u & 0 & u_0 \\ 0 & k_v & v_0 \\ 0 & 0 & 1 \end{bmatrix}}_{\mathbf{K}} \underbrace{\begin{bmatrix} \bar{x} \\ \bar{y} \\ 1 \end{bmatrix}}_{\bar{\mathbf{m}}}, \tag{8.5}$$

where scale factors (k_u, k_v) and a translation of the origin (u_0, v_0) are considered. In this model, we assume the skew factor s can be neglected. In this simplified linear model,

$$\mathbf{w} = \mathbf{f}(\bar{\mathbf{m}}) = \mathbf{K}\bar{\mathbf{m}} \quad \text{and} \quad \bar{\mathbf{m}} = \mathbf{g}(\mathbf{w}) = \mathbf{K}^{-1}\mathbf{w}. \tag{8.6}$$

If the transformation $\bar{\mathbf{m}} \rightarrow \mathbf{w}$ is non-linear, a non-linear model for $\mathbf{f}$ and for $\mathbf{g}$ must be used (see examples in Sects. 3.3.2 and 3.3.3). In this section, we will assume a linear model only. Thus, a point $\mathbf{M}$ in object coordinate system is viewed as point $\mathbf{w}$ in image coordinate system as

$$\lambda \begin{bmatrix} u \\ v \\ 1 \end{bmatrix} = \begin{bmatrix} k_u & 0 & u_0 \\ 0 & k_v & v_0 \\ 0 & 0 & 1 \end{bmatrix} \begin{bmatrix} f & 0 & 0 & 0 \\ 0 & f & 0 & 0 \\ 0 & 0 & 1 & 0 \end{bmatrix} \begin{bmatrix} \mathbf{R} & \mathbf{t} \\ \mathbf{0} & 1 \end{bmatrix} \begin{bmatrix} X \\ Y \\ Z \\ 1 \end{bmatrix}, \tag{8.7}$$

or using matrix notation:

$$\lambda \mathbf{w} = \mathbf{KPHM}. \tag{8.8}$$

If we have a pixel w in image coordinate system given by $\mathbf{w} = [u\ v\ 1]^{\mathsf{T}}$, and we want to estimate the X-ray beam $\langle C, m \rangle$ that defines w, we have to find the coordinates $\bar{\mathbf{m}} = [\bar{x}\ 1]^{\mathsf{T}}$ in projection coordinate systems using back transformation $\mathbf{g}$, i.e., $\bar{\mathbf{m}} = \mathbf{g}(\mathbf{w}) = \mathbf{K}^{-1}\mathbf{w}$ in linear case. Thus, the X-ray beam is defined by points $(\bar{X}, \bar{Y}, \bar{Z})$ that fulfill:

$$\bar{X} = \bar{x}\bar{Z}/f \quad \text{and} \quad \bar{Y} = \bar{y}\bar{Z}/f. \tag{8.9}$$

Equations (8.7) and (8.9) will be used by the simulation in the following sections.

8.2.2 X-ray Imaging

As we have already learned in Sect. 1.5, the intensity of X-ray penetrating radiation is modified by its passage through material and by discontinuities in the material. An example of this phenomenon is illustrated in Figs. 1.6 and 1.14.

Two properties of the X-rays are used to model the X-ray imaging process: *(i)* X-rays are differentially absorbed and *(ii)* X-rays travel in straight lines. The absorption can be macroscopically modeled using the exponential attenuation law for X-rays (1.2):

$$\varphi = \varphi_0 e^{-\mu x}, \tag{8.10}$$

where φ_0 is the incident intensity of radiation, φ the transmitted intensity, x thickness of the specimen, and μ is a constant known as the *linear absorption coefficient* of the material under test with dimension cm^{-1}. Coefficient μ depends on the material and the X-ray energy. As an example, Fig. 1.6 illustrates the linear absorption coefficient for aluminum plotted against X-ray energy. Typically, X-ray testing of aluminum castings uses energy values between 50 keV and 150 keV [11]. Coefficient μ can be modeled as a fourth degree polynom [17]:

$$\mu \approx \sum_{i=0}^{4} \theta_i E^i \quad \text{for } 50 \, \text{keV} \le E \le 150 \, \text{keV} \tag{8.11}$$

with

$$\theta = (6.00, -0.210, -0.00304, -1.97 \times 10^{-5}, 4.72 \times 10^{-8}).$$

A flaw such as a cavity can be simulated as a material with no absorption. In Fig. 1.6 this simulation is shown schematically. An X-ray beam penetrates an object which has a cavity with thickness d. In this case, from (8.10) the transmitted radiation φ is given by

$$\varphi = \varphi_0 e^{-\mu(x-d)}, \tag{8.12}$$

where we assume that the absorption coefficient of the cavity is zero. If the flaw is an incrusted material, its absorption coefficient μ_d must be considered.

In the example of Fig. 8.2, we have three materials with different linear absorption coefficients μ_1, μ_2, and μ_3. The thickness in direction of the X-ray beam is x_1, x_2, and x_3 for each material. It is worth noting that the thickness depends on the projection beam $\langle C, m \rangle$, i.e., for different locations of m, different thicknesses will be obtained. A simplified model can be used for different thicknesses and materials (1.4):

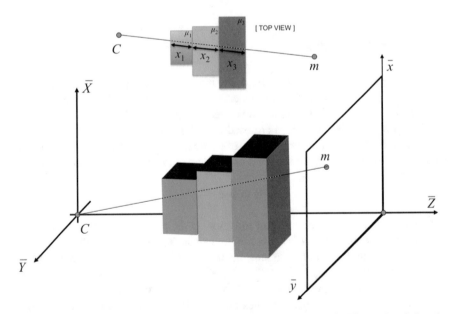

Fig. 8.2 Example of an X-ray beam that passes through three materials. The total path length through material i in direction of the beam is x_i, and the linear absorption coefficient of each material is μ_i for $i = 1, 2, 3$

$$\varphi = \varphi_0 \exp\left(-\sum_i \mu_i x_i\right). \tag{8.13}$$

Nevertheless, it is worth mentioning that μ and φ_0 depends on the energy E. In addition, if we want to compute the gray value of a pixel (u, v) of a simulated X-ray image, as a pixel is rather a square than a point, we must take into account the solid angle that corresponds to the pixel observed from the source point.

$$\varphi(R) = \varphi_0(E)\Delta\Omega \exp\left(-\sum_i \mu_i(E)x_i\right), \tag{8.14}$$

where $\varphi_0(E)$ is the incident radiation intensity of energy E, $\Delta\Omega$ is the solid angle that corresponds to region R of the image (e.g., pixel (u, v)) observed from the source point, $\mu_i(E)$ designates the attenuation coefficient associated with the material i at energy E, and x_i the total path length through material i. The X-ray source can be modeled as a raster of point sources, rays from every source point are traced to all pixels of the simulated image. The final simulated image will be an addition of each single simulation, one for each energy and each source point [8].

Finally, a linear transformation from incident energy to gray value is considered:

$$I = A\varphi + B, \tag{8.15}$$

where A and B are the linear parameters of I.

8.3 Basic General Simulation

In this section, we will see a basic approach to simulate an X-ray image of a 3D object based on *voxels*, i.e., the volume of the object is discretized in very small volume elements. In case, the volume is defined as a *polygon mesh*, the mesh can be *voxelized* [4].

A simple way to simulate an X-ray image of a 3D object is by modeling the object as a set of *voxels* as illustrated in Fig. 8.3. Thus, each voxel has a 3D location (X, Y, Z) and can have a linear absorption coefficient $V = \mu$. The value '0' for a voxel means that the voxel does not belong to the object (see cyan voxels in Fig. 8.3). In case the whole object is of the same material, e.g., an aluminum wheel, the value of a voxel can be a binary value: '0' for a voxel that does not belong to the object and '1' otherwise (see red voxels in Fig. 8.3).

In this approach, we assume that there are P voxels that belong to the object. The kth voxel, for $k = 1 \ldots P$, is defined by its linear absorption coefficient $V_k = \mu_k > 0$ and its location in object coordinate system as $\mathbf{M}_k = [X_k\ Y_k\ Z_k\ 1]^T$ in homogeneous coordinates. Using (8.8), we can obtain $\mathbf{w}_k = [u_k\ v_k\ 1]^T$, the coordinates in image coordinate systems of each projected voxel

Fig. 8.3 Object modeling using voxels. In these examples (a sphere and two cylinders) there are 15^3 voxels. The radius of each object is 5. The red voxels belong to the object

$$\lambda_k \mathbf{w}_k = \mathbf{KPHM}_k. \tag{8.16}$$

A great advantage of using homogeneous coordinates is that the projection of all points $\mathbf{M}_k$ can be done with only one multiplication: $\mathbf{W} = \mathbf{KPHM}$, where $\mathbf{M}$ is a $4 \times P$ matrix $\mathbf{M} = [\mathbf{M}_1 \ \mathbf{M}_2 \cdots \mathbf{M}_P]$ and $\mathbf{W}$ is a $3 \times P$ matrix $\mathbf{W} = [\lambda_1 \mathbf{w}_1 \ \lambda_2 \mathbf{w}_2 \cdots \lambda_P \mathbf{w}_P]$.

According to (8.13), an X-ray beam passes through different materials with different levels of thickness. In our model, each voxel can be considered as an element with a linear absorption coefficient μ_i and a thickness x_i (in direction of the X-ray beam). It is simple to accumulate in a region R of the image the contribution of all voxels that are in the corresponding X-ray beam as illustrated in Fig. 8.4. In this example, we show the voxels that belong to a spherical object in red, and those voxels that contribute to region R in blue. For region R of the image we can compute $q(R) = \sum_i \mu_i x_i$. Finally, the gray value of this region is modeled using $q(R)$ and Eqs. (8.13) and (8.15) as

$$I(R) = A\varphi_0 e^{-q(R)} + B. \tag{8.17}$$

There are two different ways to obtain the simulated X-ray image $\mathbf{I}$:

- **From pixels to voxels:** In order to simulate $\mathbf{I}$ of $N \times M$ pixels, we can estimate the intensity of each pixel (u, v), for $u = 1 \ldots N$ and $v = 1 \ldots M$ as follows:

  1. Each pixel (u, v) defines a point $(\bar{x}, \bar{y})$ in the projection coordinate system as explained in Sect. 8.2.1.
  2. Point $(\bar{x}, \bar{y})$ defines a specific X-ray beam according to (8.9). If there is an intersection of the X-ray beam with the 3D object, follow the next steps.
  3. The beam passes through n voxels of the object, that means there are corresponding linear absorption coefficients of each voxel (μ_i) and thickness (x_i) for $i = 1 \ldots n$. The absorption linear coefficient μ_i can be obtained from the corresponding voxel value. The thickness x_i can be estimated as the line seg-

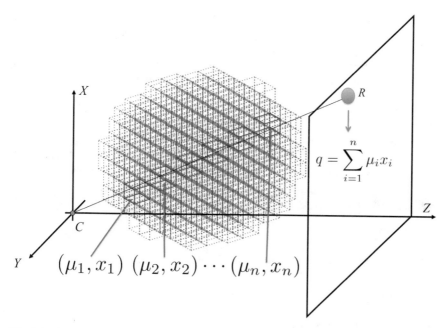

Fig. 8.4 The contribution of all voxels aligned to the X-ray beam in a region R can be modeled as $q(R) = \sum_i \mu_i x_i$. The example shows the voxels of a sphere that are in the X-ray beam

ment length of the intersection of the corresponding X-ray beam that passes through the voxel with the cube defined by the voxel.

4. The contribution $q(u, v) = \sum_i \mu_i x_i$ is computed.
5. Using (8.17) the gray value for each pixel can be estimated. In this approach, the region R corresponds to the area defined by pixel (u, v).

- **From voxels to pixels:** In order to simulate image **I**, we first define an image **Q** as a matrix with the same size of **I**, i.e., $N \times M$ pixels. All pixels of **Q** are initialized to zero. Afterwards, we can deal with the projection of each voxel k, for $k = 1 \ldots P$ as follows:

1. The kth voxel located at $\mathbf{M}_k$ is projected using (8.16), and coordinates (u_k, v_k) in the image coordinate system are obtained.
2. The contribution of kth voxel to our image is $q_k = \mu_k x_k$. The absorption linear coefficient μ_k can be obtained from the voxel value V_k. The thickness x_k can be estimated as the line segment length of the intersection of the corresponding X-ray beam that passes through the center of the voxel with the cube defined by the voxel.
3. The value q_k is added in those pixels (u, v) of image **Q** that are neighbors to (u_k, v_k).

Finally, using (8.17) the gray value for each pixel can be estimated. In this approach, the region R corresponds to the area defined by pixel (u, v).

In order to show a simple simulation of an X-ray image of a 3D object, we give some details of the second approach in the following example.

Python Example 8.1: In this example, we simulate the X-ray image of a homogeneous material using voxels. The implementation corresponds to the method 'from voxels to pixels' outlined in this section. The binary 3D matrix V, stored in `voxels_model400.npy`, contains $400 \times 400 \times 400$ voxels. Here, a voxel equals 1 (or 0) means that the voxel belongs (or does not belong) to the 3D object.

Listing 8.1 : Simulation of an X-ray image of 3D object

```python
import numpy as np
import matplotlib.pylab as plt
from pyxvis.simulation.xsim import voxels_simulation
from pyxvis.geometry.projective import rotation_matrix_3d
from pyxvis.processing.images import linimg

# Binary 3D matrix containing the voxels of a 3D object
V  = np.load('../data/voxels_model400.npy')

# Transformation (x,y)->(u,v)
(u_0,v_0,a_x,a_y) = (235,305,1.1,1.1)
K  = np.array([[a_x, 0, u_0], [ 0, a_y, v_0], [0,0,1]])

# Transformation (Xb,Yb,Zb)->(u,v)
f  = 1500  # focal length
P  = np.array([[f, 0, 0, 0], [0, f, 0, 0], [0,0,1,0]])

# Transformation (X,Y,Z)->(Xb,Yb,Zb)
R  = rotation_matrix_3d(0.5,0.1,0.6)
t  = np.array([-120, -120, 1000])
H  = np.vstack([np.hstack([R, t[:, np.newaxis]]), np.array([0, 0, 0, 1])])

# Transformation (X,Y,Z) -> (u,v)
Pt = np.matmul(K,np.matmul(P,H))

# Simulation of projection (Q) and X-ray image (X)
Q  = voxels_simulation(400,400,V,7,Pt)
X  = linimg(np.exp(-0.0001*Q))

# Output
fig1, ax = plt.subplots(1, 1, figsize=(16, 8))
ax.imshow(X, cmap='gray'), ax.axis('off')
plt.show()
```

The output of this code is illustrated in Fig. 8.5, where eight different positions are shown. The eight positions were obtained varying the rotation angles of matrix R. In this example, the X-ray image was simulated using command voxel_simulation of pyxvis Library. In this implementation we assume that the thickness of a voxel (x_k) is always 1. This is not true, however, for homogenous material, when μ_k is constant, $x_k = 1$ is a good estimation of the average value. The weighted distribution explained in step 3 is implemented as shown in Fig. 8.6. Also the reader can simulate

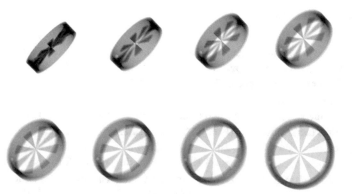

Fig. 8.5 Simulation of a wheel in eight different positions. [→ Example 8.1 🐍]

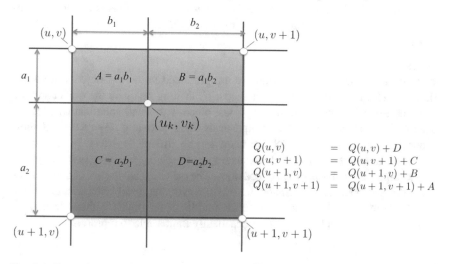

$$Q(u,v) = Q(u,v) + D$$
$$Q(u,v+1) = Q(u,v+1) + C$$
$$Q(u+1,v) = Q(u+1,v) + B$$
$$Q(u+1,v+1) = Q(u+1,v+1) + A$$

Fig. 8.6 Since pixel (u_k, v_k) does not exist, it is impossible to add the contribution $\mu_k x_k$ to this pixel. For this reason, the contribution is distributed in its four neighbor pixels according to their opposite areas A, B, C, D (note that $A + B + C + D = 1$). In our simplified model, $x_k = 1$ and μ_k is constant, that means that the contribution of each voxel is constant. [→ Example 8.1 ◀]

an X-ray image using a file in the STL format (a standard for a polygon mesh)[1] and converting it into voxels.[2] □

---

[1] http://en.wikipedia.org/wiki/STL_(file_format).
[2] See, for example, https://github.com/cpederkoff/stl-to-voxel.

8.4   Flaw Simulation

Generally, the automatic defect recognition consists of a binary classification, where a decision is performed about whether or not an initially identified hypothetical defect in an image is in fact a defect. Unfortunately, in real automatic flaw detection problems there are a reduced number of flaws in comparison with the large number of non-flaws. This skewed class distribution seriously limits the application of classification techniques [6]. Usually, the performance of an inspection method can be assessed on a few images, and an evaluation of a broader and more representative database is necessary. A good way of assessing the performance of a method for inspecting castings is to examine simulated data. This evaluation allows one the possibility of tuning the parameters of the inspection method and of testing how well the method works in critical cases.

Among the NDT community there are two groups of methods to obtain this simulated data: invasive and non-invasive methods. Table 8.1 summarizes their most important properties.

Invasive Methods

In the invasive methods, discontinuities are produced in the test object artificially. There are two published invasive methods: *(i)* drilling holes on the object surface [20] (see Fig. 8.7), and *(ii)* designing a test piece with small spherical cavities [1] (see Fig. 8.8). Usually, the first technique drills small holes (e.g., $\emptyset = 1.0 \sim 4.0$ mm) in positions of the casting which are known to be difficult to detect. In the second technique, a sphere is produced, for example, by gluing together two aluminum pieces containing half-spherical concavities. The principal advantage of these methods is that the discontinuity image is real. However, the disadvantages are: *(i)* it is impossible to introduce concavities in the middle of the object without destroying it, and *(ii)* concavities like cracks are practically impossible to reproduce.

Table 8.1 Methods for simulation of defects

	Method	Description	Advantages	Disadvantages
i n v a s i v e	Drilling holes	It drills holes on the surface of the test object	· Real X-ray image with real defects	· Cracks cannot be produced · X-ray imaging system is required
	Spherical cavities	It produces defects inside of the test object by putting together two parts with cavities	· Real X-ray image with real defects	· It destroys the test object · Cracks cannot be produced · X-ray imaging system is required
n o n i n v a s i v e	Mask superimposition	It modifies the original gray value of the image by multiplying it with a factor.	· Real X-ray image with simulated defects. · Easy implementation.	· Simulated defects differ significantly from the real ones. · X-ray imaging system is required
	Full-CAD	It simulates the X-ray imaging process by projecting a CAD model including a defect.	· No real X-ray imaging system is required · Defects and object can be modelled in 3D	· No real X-ray image of the test object. · Sophisticated computer package · Time consuming
	Flaw-CAD	It modifies the original gray value of the image by superimposing the projection of a CAD model of a flaw.	· Real X-ray image with simulated defects · No time consuming · Defects can be modelled in 3D	· X-ray imaging system is required

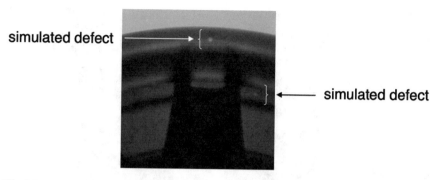

simulated defect

simulated defect

Fig. 8.7 Two defects generated using drilling holes

simulated defect

simulated defect

Fig. 8.8 Two generated defects using spherical cavities [1]

Non-invasive Methods

In the non-invasive methods, X-ray images are generated or modified without altering the test object. There are three widespread approaches that produce this simulated data [16]: (i) mask superimposition, (ii) CAD models for casting and flaw and (iii) CAD models for flaws only. In this section, they will be described in further detail.

8.4.1 Mask Superimposition

The first technique attempts to simulate flaws by superimposing masks with different gray values onto real X-ray images [9, 11, 12]. This approach is quite simple, as it neither requires a complex 3D model of the object under test nor of the flaw. It also provides a real X-ray image with real disturbances, albeit with simulated flaws.

In this technique, the original gray value I_o of a pixel (u, v) of an X-ray image is altered by

$$I_n(u, v) = I_o(u, v) (1 + M(u - u_0, v - v_0)) \tag{8.18}$$

with $I_n(u, v)$ the new gray value and M the mask that is centered on pixel (u_0, v_0), where $M(i, j)$ is defined in the interval $-\frac{n}{2} \le i \le \frac{n}{2}$ and $-\frac{m}{2} \le j \le \frac{m}{2}$. Three typical masks are shown in Fig. 8.9.

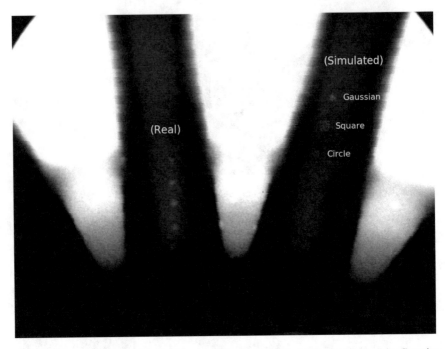

Fig. 8.9 Flaw simulation using Gaussian mask, square, and circle. As we can see, the Gaussian mask achieves the best simulation. [→ Example 8.2 🐍]

Python Example 8.2: In this example, we simulate three different flaws in an aluminum casting using Gaussian, square, and circle masks.

Listing 8.2 : Simulation of a defects using superimposed masks

```python
import numpy as np
import matplotlib.pylab as plt
from pyxvis.simulation.xsim import mask_simulation
from pyxvis.processing.helpers.kfunctions import gaussian_kernel
from pyxvis.io import gdxraydb

image_set = gdxraydb.Castings()
I = np.double(image_set.load_image(21,25)) # wheel image

p1 = [150,580] # Location of 1st defect
p2 = [200,565] # Location of 2nd defect
p3 = [250,550] # Location of 3rd defect

h1 = gaussian_kernel(35,4)                # Gaussian Mask
h1 = h1/np.max(h1)*0.9
J = mask_simulation(I,h1,p1[0],p1[1]) # Simulation
h2 = np.ones((17,17))*0.4                 # Square Mask
J = mask_simulation(J,h2,p2[0],p2[1]) # Simulation
h3 = np.zeros(h1.shape)                    # Circle mask
h3[h1>0.25] = 0.4
J = mask_simulation(J,h3,p3[0],p3[1]) # Simulation
```

```
# Output
fig1, ax = plt.subplots(1, 2, figsize=(16, 8))
ax[0].imshow(I, cmap='gray'), ax[0].axis('off')
ax[1].imshow(J, cmap='gray'), ax[1].axis('off')
ax[1].text(p1[1]+20,p1[0]+5, 'Gaussian', fontsize=8,color='white')
ax[1].text(p2[1]+20,p2[0]+5, 'Square', fontsize=8,color='white')
ax[1].text(p3[1]+20,p3[0]+5, 'Circle', fontsize=8,color='white')
ax[1].text(250,210, '(Real)', fontsize=10,color='white')
ax[1].text(565,92, '(Simulated)', fontsize=10,color='white')
plt.show()
```

The output of this code is illustrated in Fig. 8.9, where three different defects are shown in. In this example, the X-ray image was simulated using command mask_simulation of **pyxvis** Library. ☐

8.4.2 CAD Models for Object and Defect

The second approach simulates the entire X-ray imaging process [10, 30]. In this approach, characteristics of the X-ray source, the geometry, and material properties of objects and their defects, as well as the imaging process itself are modeled and simulated independently. Complex objects and defect shapes can be simulated using CAD models. In Sect. 8.7, a simulation software (aRTist) that can generate these simulated images is presented.

The principle of the simulation is shown in Fig. 8.10. The X-ray may intersect different parts of the object. The intersection points between the modeled object with the corresponding X-ray beam that is projected into pixel (u, v) are calculated for each pixel (u, v) of the simulated image as explained in Sect. 8.3.

Some complex 3D flaw shapes are reported in [30]. The defect model is coupled with a CAD interface yielding 3D triangulated objects. Other kinds of flaws like cracks can also be obtained using this simulation technique.

Although this approach offers excellent flexibility for setting the objects and flaws to be tested, it has three disadvantages for the evaluation of the inspection methods' performance: *(i)* the X-ray image of the object under test is simulated (it would be better if we could count on real images with simulated flaws); *(ii)* the simulation approach is only available when using a sophisticated computer package; *(iii)* the computing time is expensive.

8.4.3 CAD Models for Defects Only

This approach simulates only the flaws and not the whole X-ray image of the object under test [17]. This method can be viewed as an improvement of the first-mentioned technique (Sect. 8.4.1) and the 3D modeling for the flaws of the second one (Sect. 8.4.2). In this approach, a 3D modeled flaw is projected and superim-

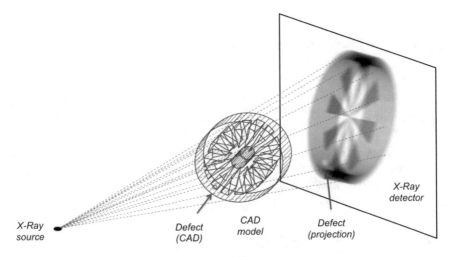

Fig. 8.10 X-ray image simulation using CAD models

posed onto real X-ray images of a homogeneous object according to the exponential attenuation law for X-rays (8.10).

As explained in Sect. 8.2.2, the gray value I of a digital X-ray image can be expressed as a linear function of the transmitted radiation φ:

$$I(x) = A\varphi(x) + B, \tag{8.19}$$

where

$$\varphi(x) = \varphi_0 e^{-\mu x}, \tag{8.20}$$

and A and B are the linear parameters of I, and x the thickness of the object under test.

Now, we investigate what happens if the penetrated object has a cavity, the thickness of which is d as shown in Fig. 1.14 and its absorption coefficient $\mu' \approx 0$. In this case, from (8.20) the transmitted radiation is given by

$$\varphi(x - d) = \varphi_0 e^{-\mu(x-d)} = \varphi(x)e^{\mu d}. \tag{8.21}$$

The gray value registered is calculated then from (8.21) and (8.19) as

$$I(x - d) = A\varphi(x)e^{\mu d} + B. \tag{8.22}$$

Substituting the value of $A\varphi(x)$ from (8.19) we see that (8.22) may be written as

$$I(x - d) = I(x)e^{\mu d} + B(1 - e^{\mu d}). \tag{8.23}$$

Parameter B can be estimated as follows: The maximal gray value (I_{max}) in an X-ray image is obtained when the thickness is zero. Additionally, the minimal gray value (I_{min}) is obtained when the thickness is x_{max}. Substituting these values in (8.19), it yields

$$\begin{cases} I_{max} = A\varphi_0 + B \\ I_{min} = A\varphi_0 e^{-\mu x_{max}} + B \end{cases} \cdot \qquad (8.24)$$

From these equations, one may compute the value for B

$$B = I_{max} - \Delta I / (1 - e^{-\mu x_{max}}), \qquad (8.25)$$

where $\Delta I = I_{max} - I_{min}$. Usually, I_{max} and I_{min} are 255 and 0 respectively. For these values, B can be written as

$$B = 255 / (1 - e^{\mu x_{max}}). \qquad (8.26)$$

This means that the gray value of the image of the cavity is

$$I(x - d) = I(x) e^{\mu d} + 255 \frac{1 - e^{\mu d}}{1 - e^{\mu x_{max}}}. \qquad (8.27)$$

Using Eq. (8.27), we can alter the original gray value of the X-ray image $I(x)$ to simulate a new image of a flaw $I(x - d)$. A 3D flaw can be modeled, projected, and superimposed onto a real radioscopic image. The new gray value of a pixel, where the 3D-flaw is projected, depends only on four parameters: (a) Original gray value $I(x)$; (b) the linear absorption coefficient of the examined material μ; (c) the length of the intersection of the 3D-flaw with the modeled X-ray beam d, that is projected into the pixel; and (d) the maximal thickness observable in the radioscopic image x_{max}.

Now, we will explain in further details how a 3D defect, namely, an ellipsoid, is projected onto an X-ray image [17]. Using this tool a simulation of an ellipsoidal flaw of any size and orientation can be made anywhere in the casting. This model can be used for flaws like blowholes and other round defects. Four examples are shown in Fig. 8.11. The simulated flaws appear to be real due to the irregularity of the gray values.

This technique presents two advantages: simulation is better than with the first technique; and with respect to the second, this technique is faster given the reduced computational complexity. However, the model used in this method has four simplifications that were not presumed in the second simulation technique: *(i)* the X-ray source is assumed as a source point; *(ii)* there is no consideration of noise in the model; *(iii)* there is no consideration of the solid angle $\Delta\Omega$ of the X-ray beam that is projected onto a pixel; and *(iv)* the spectrum of the radiation source is monochromatic.

In our approach we follow the geometric model illustrated in Fig. 8.12. This model is very similar to the geometric model we learned in Sect. 8.2.1, however,

Fig. 8.11 Simulated ellipsoidal flaws using CAD models of a defect only. See details in Table 8.2. 3D profile of a yellow square is shown in Fig. 8.14

it includes a new coordinate system (X', Y', Z') attached to the center of the ellipsoid that is modeled as

$$\frac{X'^2}{a^2} + \frac{Y'^2}{b^2} + \frac{Z'^2}{c^2} = 1, \tag{8.28}$$

where a, b, and c are the half-axes of the ellipsoid as shown in Fig. 8.12. The location of the ellipsoid relative to the object coordinate system is defined by a 3×3 rotation matrix $\mathbf{R}_e$ and a 3×1 translation vector $\mathbf{t}_e$. They can be arranged in a 4×4 matrix $\mathbf{H}_e$ as in Eq. (8.2). Using (8.1), the coordinates in the ellipsoid coordinate system (X', Y', Z') can be expressed in the world coordinate system $(\bar{X}, \bar{Y}, \bar{Z})$ by

$$\bar{\mathbf{M}} = \mathbf{H}\mathbf{H}_e\mathbf{M}' \tag{8.29}$$

with $\mathbf{M}' = [X' \ Y' \ Z' \ 1]^T$ and $\bar{\mathbf{M}} = [\bar{X} \ \bar{Y} \ \bar{Z} \ 1]^T$. Now, we can write the ellipsoid in world coordinate system from (8.28) and (8.29) as

$$\begin{aligned}
(s_{11}\bar{X} + s_{12}\bar{Y} + s_{13}\bar{Z} + s_{14})^2/a^2 + \\
(s_{21}\bar{X} + s_{22}\bar{Y} + s_{23}\bar{Z} + s_{24})^2/b^2 + \\
(s_{31}\bar{X} + s_{32}\bar{Y} + s_{33}\bar{Z} + s_{34})^2/c^2 = 1
\end{aligned} \tag{8.30}$$

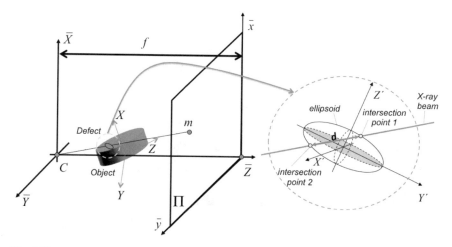

Fig. 8.12 Ellipsoid used by modeling a 3D flaw in coordinate system (X', Y', Z'). The two intersections of an X-ray beam with the surface of the ellipsoid define distance d

where s_{ij} are the elements of the 4×4 matrix $\mathbf{S} = [\mathbf{HH}_e]^{-1}$.

Suppose we have a pixel (u, v) of the X-ray image and we want to know if the X-ray beam, which produces a gray value in this pixel, intersects the modeled ellipsoid. Using $\mathbf{g}$, the inverse function of $\mathbf{f}$ (see (8.6)), we can calculate the corresponding coordinates of (u, v) in the projection coordinate systems $(\bar{x}, \bar{y})$:

$$\bar{\mathbf{m}} = \mathbf{g}(\mathbf{u}) \tag{8.31}$$

with $\mathbf{u} = [u \ v \ 1]^T$ and $\bar{\mathbf{m}} = [\bar{x} \ \bar{y} \ 1]^T$. Remember that for a linear perspective projection with no distortion, $\bar{\mathbf{m}} = \mathbf{K}^{-1}\mathbf{w}$, with $\mathbf{K}$ defined in (8.5). The X-ray beam in the world coordinate system is defined from (8.3) by

$$\begin{cases} \bar{X} = x\bar{Z}/f \\ \bar{Y} = y\bar{Z}/f \end{cases} . \tag{8.32}$$

The intersection of the X-ray beam with the ellipsoid is shown in Fig. 8.12. A intersection point must satisfy (8.30) and (8.32) simultaneously. Substituting $\bar{X}$ and $\bar{Y}$ from (8.32) in (8.30) and after some slight rearranging we obtain

$$A\bar{Z}^2 + B\bar{Z} + C = 0 \tag{8.33}$$

with

$$A = \frac{r_1^2}{a^2} + \frac{r_2^2}{b^2} + \frac{r_3^2}{c^2},$$

$$B = 2 \left(\frac{r_1 s_{14}}{a^2} + \frac{r_2 s_{24}}{b^2} + \frac{r_3 s_{34}}{c^2} \right),$$

$$C = \frac{h_{14}^2}{a^2} + \frac{h_{24}^2}{b^2} + \frac{h_{34}^2}{c^2} - 1 \quad \text{and}$$

$$r_i = s_{i1} \frac{x}{f} + s_{i2} \frac{y}{f} + s_{i3} \quad \text{for} \quad i = 1, 2, 3.$$

If $B^2 - 4AC > 0$ we obtain two intersection points of the X-ray beam with the ellipsoid given by

$$\bar{X}_{1,2} = \frac{\bar{Z}_{1,2}}{f} x$$

$$\bar{Y}_{1,2} = \frac{\bar{Z}_{1,2}}{f} y$$

$$\bar{Z}_{1,2} = \frac{-B \pm \sqrt{B^2 - 4AC}}{2A}$$

The length of the X-ray beam that penetrates into the ellipsoid can be calculated as

$$d = \sqrt{(\bar{X}_1 - \bar{X}_2)^2 + (\bar{Y}_1 - \bar{Y}_2)^2 + (\bar{Z}_1 - \bar{Z}_2)^2}, \tag{8.34}$$

that can be written as

$$d = \frac{\sqrt{B^2 - 4AC}}{A} \sqrt{\frac{x^2}{f^2} + \frac{y^2}{f^2} + 1}. \tag{8.35}$$

The algorithm to simulate a flaw can be resumed as follows:

1. Calibration: Estimate the parameters of the mapping function $3D \rightarrow 2D$ (focal length f, matrix $\mathbf{H}$, and function $\mathbf{f}$).
2. Setting of X-ray imaging parameters: Define μ and x_{max} according to the energy used by the X-ray source.
3. Definition of the 3D flaw: Define the size of the flaw (parameters a, b, and c) and the location of the flaw in the object (matrix $\mathbf{H}_e$).
4. Location of the superimposed area: Find the pixels (u, v) where the modeled 3D flaw is projected.[3]
5. Computation of intersection length d: For each determined pixel (u, v) find the length of the intersection between the X-ray beam and ellipsoid given by Eq. (8.35).

---

[3] These pixels are defined where $B^2 - 4AC > 0$ in Eq. (8.33).

6. Change of the gray value: For each determined pixel (u, v) change the original gray value using (8.27).

Python Example 8.3: In this example, we simulate a defect as an ellipsoid using the method outlined in this section.

Listing 8.3 : Simulation of a defect in an aluminum casting

```python
import numpy as np
import matplotlib.pylab as plt
from pyxvis.simulation.xsim import ellipsoid_simulation
from pyxvis.io import gdxraydb
from pyxvis.geometry.projective import rotation_matrix_3d

image_set = gdxraydb.Castings()
I = np.double(image_set.load_image(21,27)) # wheel image

# Transformation (X,Y,Z)-->(Xb,Yb,Zb)
R1 = rotation_matrix_3d(0,0,0)
t1 = np.array([-36, 40, 1000])
S = np.vstack([np.hstack([R1, t1[:, np.newaxis]]), np.array([0, 0, 0, 1])])

# Transformation (Xp,Yp,Zp)-->(X,Y,Z)
R2 = rotation_matrix_3d(0,0,np.pi/3)
t2 = np.array([0,0,0])
Se = np.vstack([np.hstack([R2, t2[:, np.newaxis]]), np.array([0, 0, 0, 1])])

# Transformation (Xp,Yp,Zp)-->(Xb,Yb,Zb)
SSe = np.matmul(S,Se)

# Transformation (x,y)-->(u,v)
K = np.array([[1.1, 0, 235], [0, 1.1, 305], [0,0,1]])

# Dimensions of the ellipsoid
abc = (5,4,3)

# Simulation
J = ellipsoid_simulation(I,K,SSe,1500,abc,0.1,400)

# Output
fig1, ax = plt.subplots(1, 2, figsize=(16, 8))
ax[0].imshow(I, cmap='gray'), ax[0].axis('off')
ax[1].imshow(J, cmap='gray'), ax[1].axis('off')
ax[1].text(328,225, '(Real)', fontsize=10,color='white')
ax[1].text(315,150, '(Simulated)', fontsize=10,color='white')
plt.show()
```

The output of this code is shown in Fig. 8.13. In this example, the simulated defects seems to be real. The defect was simulated using command ellipsoid_simulation of pyxvis Library. □

In the following, the results of the simulation of flaws in cast aluminum wheels using our approach are presented. The dimensions of the wheels used in our experiments were approximately 48 cm in diameter and 20 cm in height. The focal length (distance between X-ray source and entrance screen of the image intensifier) was 90 cm. The projection model of the X-ray imaging system was calibrated using a hyperbolic model [15, 19].

In Fig. 8.11, experimental results on four X-ray images are shown. The values used to simulate the flaws in each image are summarized in Table 8.2. We can com-

Table 8.2 Values used in the simulations of Fig. 8.11

Image No.	E (keV)	μ (1/cm)	x_{max} (cm)	a (mm)	b (mm)	c (mm)
1	54	0.8426	4.0	8	2	4
2	58	0.7569	3.8	4	2	1.5
3	50	0.9500	4.5	4	2	1.7
4	57	0.7765	3.85	6	3	2.5

pare real and simulated flaws. It was shown that the simulation results are almost identical to real flaws. In Fig. 8.14 a 3D plot of the gray values in the vicinity of the flaws shown in last X-ray image of Fig. 8.11 is illustrated. Due to the irregularity of the gray values of the simulated flaw, it seems to be real.

In defect detection, it is very common that the class of defects is underrepresented with a low number of samples. For this reason, it is very convenient to increase the number of samples by adding some simulated defects. This data augmentation strategy was used in [18], where ellipsoidal defects from different sizes and orientations were superimpose onto real X-ray images in many locations (some examples are

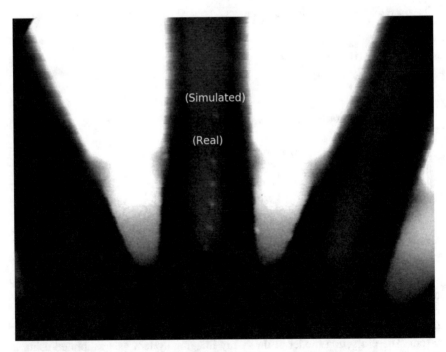

Fig. 8.13 Comparison of real defects with a simulated one (see red square) using proposed method. [→ Example 8.3 🐍]

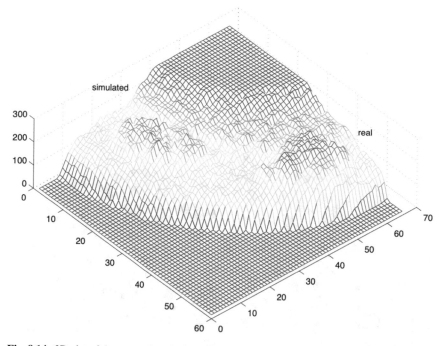

Fig. 8.14 3D plot of the gray values in the vicinity of flaws of the last X-ray image of Fig. 8.11

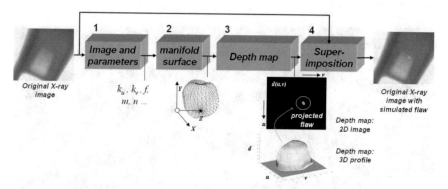

Fig. 8.15 Flaw simulation process using complex CAD models of the 3D defect

shown in Fig. 8.18. In this case, a CNN model, called Xnet-II, with 30 layers and more than 1.350.000 parameters was trained using a dataset with around 640.000 patches containing 50% of ellipsoidal defects and 50% of real background captured from different casting types.

Other complex defect shapes can be simulated using CAD models [21]. This general approach follows the block diagram of Fig. 8.15, where a 3D defect needs to be modeled as a manifold 3D mesh as illustrated in Fig. 8.16. Crack simulation can be

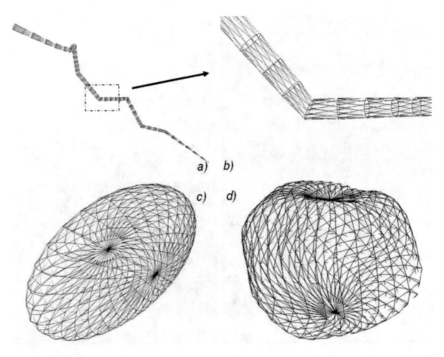

Fig. 8.16 Manifold surfaces from the 3D modeling software: **a** crack, **b** zoom of **a**, **c** ellipsoid, and **d** amorphous surface

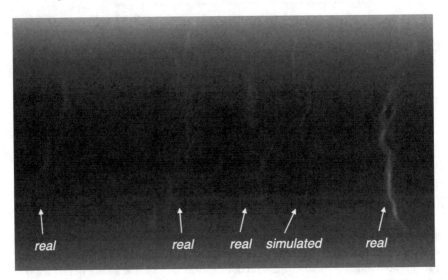

Fig. 8.17 Simulated and real cracks

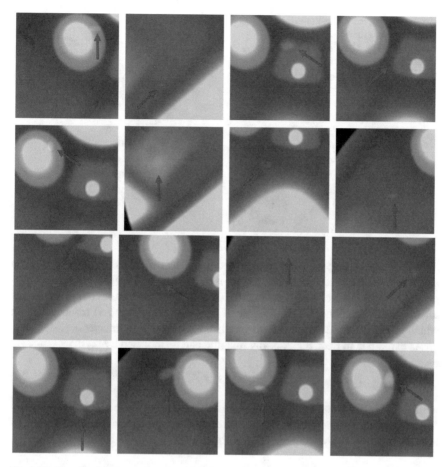

Fig. 8.18 Simulated defects using ellipsoidal model (see red arrows)

obtained by superimposing a depth map computed from a single manifold (see, for example, Fig. 8.16a). However, a real crack corresponds to a more complex 3D representation. For this reason, we simulated another crack by superimposing several single cracks onto a real X-ray image. An example of this simulation is illustrated in Fig. 8.17. Due to the irregularity of the gray values of the simulated flaw, it can be seen that both real and simulated flaws show similar patterns (Fig. 8.18).

8.5 Superimposition Using Multiplication of Images

In this section, we explain a method that can be used to simulate new X-ray images for baggage inspection [22], where simulated images can be used in training pro-

Fig. 8.19 Setup of an X-ray
imaging system, the X-ray
source irradiates the object
(a bag containing a handgun)
and produces an X-ray image

grams for human inspectors, or can be used to enhance datasets for computer vision
algorithms. The key idea of this approach is to build new X-ray images by superimposing X-ray images of objects of interest onto X-ray images of clutter. In our experiments, we simulate new X-ray images of handguns, shuriken, and razor blades, in
which it is impossible to distinguish simulated and real X-ray images.

As explained in Sect. 8.2.2, X-ray imaging can be modeled by

$$\varphi(d) = \varphi_0 e^{-\mu d} \tag{8.36}$$

with μ absorption coefficient, d thickness of the irradiated matter, φ_0 incident energy
flux density, and φ energy flux density after passage through matter with the thickness of d. As we can see in Fig. 8.2, for n materials

$$\varphi = \varphi_0 \exp\left(-\sum_{i=1}^{n} \mu_i d_i\right). \tag{8.37}$$

Finally, the grayvalue of a pixel can be linearly modeled as

$$I = A\varphi + B. \tag{8.38}$$

where A and B are constant parameters of the model.

Following the models (8.37) for the energy flux density and (8.38) for the digital
image, it is possible to model the X-ray image of the foreground (I_f), e.g., a handgun, and the background (I_b), e.g., a cluttered bag, as illustrated in Figs. 8.19 and
8.20. Thus,

$$I_f = A\varphi_f + B \qquad I_b = A\varphi_b + B, \tag{8.39}$$

where

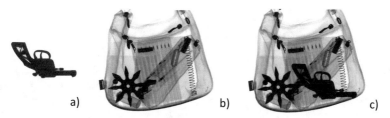

Fig. 8.20 Superimposition of an X-ray image of a handgun onto an X-ray image of a cluttered bag: **a** I_f: Foreground (threat object). **b** I_b: Background (cluttered bag). **c** I_t: Total (bag with threat object)

$$\varphi_f = \varphi_0 e^{-\mu_f d_f} \qquad \varphi_b = \varphi_0 e^{-\mu_b d_b} \tag{8.40}$$

in this case μ_f and μ_b are the absorption coefficients of the foreground and background respectively. It is worth mentioning, that $\mu_b d_b$ represents $\sum_j \mu_j d_j$ considering all cluttered objects j that lie on the X-ray beam shown in Fig. 8.2. The total X-ray image, called I_t, can be modeled as

$$\varphi_t = \varphi_0 e^{-\mu_f d_f} e^{-\mu_b d_b}, \tag{8.41}$$

$$I_t = A\varphi_t + B = Ce^{-\mu_f d_f} e^{-\mu_b d_b} + B, \tag{8.42}$$

where $C = A\varphi_0$. It is clear, that (8.42) can be used to simulate new X-ray images from I_f and I_b: replacing (8.40) in (8.39), we obtain

$$e^{-\mu_f d_f} = \frac{I_f - B}{C} \qquad e^{-\mu_b d_b} = \frac{I_b - B}{C}. \tag{8.43}$$

From (8.43) and (8.42), it yields

$$\frac{I_t - B}{C} = \frac{I_f - B}{C} \cdot \frac{I_b - B}{C}. \tag{8.44}$$

We can normalize the X-ray images by subtracting B and dividing by C, e.g., , $J_t = (I_t - B)/C$. Thus, using the normalized images for total, foreground and background images, we obtain

$$J_t = J_f \cdot J_b. \tag{8.45}$$

Easily, we can compute the total image by

$$I_t = C \cdot J_f \cdot J_b + B. \tag{8.46}$$

I_f

I_b

I_t

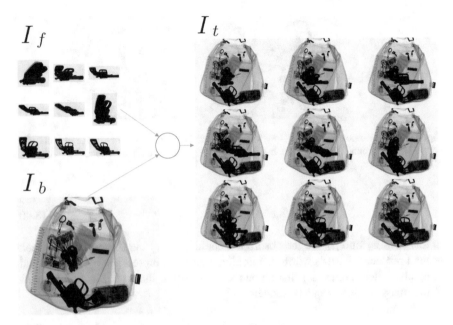

Fig. 8.21 Simulation of a handgun superimposed onto an X-ray image of a bag. The original X-ray image (I_b) has only one shuriken, whereas the simulated X-ray image (I_t) has an additional handgun in different poses (see the handgun in the middle of the image)

Indeed, image I_t in Fig. 8.20c was simulated from I_f and I_b in Fig. 8.20a, b respectively using (8.46). The simple simulation approach is summarized in Algorithm 2. It is worth mentioning that parameters, B and C can be easily estimated using a calibration approach of Sect. 8.4.3 as follows: the maximal grayvalue I_0 in an X-ray image is obtained when there is no irradiated object, i.e., it can be modeled as an object with no thickness ($d_0 = 0$). Additionally, the minimal grayvalue I_1 is obtained when the irradiated object is so thick that the X-rays are completely absorbed, i.e., it can be modeled as an object with known μ of thickness d_1. Substituting these values in (8.38) with $C = A\varphi_0$, it yields

$$\begin{cases} I_0 = A\varphi_0 e^{-\mu d_0} + B = Ce^{-\mu d_0} + B = C + B \\ I_1 = A\varphi_0 e^{-\mu d_1} + B = Ce^{-\mu d_1} + B \end{cases}. \tag{8.47}$$

From these equations, one may compute the value for B and C as

$$\begin{aligned} B &= k(-I_0 e^{-\mu d_1} + I_1), \\ C &= k(I_0 - I_1) \end{aligned}, \tag{8.48}$$

where $k = 1/(1 - e^{-\mu x_1})$.

Algorithm 2 Superimposition of X-ray images

Input: Foreground X-ray image I_f
Input: Background X-ray image I_b
Input: Parameters B and C
  1: $J_f = (I_f - B)/C$
  2: $J_b = (I_b - B)/C$
Output: $I_t = C \cdot J_f \cdot J_b + B$

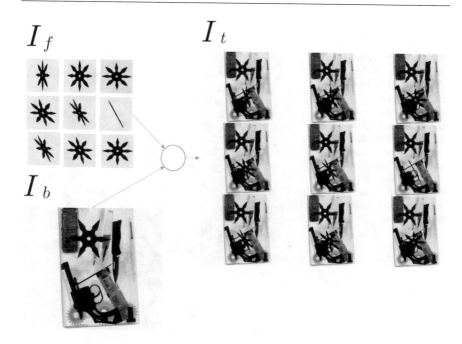

Fig. 8.22 Simulation of a shuriken superimposed onto an X-ray image of a bag. The original X-ray image (I_b) has only one shuriken, whereas the simulated X-ray image (I_t) has an additional shuriken in different poses (see the shuriken partial occluded by the gun)

Simulated images can be used in training programs for human inspectors, or can be used to enhance datasets for computer vision algorithms. The idea is simple, we have to acquire X-ray images of objects that are completely isolated and then we can superimposed them onto X-ray images of cluttered bags. In order to acquire isolated X-ray images, the threat object can be located inside a sphere of expanded polystyrene (EPS) due to its low X-ray absorption coefficient as suggested in [25]. In GDXray+ we have those kinds of images, where a threat object is irradiated from different points of views. Thus, the threat object can be superimposed in many different poses.

In order to illustrate the similarity between original and simulated X-ray images, we show experiments where the original X-ray image has only one threat object and the simulated image has the original threat object and the superimposed threat object, so in the same image we can compare both of them. We tested with the following threat objects: handguns, razor blades, and shuriken (ninja stars) in nine different poses. The results are given in Figs. 8.21 and 8.22 respectively. In our results, the reader can see both threat objects—simulated and original—and can conclude that both objects are so similar that it is impossible to say which one is simulated and which one is the original.

Fig. 8.23 Simulation of three threat objects onto an X-ray image of a backpack. The original X-ray image (Top) has only one handgun, whereas the simulated X-ray image (Bottom) contains the original handgun and the three superimposed three objects. The location of the bounding boxes can be easily defined because they correspond to the location where the foreground objects have been superimposed. [→ Example 8.4 🐍]

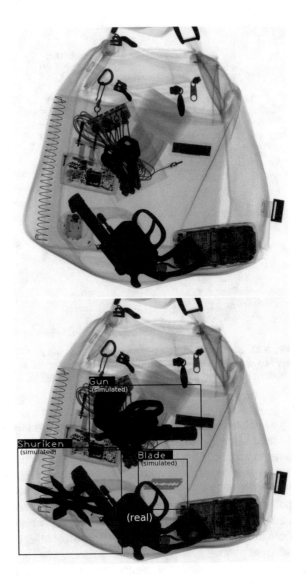

Python Example 8.4: In this example, we superimpose a gun, a shuriken, and a razor blade onto an X-ray image of a backpack using Algorithm 2.

Listing 8.4 : Superimposition of X-ray images

```python
import numpy as np
import matplotlib.pylab as plt
from pyxvis.simulation.xsim import superimpose_xray_images, draw_bounding_box

import gdxraydb

image_set = gdxraydb.Baggages()
Ib  = np.double(image_set.load_image(46, 2)) # background image

If1 = np.double(image_set.load_image(49, 2)) # foreground: Gun
p1  = [700,700]
It  = superimpose_xray_images(Ib, If1, p1[0], p1[1])

If2 = np.double(image_set.load_image(50, 4)) # foreground: Shuriken
p2  = [1200,100]
It  = superimpose_xray_images(It, If2,p2[0], p2[1])

If3 = np.double(image_set.load_image(51, 2)) # foreground: Razor Blade
p3  = [1300,1100]
It  = superimpose_xray_images(It, If3,p3[0],p3[1])

It = draw_bounding_box(It,p1[0],p1[1],If1.shape[0],If1.shape[1],'Gun')
It = draw_bounding_box(It,p2[0],p2[1],If2.shape[0],If2.shape[1],'Shuriken')
It = draw_bounding_box(It,p3[0],p3[1],If3.shape[0],If3.shape[1],'Blade')

fig1, ax = plt.subplots(1, 2, figsize=(16, 8))
ax[0].imshow(Ib, cmap='gray'), ax[0].axis('off')
ax[1].imshow(It, cmap='gray'), ax[1].axis('off')
ax[1].text(p1[1]+20,p1[0]+50, '(simulated)', fontsize=8,color='white')
ax[1].text(p2[1]+20,p2[0]+50, '(simulated)', fontsize=8,color='black')
ax[1].text(p3[1]+20,p3[0]+50, '(simulated)', fontsize=8,color='black')
ax[1].text(1000,1800, '(real)', fontsize=12,color='white')
plt.show()
```

The output of this code is shown in Fig. 8.23. The reader can observe the use of function superimpose_xray_images to superimpose a foreground image (e.g., If1) onto the background image Ib. □

Simulated images can be used in training programs for human inspectors, or can be used to enhance datasets for computer vision algorithms. As we can see in Fig. 8.23, the location of the bounding boxes can be easily and precisely defined because we know exactly where the foreground object has been located. This means, that the definition of the ground truth (of the superimposed objects) in the enhanced dataset is straightforward. This simulation model was used in [26] as data augmentation strategy to enhance the training dataset, where GDXray+ series B0083—X-ray images of backpacks with no threat objects—was used as background images, and series B0049-51—X-ray images of threat objects—was used as foreground images.[4]

---

[4]See details of implementation in https://github.com/dlsaavedra/Detector_GDXray, and some examples in Sect. 7.7.6.

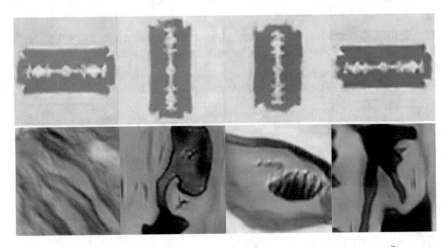

Fig. 8.24 Simulation of razor blades and background using GAN[6]. [→ Example 7.6 🐍]

8.6   Simulation of X-ray Images Using GAN

Generative Adversarial Networks (GANs) have been used successfully in the last years to generate realistic synthetic data [5, 7, 14]. In X-ray testing, we use GAN to simulate X-ray images, for example, as data augmentation in training data to increase the number of samples of some underrepresented class, or as new data in a training course for human inspectors. Some applications of simulated X-ray images using GAN can be found in [18] for the simulation of casting defects[5] and in [26, 31, 33, 34] for the simulation of threat objects. The simulated X-ray images using GAN are very realistic as we can see in Figs. 7.25 and 7.26 for defects and shuriken respectively. The reader is referred to Sect. 7.6 that gives further details of GAN in X-ray testing, where the theory is explained and some examples in Python are shown. Using Python code of Example 7.6, we generate two extra simulations that are shown in Fig. 8.24.[6]

---

[5]GAN solutions have been used in other kinds of defects, see, for example, [23].

[6]For the simulation of razor blades and background, we used training datasets `blade_64 × 64.npy` and `back_64 × 64.npy` respectively that can be download from https://domingomery.ing.puc.cl/material/. [→ Example 7.6 🐍].

8.7 Simulation with aRTist

The *analytical Radiographic Testing inspection simulation tool* (aRTist) is a simulation software that has been developed by BAM[7] for quantitative description of X-ray testing. It can model the radiation source, attenuation of radiation, X-ray films, and digital detectors using an interactive virtual scene with CAD interface [2, 3].

In aRTist, the simulation model consists of four components:

- Radiation source: It is modeled as an X-ray beam and its energy spectrum (that depends on the interaction X-ray—penetrated material). The focal spot is defined as a raster of point sources.
- Interaction of radiation with material: It is modeled considering the photoelectric effect, coherent and incoherent scattering, and pair production (for photon energies larger than 1 MeV), electron binding effects, and X-ray fluorescence. Additionally, it calculates X-ray spectra based on the interaction cross sections for Bremsstrahlung generation.
- Detection of radiation: The simulation of the X-ray image is performed by tracing beams from source points to detector pixels. Different types of X-ray films and X-ray detectors are also modeled Transmission functions like the characteristic film curves for different types of film classes are used to describe the properties of different detectors. Blurriness can be simulated by Gaussian kernels. The simulation of noise can be added to generate more realistic X-ray images.
- Geometry of the object under test: The geometry is provided in STL format in footnote 1 that is used to model boundary representations of volumes of homogeneous materials. In aRTist, facetted (triangulated) boundaries are used. Curves are approximated to many planes.

Some examples are illustrated in Figs. 8.25 and 8.26. The simulations are very realistic and can be easily used in many purposes of X-ray testing, e.g., evaluation of X-ray systems, testing of computer vision algorithms, training of human operators, etc.

8.8 Summary

To evaluate the performance of computer vision techniques, it is convenient to examine simulated data. This offers the possibility of tuning the parameters of the computer vision algorithm and to testing how it works in critical cases.

A simulation tool should model the physics of the X-ray formation (generation, interaction, and detection) and handle complex 3D objects efficiently. State of the art of computer modeling of X-ray testing methods are able to simulate different X-ray

---

[7]BAM, Bundesanstalt für Materialprüfung, is a senior scientific and technical federal institute with responsibility to the Federal Ministry for Economic Affairs and Energy of Germany (see https://www.bam.de). For aRTist see http://artist.bam.de.

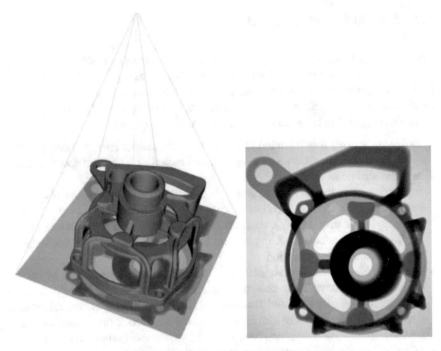

Fig. 8.25 Simulation of an X-ray image of a casting using aRTist: virtual model and simulated X-ray image. Courtesy of Carsten Bellon of BAM, Germany

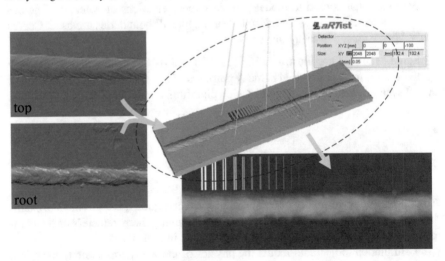

Fig. 8.26 Simulation of an X-ray image of a weld using aRTist: virtual model and simulated X-ray image. Courtesy of Carsten Bellon of BAM, Germany

spectrum and X-ray source size, varied photon–matter interactions, and several X-ray detector responses. Thus, a computer simulator for X-ray testing should include the following modules: X-ray source model, ray-tracing engine, material database, straight line attenuation model and detector model.

In this chapter, we reviewed some basic concepts of simulation of X-ray images. We gave simple geometric and imaging models that can be used in the simulation. We explained the basic simulation principles and we addressed some techniques to simulate defects (that can be used to assess the performance of a computer vision method for automated defect recognition). Afterwards, the chapter gives an overview of the use of Generative Adversarial Networks (GANs) in the simulation of realistic X-ray images. Finally, we present 'aRTist', a simulation software that can be used to generate very realistic X-ray images. The chapter has some Python examples that the reader can run and follow. Examples of simulated defects in castings and welds, and simulated threat objects are also given.

References

1. Bavendiek, K.: Prüfkörper für die automatischen überprüfung der Bildqualität und der Messung der Erkennungssicherheit bei ADR Systemen. In: German Conference on Nondestructive Testing. Berlin (2001). (in German)
2. Bellon, C., Deresch, A., Jaenisch, G.R.: Radiography simulation with artist-combining analytical and monte carlo methods. In: Proceedings of International Symposium on Digital Industrial Radiology and Computed Tomography (DIR2015), Ghent, Belgium (2015)
3. Bellon, C., Jaenisch, G.R.: Artist–analytical RT inspection simulation tool. In: International Symposium on Digital industrial Radiology and Computed Tomography (DIR2007), June 25–27, 2007, Lyon, France, pp. 25–27 (2007)
4. Botsch, M., Kobbelt, L., Pauly, M., Alliez, P., Lévy, B.: Polygon Mesh Processing. CRC Press, Boca Raton (2010)
5. Brock, A., Donahue, J., Simonyan, K.: Large scale gan training for high fidelity natural image synthesis. In: The International Conference on Learning Representations (ICLR 2019), pp. 1–35 (2019)
6. Carvajal, K., Chacón, M., Mery, D., Acuna, G.: Neural network method for failure detection with skewed class distribution. Insight **46**(7), 399–402 (2004)
7. Creswell, A., White, T., Dumoulin, V., Arulkumaran, K., Sengupta, B., Bharath, A.A.: Generative adversarial networks: an overview. IEEE Signal Process. Mag. **35**(1), 53–65 (2018)
8. Duvauchelle, P., Freud, N., Kaftandjian, V., Babot, D.: A computer code to simulate X-ray imaging techniques. Nucl. Instrum. Methods Phys. Res. B **2000**(170), 245–258 (2000)
9. Filbert, D., Klatte, R., Heinrich, W., Purschke, M.: Computer aided inspection of castings. In: IEEE-IAS Annual Meeting, Atlanta, USA, pp. 1087–1095 (1987)
10. Freud, N., Duvauchelle, P., Babot, D.: Simulation of X-ray NDT imaging techniques. In: Proceedings of the 15th World Conference on Non-Destructive Testing (WCNDT–2000), Rome (2000)
11. Hecker, H.: A new method to process X-ray images in the automated inspection of castings. Ph.D. thesis, Institute for Measurement and Automation, Faculty of Electrical Engineering, Technical University of Berlin (1995). (in German)
12. Heinrich, W.: Automated inspection of castings using X-ray testing. Ph.D. thesis, Institute for Measurement and Automation, Faculty of Electrical Engineering, Technical University of Berlin (1988). (in German)

13. Huang, Q., Wu, Y., Baruch, J., Jiang, P., Peng, Y.: A template model for defect simulation for evaluating nondestructive testing in X-radiography. IEEE Trans. Syst. Man Cybern. Part A: Syst. Hum. **39**(2), 466–475 (2009)

14. Karras, T., Laine, S., Aila, T.: A style-based generator architecture for generative adversarial networks. In: Proceedings of the IEEE Conference on Computer Vision and Pattern Recognition, pp. 4401–4410 (2019)

15. Mery, D.: Automated Flaw Detection in Castings from Digital Radioscopic Image Sequences. Verlag Dr. Köster, Berlin (2001). (Ph.D. Thesis in German)

16. Mery, D.: Flaw simulation in castings inspection by radioscopy. Insight **43**(10), 664–668 (2001)

17. Mery, D.: A new algorithm for flaw simulation in castings by superimposing projections of 3D models onto X-ray images. In: Proceedings of the XXI International Conference of the Chilean Computer Science Society (SCCC-2001), pp. 193–202. IEEE Computer Society Press, Punta Arenas (2001)

18. Mery, D.: Aluminum casting inspection using deep learning: a method based on convolutional neural networks. J. Nondestruct. Eval. **39**(1), 12 (2020)

19. Mery, D., Filbert, D.: The epipolar geometry in the radioscopy: theory and application. at - Automatisierungstechnik **48**(12), 588–596 (2000). (in German)

20. Mery, D., Filbert, D.: Automated flaw detection in aluminum castings based on the tracking of potential defects in a radioscopic image sequence. IEEE Trans. Robot. Autom. **18**(6), 890–901 (2002)

21. Mery, D., Hahn, D., Hitschfeld, N.: Simulation of defects in aluminum castings using cad models of flaws and real X-ray images. Insight **47**(10), 618–624 (2005)

22. Mery, D., Katsaggelos, A.K.: A logarithmic X-ray imaging model for baggage inspection: simulation and object detection. In: Proceedings of the IEEE Conference on Computer Vision and Pattern Recognition Workshops, pp. 57–65 (2017)

23. Niu, S., Li, B., Wang, X., Lin, H.: Defect image sample generation with GAN for improving defect recognition. IEEE Trans. Autom. Sci. Eng. (2020)

24. Rebuffel, V., Tabary, J., Tartare, M., Brambilla, A., Verger, L.: SINDBAD: a simulation software tool for multi-energy X-ray imaging. In: Proceedings of 11th European Conference on Non Destructive Testing, Prague (2014)

25. Riffo, V., Mery, D.: Automated detection of threat objects using adapted implicit shape model. IEEE Trans. Syst. Man Cybern.: Syst. **46**(4), 472–482 (2016)

26. Saavedra, D., Banerjee, S., Mery, D.: Detection of threat objects in baggage inspection with X-ray images using deep learning. Neural Comput. Appl. pp. 1–17. Springer (2020)

27. Salvat, F., Fernández-Varea, J.M., Sempau Roma, J.: Penelope-2008: A code system for monte carlo simulation of electron and photon transport. In: Workshop Proceedings, Barcelona, 30 June–3 July 2, 2008. OECD (2009)

28. Schumm, A., Duvauchelle, P., Kaftandjian, V., Jaenisch, R., Bellon, C., Tabary, J., Mathy, F., Legoupil, S.: Modelling of radiographic inspections. In: Nondestructive Testing of Materials and Structures, pp. 697–702. Springer (2013)

29. Tabary, J., Hugonnard, P., Mathy, F.: SINDBAD: a realistic multi-purpose and scalable X-ray simulation tool for NDT applications. In: International Symposium on DIR and CT, Lyon, vol. 1, pp. 1–10 (2007)

30. Tillack, G.R., Nockemann, C., Bellon, C.: X-ray modelling for industrial applications. NDT & E Int. **33**(1), 481–488 (2000)

31. Yang, J., Zhao, Z., Zhang, H., Shi, Y.: Data augmentation for X-ray prohibited item images using generative adversarial networks. IEEE Access **7**, 28894–28902 (2019)

32. Yao, M., Duvauchelle, P., Kaftandjian, V., Peterzol-Parmentier, A., Schumm, A.: X-ray imaging plate performance investigation based on a Monte Carlo simulation tool. Spectrochim. Acta Part B: At. Spectrosc. **103**, 84–91 (2015)

33. Zhao, Z., Zhang, H., Yang, J.: A gan-based image generation method for X-ray security prohibited items. In: Chinese Conference on Pattern Recognition and Computer Vision (PRCV), pp. 420–430. Springer (2018)

34. Zhu, Y., Zhang, Y., Zhang, H., Yang, J., Zhao, Z.: Data augmentation of X-ray images in baggage inspection based on generative adversarial networks. IEEE Access **8**, 86536–86544 (2020)

Chapter 9
Applications in X-ray Testing

Abstract In this chapter, relevant applications on X-ray testing are described. We cover X-ray testing in (i) castings, (ii) welds, (iii) baggage, (iv) natural products, and (v) others (like cargos and electronic circuits). For each application, the state of the art is presented. Approaches in each application are summarized showing how they use computer vision techniques. A detailed approach is shown in each application and some examples using Python are given in order to illustrate the performance of the methods.

Cover Image: 3D representation of the X-ray image of a wheel (X-ray image C0023_0001 *colored with 'sinmap' colormap).*

© Springer Nature Switzerland AG 2021
D. Mery and C. Pieringer, *Computer Vision for X-Ray Testing*,
https://doi.org/10.1007/978-3-030-56769-9_9

9.1   Introduction

In this chapter, we review some relevant applications in X-ray testing such as (i) castings, (ii) welds, (iii) baggage, (iv) natural products, and (v) others (like cargos and electronic circuits). For the first four application applications, in which the authors have been undertaking research over the last decades, we will present a description, the state of the art, a detailed approach and an example in Python. For the last application, different techniques are mentioned.

9.2   Castings

Light-alloy castings produced for the automotive industry, such as wheel rims, steering knuckles, and steering gear boxes are considered important components for overall roadworthiness. Non-homogeneous regions can be formed within the work piece in the production process. These are manifested, for example, by bubble-shaped voids, fractures, inclusions, or slag formation. To ensure the safety of construction, it is necessary to check every part thoroughly using X-ray testing. In casting inspection, automated X-ray systems have not only raised quality, through repeated objective inspections and improved processes, but have also increased productivity and consistency by reducing labor costs. Some examples are illustrated in Fig. 9.1.

9.2.1   State of the Art

Different methods for the automated detection of casting discontinuities using computer vision have been described in the literature over more than thirty years [22, 42]. In the past, the published approaches to detecting were divided into three groups [100]:

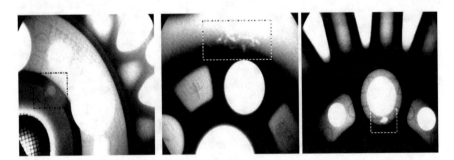

Fig. 9.1 Real defects in X-ray images of wheels

- Reference methods: In reference methods, it is necessary to take still images at selected programmed inspection positions. A test image is then compared with the reference image. If a significant difference is identified, the test piece is classified as defective.
- Methods without apriori knowledge of the structure: These approaches using pattern recognition, expert systems, artificial neural networks, general filters or multiple-views analyzes to make them independent of the position and structure of the test piece.
- Computed tomography: These approaches use computed tomography to make a reconstruction of the cast piece and thereby detect discontinuities.

Nowadays, computed tomography and multiple views for the inspection of castings are rarely used. It is clear that the methods that achieve the best performance are based on deep learning using single views. Deep learning has been successfully used in image and video recognition (see, for example, [20, 77, 155]), and it has been established as the state of the art in many areas of computer vision. The key idea of deep learning, as we show in Chap. 7, is to replace *handcrafted* features with features that are *learned* efficiently using a hierarchical feature extraction approach.

Selected approaches are summarized in Table 9.1. In this table, we follow the 3$\mathbb{X}$-strategy outlined in Sect. 1.8, in which we distinguish (i) the X-ray energy used to generate the X-ray images (monochromatic, dual-, or multi-energy), (ii) the number of views used by the algorithms (single-view, multi-views, or computed tomography) and complexity of the algorithms (simple, medium, and complex—here, deep learning methods–). In this area, the automated systems are very effective, because the inspection task is fast and obtains a high performance.

9.2.2 An Application

In this section, we present a method for the automated detection of flaws based on *tracking principle* in an X-ray image sequence, i.e., first, it identifies potential defects in each image of the sequence, and second, it matches and tracks these from image to image. The key idea is to consider as false alarms those potential defects which cannot be tracked in the sequence [107]. The method for automated flaw detection presented here has basically two steps (see Fig. 9.2): *identification* and *tracking of potential flaws*. These will be described in this section.

Identification of Potential Flaws

A digital X-ray image sequence of the object test is acquired (see, for example, series C0001 of GDXray+). In order to ensure the tracking of flaws in the X-ray images, similar projections of the specimen must be achieved along the sequence. For this reason, the sequence consists of X-ray images taken by the rotation of the casting at small intervals (e.g., 5^0). Since many images are captured, the time of the data acquisition is reduced by taking the images without frame averaging. The position of

Table 9.1 State of art in inspection of castings

Authors	Year	Ref	$\mathbb{X}_1^*$ energies 1 2 3	$\mathbb{X}_2^*$ views 1 2 3	$\mathbb{X}_3^*$ algorithms 1 2 3
Bandara et al.	2020	[13]	⊞ ⊞ □	□ □ ⊞	⊞ □ □
Carrasco and Mery	2011	[24]	⊞ □ □	⊞ ⊞ □	⊞ ⊞ □
Cogranne and Retraint	2014	[27]	⊞ □ □	⊞ □ □	⊞ ⊞ □
Du et al.	2019	[34]	⊞ □ □	⊞ □ □	□ □ ⊞
Ferguson et al.	2017	[40]	⊞ □ □	⊞ □ □	□ □ ⊞
Ferguson et al.	2017	[41]	⊞ □ □	⊞ □ □	□ □ ⊞
Jin et al.	2020	[66]	⊞ □ □	⊞ □ □	⊞ ⊞ □
Kamalakannan and Rajamanickam	2017	[68]	⊞ □ □	⊞ □ □	⊞ ⊞ □
Li et al.	2006	[81]	⊞ □ □	⊞ □ □	⊞ ⊞ □
Li et al.	2015	[80]	⊞ □ □	⊞ □ □	⊞ ⊞ □
Li et al.	2019	[79]	⊞ □ □	□ □ ⊞	⊞ □ □
Lin et al.	2018	[86]	⊞ □ □	⊞ □ □	□ □ □
Mery and Filbert	2002	[107]	⊞ □ □	⊞ ⊞ □	⊞ ⊞ □
Mery et al.	2013	[114]	⊞ □ □	⊞ ⊞ □	⊞ ⊞ □
Mery	2015	[103]	⊞ □ □	⊞ ⊞ □	⊞ ⊞ □
Mery and Arteta	2017	[105]	⊞ □ □	⊞ □ □	⊞ ⊞ ⊞
Mery	2020	[104]	⊞ □ □	⊞ □ □	□ □ ⊞
Pieringer and Mery	2010	[136]	⊞ □ □	⊞ ⊞ □	⊞ ⊞ □
Pizarro et al.	2008	[137]	⊞ □ □	⊞ ⊞ □	⊞ ⊞ □
Ramirez and Allende	2013	[138]	⊞ □ □	⊞ □ □	⊞ ⊞ □
Ren et al.	2019	[139]	⊞ □ □	⊞ □ □	□ □ ⊞
Tang et al.	2019	[163]	⊞ □ □	⊞ □ □	□ □ ⊞
Tang et al.	2009	[162]	⊞ □ □	⊞ □ □	⊞ ⊞ □
Yahaghi et al.	2020	[179]	⊞ □ □	⊞ □ □	⊞ ⊞ □
Yong et al.	2016	[182]	⊞ □ □	⊞ □ □	⊞ ⊞ □
Zhao et al.	2014	[189]	⊞ □ □	⊞ □ □	⊞ ⊞ □
Zhao et al.	2015	[190]	⊞ □ □	⊞ □ □	⊞ ⊞ □
Zhang et al.	2018	[187]	⊞ □ □	⊞ □ □	⊞ ⊞ □
		*1	Mono	Mono	Simple
		2	Dual	Multi	Medium
		3	Multi	CT	Complex

□ not used, ⊞ used

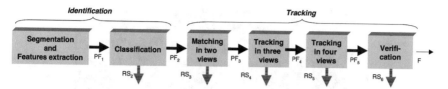

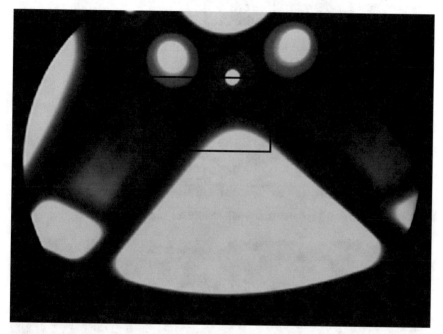

Fig. 9.2 Automated flaw detection in aluminum castings based on the tracking of potential defects in an X-ray image sequence: PF = potential flaws, RS = potential flaws classified as regular structures, F = detected flaws [107]

Fig. 9.3 X-ray image C0001_0030 of an aluminum wheel (see zoom in Fig. 9.4)

the casting, provided on-line by the manipulator is registered at each X-ray image to calculate the perspective projection matrix $\mathbf{P}_p$ (for details see Sect. 3.3.4 and Example 3.5). An X-ray image sequence is shown in Fig. 9.5.

The detection of potential flaws identifies regions in X-ray images that may correspond to real defects. This process takes place in each X-ray image of the sequence without considering information about the correspondence between them. Two general characteristics of the defects are used for identification purposes: (i) a flaw can be considered as a connected subset of the image, and (ii) the gray level difference between a flaw and its neighborhood is significant. However, as the signal-to-noise ratio in our X-ray images is low, the flaws signal is slightly greater than the background noise, as illustrated in Fig. 9.4. In our experiments, the mean gray level of the flaw signal (without background) was between 2.4 and 28.8 gray values with a

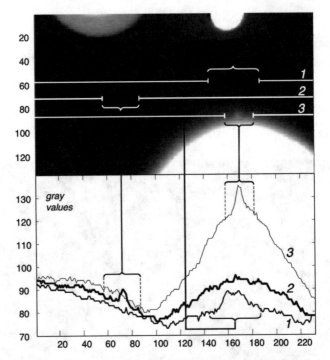

Fig. 9.4 Zoom of Fig. 9.3 and gray level profile along three rows crossing defects

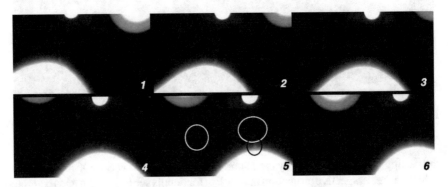

Fig. 9.5 X-ray image sequence with three flaws (image 5 is shown in Fig. 9.4)

standard deviation of 6.1. Analyzing a homogeneous background in different areas of interest of normal parts, we found that the noise signal was within ±13 gray values with a standard deviation of 2.5. For this reason, the identification of real defects with poor contrast can also involve the detection of false alarms.

According to the mentioned characteristics of the real flaws, our method of identification has the following two steps (see Fig. 9.6):

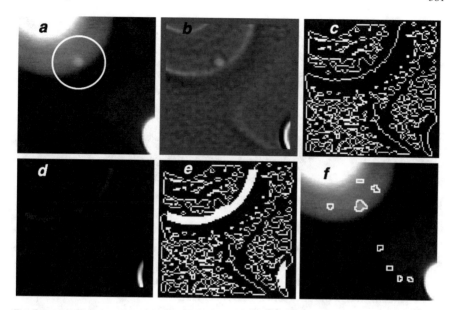

Fig. 9.6 Identification of potential flaws: **a** X-ray image with a small flaw at an edge of a regular structure, **b** Laplacian-filtered image with $\sigma = 1.25$ pixels (kernel size = 11 × 11), **c** zero-crossing image, **d** gradient image, **e** edge detection after adding high gradient pixels, and **f** potential flaws

<u>Edge Detection:</u> A Laplacian-of-Gaussian (LoG) kernel and a zero-crossing algorithm [37] are used to detect the edges of the X-ray images. The LoG-operator involves a Gaussian low-pass filter which is a good choice for the pre-smoothing of our noisy images. The resulting binary edge image should produce at real flaws closed and connected contours which demarcate *regions*. However, a flaw may not be perfectly enclosed if it is located at an edge of a regular structure as shown in Fig. 9.6c. In order to complete the remaining edges of these flaws, a thickening of the edges of the regular structure is performed as follows: (a) the gradient image[1] of the original image is computed (see Fig. 9.6d); (b) by thresholding the gradient image at a high gray level a new binary image is obtained; and (c) the resulting image is added to the zero-crossing image (see Fig. 9.6e).

<u>Segmentation and Classification of Potential Flaws:</u> Afterwards, each closed region is segmented and classified as a potential flaw if (a) its mean gray level is 2.5% greater than the mean gray level of its surroundings (to ensure the detection of the flaws with a poor contrast); and (b) its area is greater than 15 pixels (very small flaws are permitted). A statistical study of the classification of potential flaws using more than 70 features can be found in [108].

---

[1]The gradient image is computed by taking the square root of the sum of the squares of the gradient in a horizontal and vertical direction. These are calculated by the convolution of the X-ray image with the first derivative (in the corresponding direction) of the Gaussian low-pass filter used in the LoG-filter.

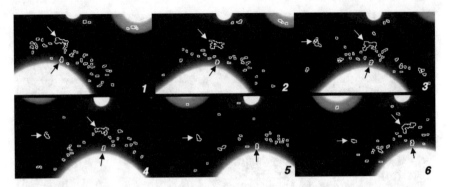

Fig. 9.7 Identification of potential flaws (the arrows indicate real flaws)

This is a very simple detector of potential flaws (see implementation in Example 5.6). However, the advantages are as follows: (a) it is a single detector (it is the same detector for each image), and (b) it is able to identify potential defects independent of the placement and the structure of the specimen.

Using this method, some real defects cannot be identified in all X-ray images in which they appear if the contrast is very poor or the flaw is not enclosed by edges. For example, in Fig. 9.7 one can observe that the biggest real flaw was identified in images 1, 2, 3, 4, and 6, but not in image 5 where only two of the three real flaws were identified (compare with Fig. 9.4). Additionally, if a flaw is overlapped by edges of the structure of the casting, not all edges of the flaw can be detected. In this case, the flaw will not be enclosed and therefore not be segmented. Furthermore, a small flaw that moves in front (or behind) a thick cross section of the casting, in which the X-rays are highly absorbed, may cause an occlusion. In our experiments, this detector identified the real flaws in four or more (not necessarily consecutive) images of the sequence.

Multiple-View Detection

In the previous step, n_1 potential regions were segmented and described in the entire image sequence $\mathbb{I}$. Each segmented region is labeled with a unique number $r \in \mathbf{T}_1 = \{1, ..., n_1\}$. In view i, there are m_i segmented regions that are arranged in a subset $\mathbf{t}_i = \{r_{i,1}, r_{i,2}, ..., r_{i,m_i}\}$, i.e., $\mathbf{T}_1 = \mathbf{t}_1 \cup \mathbf{t}_2 \cup ... \mathbf{t}_m$.

The matching and tracking algorithms combine all regions to generate consistent tracks of the object's parts of interest across the image sequence. The algorithm has the following steps:

Matching in Two Views: All regions in view i that have corresponding regions in the next p views are searched, i.e., regions $r_1 \in \mathbf{t}_i$ that have corresponding regions $r_2 \in \mathbf{t}_j$ for $i = 1, ..., m - 1$ and $j = i + 1, ..., \min(i + p, m)$. In our experiments, we use $p = 3$ to reduce the computational cost. The matched regions (r_1, r_2) are those that meet *similarity* and *location* constraints. The similarity constraint means that corresponding descriptors $\mathbf{y}_{r_1}$ and $\mathbf{y}_{r_2}$ must be similar enough such that

$$||\mathbf{y}_{r_1} - \mathbf{y}_{r_2}|| < \varepsilon_1. \tag{9.1}$$

The location constraint means that the corresponding locations of the regions must meet the epipolar constraint. In this case, the Sampson distance between $\mathbf{x}_{r_1}$ and $\mathbf{x}_{r_2}$ is used, i.e., the first-order geometric error of the epipolar constraint must be small enough such that:

$$|\mathbf{x}_{r_2}^\mathsf{T} \mathbf{F}_{ij} \mathbf{x}_{r_1}| \left(\frac{1}{\sqrt{a_1^2 + a_2^2}} + \frac{1}{\sqrt{b_1^2 + b_2^2}} \right) < \varepsilon_2, \tag{9.2}$$

with $\mathbf{F}_{ij}\mathbf{x}_{r_1} = [a_1\, a_2\, a_3]^\mathsf{T}$ and $\mathbf{F}_{ij}^\mathsf{T}\mathbf{x}_{r_2} = [b_1\, b_2\, b_3]^\mathsf{T}$. In this case, $\mathbf{F}_{ij}$ is the fundamental matrix between views i and j calculated from projection matrices $\mathbf{P}_i$ and $\mathbf{P}_j$ [56] (see Sect. 3.5.1). In addition, the location constraint used is as follows:

$$||\mathbf{x}_{r_1} - \mathbf{x}_{r_2}|| < \rho(j - i), \tag{9.3}$$

because the translation of corresponding points in these sequences is smaller than ρ pixels in consecutive frames.

If we have 3D information about the space where our test object should be, it is worth to evaluating whether the 3D point reconstructed from the centers of mass of the regions must belong to the space occupied by the casting. From $\mathbf{m}_p^a$ and $\mathbf{m}_q^b$ the corresponding 3D point $\hat{\mathbf{M}}$ is estimated using the linear approach of Hartley in [56]. For two views this approach is faster than the least squares technique. It is necessary to examine if $\hat{\mathbf{M}}$ resides in the volume of the casting, the dimensions of which are usually known a priori (e.g., a wheel is assumed to be a cylinder)[2].

Finally, a new matrix $\mathbf{T}_2$ sized $n_2 \times 2$ is obtained with all matched duplets (r_1, r_2), one per row. If a region is found to have no matches, it is eliminated. Multiple matching, i.e., a region that is matched with more than one region, is allowed. Using this method, problems like non-segmented regions or occluded regions in the sequence can be solved by tracking if a region is not segmented in consecutive views.

Matching in 3 Views: Based on the matched regions stored in matrix $\mathbf{T}_2$, we look for triplets (r_1, r_2, r_3), with $r_1 \in \mathbf{t}_i, r_2 \in \mathbf{t}_j, r_3 \in \mathbf{t}_k$ for views i, j, and k. We know that a row a in matrix $\mathbf{T}_2$ has a matched duplet $[T_2(a, 1)\, T_2(a, 2)] = [r_1\, r_2]$. We then look for rows b in $\mathbf{T}_2$ in which the first element is equal to r_2, i.e., $[T_2(b, 1)\, T_2(b, 2)] = [r_2\, r_3]$. Thus, a matched triplet (r_1, r_2, r_3) is found if the regions r_1, r_2, and r_3 meet the trifocal constrain:

$$||\hat{\mathbf{x}}_{r_3} - \mathbf{x}_{r_3}|| < \varepsilon_3, \tag{9.4}$$

---

[2]It is possible to use a CAD model of the casting to evaluate this criterion more precisely. Using this model we could discriminate a small hole of the regular structure that is identified as a potential flaw. Additionally, the CAD model can be used to inspect the casting geometry, as shown in [129].

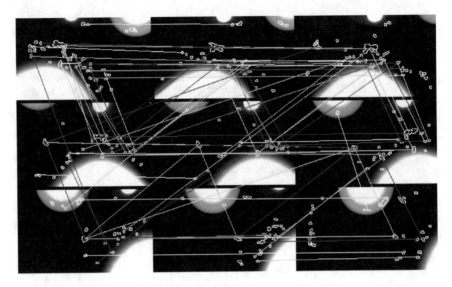

Fig. 9.8 Matching of potential flaws in two views

This means that $\mathbf{x}_{r_3}$ must be similar enough to the re-projected point $\hat{\mathbf{x}}_{r_3}$ computed from the points in views i and j ($\mathbf{x}_{r_1}$ and $\mathbf{x}_{r_2}$), and the trifocal tensors T_i^{jk} of views i, j, k calculated from projection matrices $\mathbf{P}_i$, $\mathbf{P}_j$, and $\mathbf{P}_k$ [56] (see (3.76)). A new matrix $\mathbf{T}_3$ sized $n_3 \times 3$ is built with all matched triplets (r_1, r_2, r_3), one per row. Regions in which the three views do not match are eliminated.

The results of our example are shown in Fig. 9.8.

Matching in More Views: For $v = 4, ..., q \leq m$ views, we can build the matrix recursively $\mathbf{T}_v$, sized $n_v \times v$, with all possible v-tuplets $(r_1, r_2, ..., r_v)$ that fulfill $[T_{v-1}(a, 1) \ldots T_{v-1}(a, v - 1)] = [r_1 \, r_2 \ldots r_{v-1}]$ and $[T_{v-1}(b, 1) \ldots T_{v-1}(b, v - 1)] = [r_2 \ldots r_{l-1} \, r_v]$, for $j, k = 1, ..., n_{v-1}$. No more geometric constraints are required because it is redundant. The final result is stored in matrix $\mathbf{T}_q$. For example, for $q = 4$ we store in matrix $\mathbf{T}_4$ the matched quadruplets (r_1, r_2, r_3, r_4) with $r_1 \in \mathbf{t}_i$, $r_2 \in \mathbf{t}_j, r_3 \in \mathbf{t}_k, r_4 \in \mathbf{t}_l$ for views i, j, k and l.

Figure 9.10 shows the tracked regions of our example that fulfill this criterion. Only two false trajectories are observed (see arrows).

As our detector cannot guarantee the identification of all real flaws in more than four views, a tracking in five views could lead to the elimination of those real flaws that were identified in only four views. However, if a potential flaw is identified in more than four views, more than one quadruplet can be detected. For this reason, these corresponding quadruplets are joined in a trajectory that contains more than four potential flaws (see trajectory with arrows in Fig. 9.10).

The matching condition for building matrix $\mathbf{T}_i, i = 3, ..., q$, is efficiently evaluated (avoiding an exhaustive search) by using a k-d tree structure [21] to search

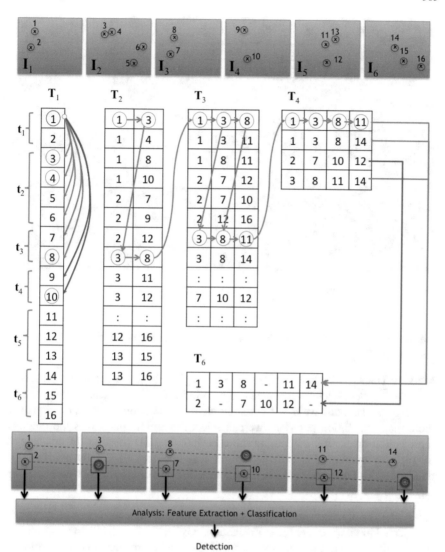

Fig. 9.9 Tracking example with $m = 6$ views. In each view there are 2, 4, 2, 2, 3, and 3 segmented regions, i.e., there are $n_1 = 16$ regions in total. For each region we seek corresponding regions in the next $p = 3$ views (see matching arrows in $\mathbf{T}_1$: region 1 with regions (3, 4, 5, 6) in view 2, regions (7, 8) in view 3, and (9, 10) in view 4). We observe that after tracking in 2, 3, and 4 views there are only two tracks in $\mathbf{T}_6$ that could be tracked in 5 and 4 views respectively. The regions that were not segmented can be recovered by reprojection (see gray circles in views 2, 4, and 6). Finally, each set of tracked regions are analyzed in order to take the final decision

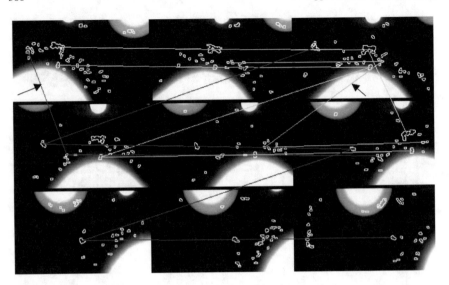

Fig. 9.10 Tracking in more views (the arrows indicate false detections)

the nearest neighbors for zero Euclidean distance between the first and last $i - 2$ columns in $\mathbf{T}_{i-1}$.

Merging Tracks: Matrix $\mathbf{T}_q$ defines tracks of regions in q views. It can be observed that some of these tracks correspond to the same region. For this reason, it is possible to merge tracks that have $q - 1$ common elements. In addition, if a new track has more than one region per view, we can select the region that shows the minimal reprojection error after computing the corresponding 3D location. In this case, a 3D reconstruction of $\hat{\mathbf{X}}$ is estimated from tracked points [56]. Finally, matrix $\mathbf{T}_m$ is obtained with all merged tracks in the m views. See an example of the whole tracking algorithm in Fig. 9.9.

Analysis: The 3D reconstructed point $\hat{\mathbf{X}}$ from each set of tracked points of $\mathbf{T}_m$ can be reprojected in views where the segmentation may have failed to obtain the complete track in all views. The reprojected points of $\hat{\mathbf{X}}$ should correspond to the centroids of the non-segmented regions. It is then possible to calculate the size of the projected region as an average of the sizes of the identified regions in the track. In each view, a small window centered in the computed centroids is defined. These corresponding small windows, referred to as *tracked part*, will be denoted as $\mathbb{W} = \{\mathbf{W}_1, ..., \mathbf{W}_m\}$. In each view a small window is defined with the estimated size in the computed centers of gravities (see Fig. 9.11). Afterwards, the corresponding windows are averaged. Thus, the attempt is made to increase the signal-to-noise ratio by the factor $\sqrt{n}$, where n is the number of averaged windows. As flaws must appear as contrasted zones relating to their environment, we can verify if the contrast of each averaged window is greater than 2.5%. With this verification it is possible to eliminate all remaining false detections. Figure 9.11 shows the detection in our sequence using

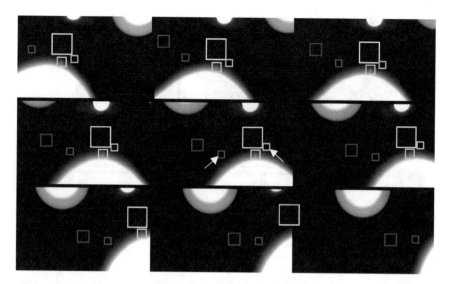

Fig. 9.11 Reconstruction and verification: the false detections (indicated by the arrows) are eliminated after the verification in all images of the sequence

this method. Our objective is then achieved: the real defects were separated from the false ones.

Experimental Results

In this section, results of automatic inspection of cast aluminum wheels using the outlined approach are presented. These results have been achieved recently on synthetic flaws and real data. The parameters of our method have been manually tuned, giving $\sigma = 1.25$ pixels (for LoG-operator), $\varepsilon_2 = 0.75$ mm, $\varepsilon_s = 0.7$, and $\varepsilon_3 = 0.9$ mm. These parameters were not changed during these experiments. A wheel was considered to be a cylinder with the following dimensions: 470 mm diameter and 200 mm height. The focal length (distance between X-ray source and entrance screen of the image intensifier) was 884 mm. The bottom of a wheel was 510 mm from the X-ray source. Thus, a pattern of 1 mm in the middle of the wheel is projected in the X-ray projection coordinate system as a pattern of 1.73 mm, and in the image coordinate system as a pattern of 2.96 pixels. The sequences of X-ray images were taken by rotation of the casting at 5^0.

The detection performance will be evaluated by computing the number of True Positives (TP) and False Positives (FP). They are respectively defined as the number of flaws that are correctly classified and the number of misclassified regular structures. The TP and FP will be normalized by the number of existing flaws (E) and the number of identified potential flaws (I). Thus, we define the following percentages: TPP = TP / E $\times 100$ and FPP = FP / I $\times 100$. Ideally, TPP = 100% and FPP = 0%.

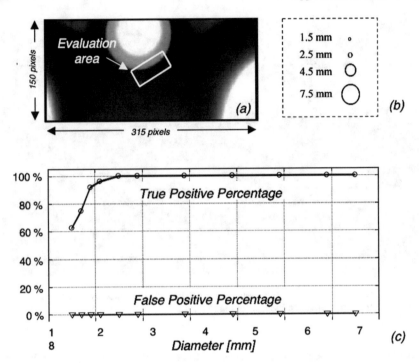

Fig. 9.12 Detection on synthetic flaws: **a** X-ray image and evaluated area, **b** flaw sizes, and **c** TPP and FPP

Synthetic Flaws: To evaluate the performance of our method in critical cases, real data in which synthetic flaws have been added were examined (see Sect. 8.4.3). A simple 3D modeled flaw (a spherical bubble) was projected and superimposed on real X-ray images of an aluminum wheel according to the law of X-ray absorption [98]. In our experiment, a flaw is simulated in 10 X-ray images of a real casting, in an area that included an edge of the structure (see Fig. 9.12a). In this area the synthetic flaw was located in 24 different positions in a regular grid manner. At each position TPP and FPP were tabulated. This test was repeated for different sizes of the flaws ($\emptyset = 1.5 \sim 7.5$ mm) which are illustrated in Fig. 9.12b. The results are shown in Fig. 9.12c. It was observed that the FPP was always zero. The TPP was 100% for $\emptyset \geq 2.5$ mm, and greater than 95% for $\emptyset \geq 2.1$ mm. However, the identification of the flaw may fail (and therefore also its detection) if it is very small and is located at the edge of the structure of the casting. In this case one may choose a smaller value of the parameter σ in the LoG operator of the edge detection, which will unfortunately increment the FPP. Other non-critical experiments, where the area of the simulation does not include an edge of the structure, have led to perfect results (TPP = 100%, FPP = 0%) for $\emptyset \geq 1.5$ mm (≥ 4.4 pixels). Usually, the minimum detectable defect size according to inspection specifications is in the order of $\emptyset = 2$ mm. In X-ray

Table 9.2 Detection of flaws on real data

Seq.	X-ray Images	Flaws in the Sequence	Flaws in the Images (E)	Identification			Detection	
				TP	FP	Total (I)	TP	FP
1	10	2	12	12	249	261	2	0
2	9	1	9	8	238	246	1	0
3	9	3	23	19	253	272	3	0
4	8	1	8	4	413	417	1	0
5	6	1	6	6	554	560	1	0
6	8	1	8	8	196	204	1	0
7	6	3	18	14	445	459	3	0
8	6	0	0	0	178	178	0	0
9	9	0	0	0	256	256	0	0
10	8	0	0	0	150	150	0	0
11	8	0	0	0	345	345	0	0
12	6	0	0	0	355	355	0	0
13	6	0	0	0	365	365	0	0
14	9	0	0	0	313	313	0	0
Total	108	12	84	71	4310	4381	12	0
Percentage				85%	98%		100%	0%

testing, smaller flaws can be detected by decreasing the distance of the object test to the X-ray source.

Real Data: Fourteen X-ray image sequences of aluminum wheels with twelve known flaws were inspected. Three of these defects were existing blow holes (with $\emptyset = 2.0 \sim 7.5$ mm). They were initially detected by a visual (human) inspection. The remaining nine flaws were produced by drilling small holes ($\emptyset = 2.0 \sim 4.0$ mm) in positions of the casting which were known to be difficult to detect. Casting flaws are present only in the first seven sequences. The results are summarized in Table 9.2, Figs. 9.13, and 9.14. In the identification of potential flaws, it was observed that the FPP was 98% (4310/4381). Nevertheless, the TPP in this experiment was good, and it was possible to identify 85% (71/84) of all projected flaws in the sequences (13 of the existing 84 flaws were not identified because the contrast was poor or they were located at edges of regular structures). It was observed that in the next steps, the FPP was reduced to nil. The detection of the real flaws was successful in all cases. The first six images of sequence 3 and its results were already illustrated in Figs. 9.5, 9.7, 9.8, 9.9, 9.10 and 9.11. The results on the other sequences with flaws are shown in Fig. 9.13.

Comparison with Other Methods: In this section, we present a comparison of our proposed algorithm with other methods that can be used to detect defects in aluminum

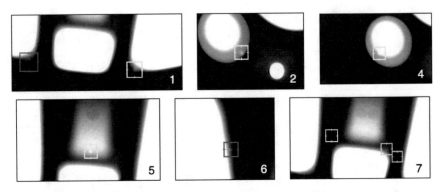

Fig. 9.13 Detected flaws in sequences 1, 2, 4, 5, 6, and 7 (sequence 3 is shown in Fig. 9.11)

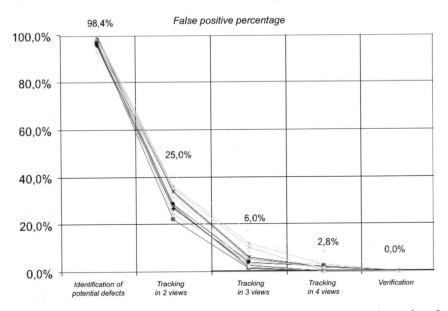

Fig. 9.14 False positive percentage on real data in the fourteen real sequences (the number of identified potential flaws corresponds to 100%). The mean of each step is given over the fourteen curves

castings. In this comparison, we evaluate the same real fourteen sequences used in the previous section. The results are summarized in Table 9.3.

Firstly, we compared the first step of our method (*identification of potential flaws*). The objective of this step is the use of a single filter, instead of a set of filters adapted to the regular structure of the specimen. We evaluated the well-known Canny filter (see, for example, [37]). As this filter detects sparse edge pixels that not necessarily produce at real flaws closed and connected contours, the TPP of this detector was unacceptable, only 4% of the real flaws were identified ('Canny I' in Table 9.3). In

Table 9.3 Comparison with other methods

Method	Identification		Detection	
	TPP	FPP	TPP	FPP
Canny I	4%	97%	0%	–
Canny II	40%	99%	17%	40%
Median I	55%	85%	33%	36%
Median II	88%	98%	92%	45%
Tracking in 3	85%	98%	100%	25%
Tracking in 5	85%	98%	83%	0%
PXV-5000	–	–	100%	0%
Proposed	85%	98%	100%	0%

order to increase the number of closed regions a dilation of the edges using a 3×3 mask was performed. Although the TPP is improved to 40% ('Canny II' in Table 9.3), many flaws were not detected in any of the images of the sequence. For this reason, only 17% of the real flaws were detected after the tracking and verification.

Another detection of potential flaws can be performed using a region-based segmentation. Median filtering is normally used to generate an error-free image, since defect structures are essentially eliminated, while design features of the test piece are normally preserved [109]. Once the error-free reference image is computed, an error difference image between original and error-free images is calculated. Casting defects are then identified when a sufficiently large gray level in the error difference image occurs. The best results were obtained using a median filter with a 11×11 mask. We evaluated two thresholds: $\theta = 6$ and $\theta = 2$—by 256 gray levels—(see 'Median I' and 'Median II' in Table 9.3). In the first case the TPP was only 55%. By decreasing the threshold value we increased the TPP to 88%, that is slightly better than our detector (85%). However, systematic false alarms were detected at the corners of the regular structures. Since these false alarms satisfy the multifocal conditions, they can be tracked in the sequence. For this reason, this detector can only be used if the median filter is adapted to the regular structures of the specimen using a priori information. Normally, a set of median filters is used for each X-ray image [42, 58, 59].

In order to evaluate the second step of our method (*tracking of potential flaws*), we tested the method by tracking the potential flaws in 3 and in 5 views, instead of 4 views (see 'Tracking in 3', 'Tracking in 5', and 'Proposed' in Table 9.3). By considering only three views we obtained so many false alarms that the verification step detected 4 false alarms (25%). In the other case, by tracking the potential flaws in five views, real flaws that were segmented in only four views of the sequences were not tracked. For this reason, only 83% of the real flaws were detected.

Finally, we inspected the test castings using a classic image processing method. In our experiments, we used the industrial software PXV-5000 [110]. The results were excellent: 100% of the real flaws were detected without false alarms. As a result of its peak detection performance, the classic image processing methods have become the most widely established in industrial applications. However, these methods suffer from the complicated configuration of the filtering, which is tailored to the test piece. In our experiments, the configuration process has taken two weeks. Nevertheless, as our method requires only a few number of parameters, the configuration could be carried out in hours.

Conclusions

A new method for automated flaw detection in aluminum castings using multiple-view geometry has been developed. Our method is very efficient because it is based on a two-step analysis: identification and tracking. The idea was to try to imitate the way a human inspector inspects X-ray images: first relevant details (potential defects) are detected, followed by tracking them in the X-ray image sequence. In this way, the false detections can be eliminated without discriminating the real flaws.

The great advantage of our first step is the use of a single filter to identify potential defects, which is independent of the structure of the specimen. Nevertheless, its disadvantages are as follows: (a) the false positive percentage is enormous; (b) the true positive percentage could be poor if the flaws to be detected are very small and located at the edge of a structure; and (c) the identification of regions is time-consuming. Contrarily, the second step is highly efficient in both discrimination of false detections and tracking of real defects, and is not time-consuming, due to the use of the multiple-view tensors.

To inspect a whole wheel our method requires approximately 100 views of 256 × 256 pixels, that can be processed in one minute. The required computing time is acceptable for practical applications because a typical inspection process takes about one minute, independently of whether it is performed manually or automatically.

We have shown that these preliminary results are promising. However, given that the performance of the method has been verified on only a few X-ray image sequences, an evaluation on a broader data base is necessary.

It is possible to combine our second step with existing defect detection technologies, which use a priori information of the regular structures of the casting to detect flaws in single images (see, for example [110]). This method could also be used in the automated flaw detection of other objects. In the adaptation of our method, one must determine the number of views in which a flaw must be tracked. If the false positive percentage by identifying potential flaws is low (or high), one may track a flaw in fewer (or more) views of the sequence. However, one must guarantee that the real flaws will be identified as potential flaws in these views.

9.2.3 An Example

In this section, an implementation that can be used for defect detection of castings in single views is presented. It consists of features that are extracted from positive class (the defects) and negative class (the background).

An example of using detection in multiple views can be found in Sect. 9.4.3.

Python Example 9.1: In this example, we show how to implement a classifier that is able to defect casting defects in single X-ray images. For this end we use series C0002 that contains small images with and without defects. In addition, for this series we have the ground truth for all defects. The strategy of this example is the strategy that we proposed in Algorithm 1, that means we extract many features the proposed algorithm searches the combination of features and a classifier that maximizes the accuracy.

Listing 9.1 : Defect detection in castings

```python
import numpy as np
import numpy as np
from pyxvis.io.data import load_features,save_features
from pyxvis.learning.evaluation import best_features_classifier
from pyxvis.features.selection import clean_norm,clean_norm_transform
from pyxvis.features.extraction import extract_features_labels


dataname = 'c32' # prefix of npy files of training and testing data
fxnew    = 1     # the features are (0) loaded or (1) extracted and saved
if fxnew:
    # features to extract
    fx       = ['basicint','gabor-ri','lbp-ri','haralick-2','fourier','hog','clp']
    # feature extraction in training images
    path     = '../images/castings/'
    X,d      = extract_features_labels(fx,path+'train','jpg')
    # feature extraction in testing images
    Xt,dt    = extract_features_labels(fx,path+'test','jpg')
    # backup of extracted features
    save_features(X,d,Xt,dt,dataname)
else:
    X,d,Xt,dt = load_features(dataname)

X,sclean,a,b = clean_norm(X)
Xt           = clean_norm_transform(Xt,sclean,a,b)
# Classifiers to evaluate
ss_cl        = ['maha','bayes-kde','svm-lin','svm-rbf','qda','lda','knn3','knn7','nn']
# Number of features to select
ff           = [3,5,10,12,15]
# Feature selectors to evaluate
ss_fs        = ['fisher','qda','svm-lin','svm-rbf']

clbest,ssbest = best_features_classifier(ss_fs,ff,ss_cl,X,d,Xt,dt,
                              'Accuracy in Castings')
print('    Selected Features: '+str((np.sort(sclean[ssbest]))))
```

Fig. 9.15 Accuracy on training and testing dataset for castings defect detection. In this example we use the strategy proposed in Algorithm 1. [→ Example 9.1]

The output of this code is the estimated accuracy:

```
----------------------------------------------------------------------------
        Best iteration: 8 (maximum of testing accuracy)
     Feature Selector: fisher with 10 features
                     : (Fisher, )
            Classifier: knn3
                     : (KNeighborsClassifier, n_neighbors=3) CrossVal with 5 folds
         Training-Acc: 0.9676
          Testing-Acc: 0.9609
     Selected Features: [ 1   2   3 16 20 24 25 26 38 72]
----------------------------------------------------------------------------
```

The accuracy of the selected classifier (knn with 3 neighbors) is 96.76% with 10 features (Fig. 9.15). In this code we used best_features_classifier of pyxvis Library. The reader can use additional series of GDXray+, that contain annotated defects in aluminum wheels.   □

9.3   Welds

In welding process, a mandatory inspection using X-ray testing is required in order to detect defects like porosity, inclusion, lack of fusion, lack of penetration, and cracks. Industrial X-ray images of welds is widely used for detecting those defects in the petroleum, chemical, nuclear, naval, aeronautics and civil construction industries, among others. An example is illustrated in Fig. 9.20.

9.3.1 State of the Art

Over the last 35 years, substantial research has been performed on automated detection and classification of welding defects in continuous welds using X-ray imaging [152, 153]. Typically, the approaches follow a classical computer vision schema: (i) image acquisition—an X-ray digital imFage is taken and stored in the computer, (ii) pre-processing—the digital image is improved in order to enhance the details, (iii) segmentation—potential welding defects are found and isolated, (iv) feature extraction/selection—significant features of the potential welding defects and their surroundings are quantified, and (v) classification—the extracted features are interpreted automatically using a priori knowledge of the welding defects in order to separate potential defects into detected welding defects or false alarms. In the last few years, some methods based on deep learning have been developed with promising results. Selected approaches are summarized in Table 9.4. In this table, we follow the $3\mathbb{X}$-strategy outlined in Sect. 1.8, in which we distinguish (i) the X-ray energy used to generate the X-ray images (monochromatic, dual-, or multi-energy), (ii) the number of views used by the algorithms (single-view, multi-views, or computed tomography) and complexity of the algorithms (simple, medium, and complex—here, deep learning methods–). As we can see there is much research on weld inspection. Achieved performance of the developed algorithms is still not high enough, thus it is not suitable for fully automated inspection.

9.3.2 An Application

In computer vision, many object detection and classification problems have been solved without classic segmentation using *sliding-windows*. Sliding-window approaches have established themselves as state of the art in computer vision problems where an object must be separated from the background (see, for example, successful applications in face detection [171] and human detection [30]). In sliding-window methodology, a detection window (see black square in Fig. 9.16) is sledded over an input image in both horizontal and vertical directions, and for each localization of the detection window a classifier decides to which class the corresponding portion of the image belongs to according to its features. In this section, an approach to detect defects based on sliding-windows in welds is presented [102].

Overview
We developed an X-ray computer vision approach to detect welding defects using this methodology yielding promising results. We will differentiate between the 'detection of defects' and the 'classification of defects' [82]. In the detection problem, the classes that exist are only two: 'defects' and 'no-defects', whereas the recognition of the type of the defects (e.g., porosity, slag, crack, lack of penetration, etc.) is known as classification of flaw types. This section describes our approach on detection only

Table 9.4 State of art in inspection of welds

Authors	Year	Ref	X_1^* energies 1 2 3	X_2^* views 1 2 3	X_3^* algorithms 1 2 3
Ajmi et al.	2018	[4]	⊠ ☐ ☐	⊠ ☐ ☐	⊠ ⊠ ☐
Anand et al.	2009	[11]	⊠ ☐ ☐	⊠ ☐ ☐	⊠ ⊠ ☐
Baniukiewicz	2014	[14]	⊠ ☐ ☐	⊠ ☐ ☐	⊠ ⊠ ☐
Gao and Yu	2014	[47]	⊠ ☐ ☐	⊠ ☐ ☐	⊠ ⊠ ☐
Hassan et al.	2012	[57]	⊠ ☐ ☐	⊠ ☐ ☐	⊠ ⊠ ⊠
Hou et al.	2018	[61]	⊠ ☐ ☐	⊠ ☐ ☐	☐ ☐ ⊠
Hou et al.	2019	[62]	⊠ ☐ ☐	⊠ ☐ ☐	☐ ☐ ⊠
Kaftandjian et al.	2003	[67]	⊠ ☐ ☐	⊠ ☐ ☐	⊠ ⊠ ☐
Kumar et al.	2014	[74]	⊠ ☐ ☐	⊠ ☐ ☐	⊠ ⊠ ☐
Kumar et al.	2014	[75]	⊠ ☐ ☐	⊠ ☐ ☐	⊠ ⊠ ☐
Liao	2008	[83]	⊠ ☐ ☐	⊠ ☐ ☐	⊠ ⊠ ☐
Liao	2009	[84]	⊠ ☐ ☐	⊠ ☐ ☐	⊠ ⊠ ☐
Lindgren	2014	[87]	⊠ ☐ ☐	⊠ ⊠ ☐	⊠ ⊠ ☐
Liu et al.	2017	[88]	⊠ ☐ ☐	⊠ ☐ ☐	☐ ☐ ⊠
Mery and Berti	2003	[106]	⊠ ☐ ☐	⊠ ☐ ☐	⊠ ⊠ ☐
Mery	2011	[102]	⊠ ☐ ☐	⊠ ☐ ☐	⊠ ⊠ ☐
Mu et al.	2011	[120]	⊠ ☐ ☐	⊠ ☐ ☐	⊠ ⊠ ☐
Muniategui et al.	2019	[121]	⊠ ☐ ☐	⊠ ☐ ☐	⊠ ⊠ ☐
Muravyov and Pogadaeva	2020	[122]	⊠ ☐ ☐	⊠ ☐ ☐	⊠ ⊠ ☐
Pan et al.	2020	[135]	⊠ ☐ ☐	⊠ ☐ ☐	⊠ ⊠ ⊠
Shao et al.	2014	[149]	⊠ ☐ ☐	⊠ ⊠ ☐	⊠ ⊠ ☐
Shi et al.	2007	[150]	⊠ ☐ ☐	⊠ ☐ ☐	⊠ ⊠ ☐
da Silva et al.	2009	[154]	⊠ ☐ ☐	⊠ ☐ ☐	⊠ ⊠ ☐
Suyama et al.	2019	[161]	⊠ ☐ ☐	⊠ ☐ ☐	☐ ☐ ⊠
Tong et al.	2012	[164]	⊠ ☐ ☐	⊠ ☐ ☐	⊠ ⊠ ☐
Vilar et al.	2009	[170]	⊠ ☐ ☐	⊠ ☐ ☐	⊠ ⊠ ☐
Wang et al.	2008	[174]	⊠ ☐ ☐	⊠ ☐ ☐	⊠ ⊠ ☐
Wang et al.	2019	[173]	⊠ ☐ ☐	⊠ ☐ ☐	⊠ ⊠ ☐
Yiron et al.	2015	[181]	⊠ ☐ ☐	⊠ ⊠ ☐	⊠ ⊠ ☐
Zapata et al.	2008	[185]	⊠ ☐ ☐	⊠ ☐ ☐	⊠ ⊠ ☐
		*1	Mono	Mono	Simple
		2	Dual	Multi	Medium
		3	Multi	CT	Complex

☐ not used, ⊠ used

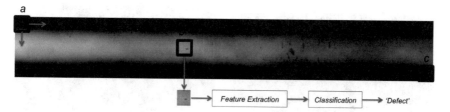

Fig. 9.16 Sliding window approach: A detection window (see black square) is sledded over the X-ray image starting at place 'a' and ending at 'c'. For each position, e.g., at 'b', features are extracted only from the sub-image defined by the square, and a classifier determines the class of this portion of the image

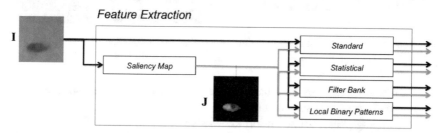

Fig. 9.17 Feature extraction: from each detection window several features are extracted (see black path). Additionally, the same features are extracted from a saliency map of the sub-window (see gray path)

and the corresponding validation experiments. The classification of defects can be developed by the reader using a similar methodology.

The key idea of this example is to use a computer vision methodology, as shown in Figs. 9.16 and 9.17, to automatically detect welding defects. In the following, feature extraction, feature selection, classification, and validation will be explained in further detail.

Feature Extraction, Selection, and Classification

Features provide information about the intensity of a sub-image. In our approach, p features per *intensity* channel were extracted. The used intensity channels in our work are only two: the grayscale X-ray image ($\mathbf{I}$) and a saliency map ($\mathbf{J}$) computed from $\mathbf{I}$, i.e., , $p \times 2$ features for two intensity channels. In order to reduce the computational time, we restricted the feature extraction for these only two channels, however, other channels, like Harris transform [55] or other saliency maps, can be used.

The saliency map $\mathbf{J}$ is obtained using a center-surround saliency mechanism based on a biologically inspired attention system [118][3]. In order to achieve faster processing, this theory proposes that the human visual system uses only a portion of the image, called *focus of attention*, to deal with complex scenes. In our approach, we

---

[3]The saliency function is implemented in saliency of **pyxvis** Library.

Fig. 9.18 X-ray images used in our experiments (series W0001 of GDXray+)

use the *off-center* saliency map that measures the different dark areas surrounded by
a bright background, as shown in Fig. 9.17.

In a training phase, using a priori knowledge of the welding defects, the detection windows are manually labeled as one of two classes: 'defects' and 'no-defect'.
The first class corresponds to those regions where the potential welding defects are
indeed welding defects. Alternatively, the second class corresponds to false alarms.
For this end, we use series W0001 and W0002 of GDXray+. In the first series,
we have the X-ray images, whereas in the second one we have the corresponding
binary images representing the ground truth. Thus, the ideal segmentation of image
W0001_00i.png is binary image W0002_00i.png, for i = 01 ... 10.. Intensity features of each channel are extracted for both classes. Features extracted from
each area of an X-ray image region are divided into four groups: basic intensity features (see Sect. 5.3.1), statistical features (see Sect. 5.3.5), Fourier and DCT features
(see Sect. 5.3.7), Gabor features (see Sect. 5.3.6), and Local Binary Patterns (see
Sect. 5.4.1). Afterwards, the extracted features are selected using feature selection
approaches (see Sect. 5.6, and several classifiers (see Sect. 6.2) were evaluated using
cross-validation (see Sect. 6.3.2). indexGabor features

Experiments
We experimented with 10 representative X-ray images (see Fig. 9.18). The average
size of the image was 1.35 mega-pixels. For each X-ray image, 250 detection windows
with detects and 250 without defects were selected, yielding $2 \times 250 \times 10 = 5000$
detection windows. Each detection window was labeled with '1' for class *defects*
and '0' for *no-defects*. The size of the detection windows were 24×24 pixels. For
each detection window 586 features were extracted. This means that 586 features
were extracted from 5000 samples (2500 with defects and 2500 without defects) .

After the feature extraction, 75% of the samples from each class were randomly
chosen to perform the feature selection. The best performance was achieved using

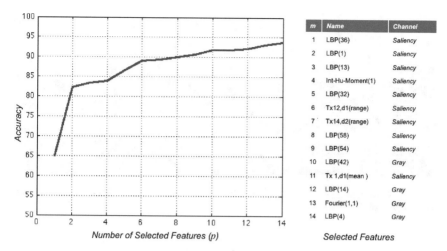

m	Name	Channel
1	LBP(36)	Saliency
2	LBP(1)	Saliency
3	LBP(13)	Saliency
4	Int-Hu-Moment(1)	Saliency
5	LBP(32)	Saliency
6	Tx12,d1(range)	Saliency
7	Tx14,d2(range)	Saliency
8	LBP(58)	Saliency
9	LBP(54)	Saliency
10	LBP(42)	Gray
11	Tx 1,d1(mean)	Saliency
12	LBP(14)	Gray
13	Fourier(1,1)	Gray
14	LBP(4)	Gray

Selected Features

Fig. 9.19 Classification performance using the first p features

Sequential Forward Selection. The best 14 features are shown in Fig. 9.19 in ascending order.

The performance of the classification using the SVM classifier and the first p selected features was validated using an average of ten cross-validation with 10 folds. The results are shown in Fig 9.19. We observe that by using 14 features, the performance was almost 94% with a 95% confidence interval between 93.0 and 94.5%.

In order to test this methodology on X-ray images, the technique was implemented using a sliding window sized 24×24 pixels that was shifted by 4 pixels. Thus, in each position a sub-window of 24×24 pixels was defined and the corresponding features were extracted. The sub-window was marked if the trained classifier detected it as a discontinuity. Using a size of 24×24 pixel and a shift of 4 pixels, an image pixel could be marked from 0 to 36 times. Finally, if a pixel of the image was marked more than 24 times, then the pixel was considered as a discontinuity. The aforementioned parameters were set using an exhaustive search. The described steps are shown in Fig. 9.20 for one X-ray image. The results on other X-ray images are shown in Fig. 9.21. From these, one can see the effectiveness of the proposed technique.

Conclusions

In this section, we presented a new approach to detecting weld defects without segmentation based on sliding-windows and novel features. The promising results outlined in our work show that we achieved a very high classification rate in the detection of welding defects using a large number of features combined with efficient feature selection and classification algorithms. The key idea of the proposed method was to select, from a large universe of features, namely 572 features, only those features that were relevant for the separation of the two classes. We tested our method

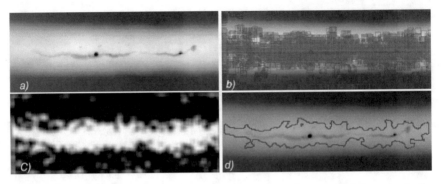

Fig. 9.20 Weld inspection using a sliding-window: **a** X-ray image, **b** detected windows, **c** activation map, **d** detection [102]

on 10 representative X-ray images yielding a performance of 94% in accuracy using only 14 features and support vector machines. It is important to note that local binary pattern features extracted from the saliency map play an important role in the performance of the classifier. The method was implemented and tested on real X-ray images showing high effectiveness.

9.3.3 An Example

In this section, we present a Python code that can be used to detect defects in welds according to sliding-windows approach explained above.

Python Example 9.2: In this example, we show how to implement—for a simple perspective—the strategy explained in the previous section using CNN. We will use one part of image W0001_0001.png as training, and another part as testing. Using a sliding windows strategy, we will extract patches of 32 × 32 pixels on the right side of the image for training and on the left side for testing. These patches are stored in file welds32x32.mat. In this example, there are around 10,000 patches for training and other 10,000 for testing. The reader can modify this dataset including more X-ray images of GDXray+ in order to achieve better results. The reader will note that this example has pedagogical purposes only. In order to develop a real application, more training images must be taken into account.

Listing 9.2 : Detection of weld defects using CNN

```
from pyxvis.learning.cnn import CNN

# execution type
type_exec   = 0 # training & testing

# pacthes' file for training and testing
patches_file  = '../data/weld32x32.mat'
```

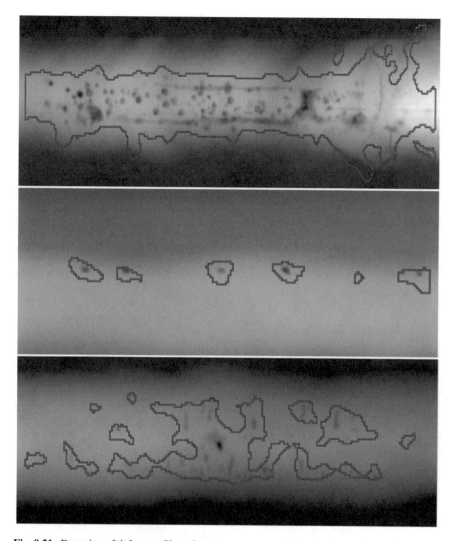

Fig. 9.21 Detection of defects on X-ray images

```
# architechture
p = [9,7,5,3]          # Conv2D mask size
d = [32,64,128,256]    # Conv2D channels
f = [64,32]            # fully connected

# training and testing
CNN(patches_file,type_exec,p,d,f)
```

The output of this code is shown in Fig. 9.22. We can see the final detection on testing image. In this example, we use CNN of pyxvis Libraryto train the convolutional neural network as explained in Sect. 7.3. The patches are extracted only in the region of

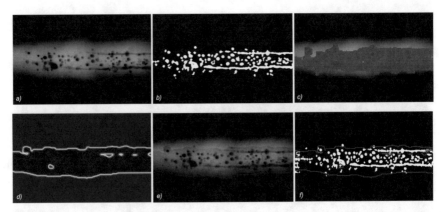

Fig. 9.22 Detection of defects on X-ray images using sliding windows. **a** Testing image. **b** Ground truth (binary image). **c** Ground truth (sliding windows patches of targe class). **d** Activation map of the detection. **e** Detection (after thresholding) and testing image **f** Boundary of the detection and binary ground truth. [→ Example 9.2 🐍]

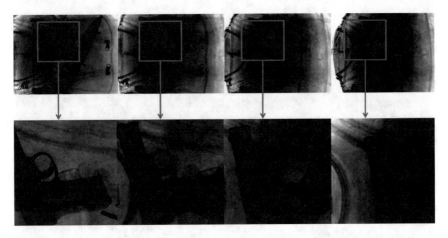

Fig. 9.23 Detection of a handgun based on the trigger identification in multiple views [112]

interest defined by the segmentation of the weld using seg_bimodal of pyxvis Library. The reader can observe the effectiveness of this strategy. However, it is clear that better results can be achieved by considering more features, classifiers, and training images.   □

9.4 Baggage

Since the September 11 attacks, automated (or semi-automated) 3D recognition using X-ray images have become a very important element in baggage screening. The inspection process, however, is complex, basically because threatening items are very difficult to detect when placed in close-packed bags, superimposed by other objects, and/or rotated showing an unrecognizable view [186]. In baggage screening, where human security plays an important role and inspection complexity is very high, human inspectors are still used. Nevertheless, during peak hours in airports, human screeners have only a few seconds to decide whether a bag contains or not a prohibited item, and detection performance is only about 80–90% [117].

9.4.1 State of the Art

Before 9/11, the X-ray analysis of luggage mainly focused on capturing the images of their content: the reader can find in [123] an interesting analysis carried out in 1989 of several aircraft attacks around the world, and the existing technologies to detect terrorist threats based on Thermal-Neutron Activation (TNA), Fast-Neutron Activation (FNA), and dual-energy X-rays (used in medicine since the early 70s). In the 90s, Explosive Detection Systems (EDS) were developed based on X-ray imaging [124], and computed tomography through elastic scatter X-ray (comparing the structure of irradiated material, against stored reference spectra for explosives and drugs) [160]. All these works were concentrated on image acquisition and simple image processing; however, they lacked advanced image analysis to improve detection performance. Nevertheless, the 9/11 attacks increased the security measures taken at airports, which in turn stimulated the interest of the scientific community in the research of areas related to security using advanced computational techniques. Over the last decade, the main contributions were: analysis of human inspection [172], pseudocoloring of X-ray images [1, 25], enhancement and segmentation of X-ray images [156], and detection of threatening items in X-ray images, based on texture features (detecting a 9mm Colt Beretta automatic (machine) pistol) [131], neural networks and fuzzy rules (yielding about 80% of performance) [89], and SVM classifier (detecting guns in real time) [126].

In baggage screening, the use of multiple-view information yields a significant improvement in performance as certain items are difficult to recognize using only one viewpoint. As reported in a study that measures the human performance in baggage screening [17], (human) multiple-view X-ray inspection leads to a higher detection performance of prohibited items under difficult conditions, however, there are no significant differences between the detection performance (single versus multiple view) for difficult-easy multiple-view conditions, i.e., two *difficult* or two *easy* views are redundant. We observed that for intricate conditions, multiple-view X-ray inspection is required.

Recently, some algorithms based on multiple X-ray views were reported in the literature. For example: synthesis of new X-ray images obtained from Kinetic Depth Effect X-ray (KDEX) images based on SIFT features in order to increase detection performance [2]; an approach for object detection in multi-view dual-energy X-ray with promising preliminary results [45]; X-ray active vision that is able to adequate the viewpoint of the target object in order to obtain better X-ray images to analyze [142]; and tracking across multiple X-ray views in order to verify the diagnoses performed using a single view [101, 103, 112, 114].

Finally, methods based on deep learning have been proposed in the last years and they have established themselves as state of the art in baggage inspection. In single views, we can mention [115] using mono-energy and [7–9, 12, 116] using dual-energy. In addition, there are some contributions based on GAN's (see Sect. 7.6) to generate synthetic X-ray images that can be used as data augmentation in the training stage [5, 146, 180]. A review of deep learning method in baggage inspection can be found in [6].

An example is illustrated in Fig. 9.23. A survey on explosives detection can be found in [157, 176]. Selected approaches are summarized in Table 9.5. In baggage screening, where human security plays an important role and inspection complexity is very high, human inspectors are still used. For intricate conditions, multiple-view X-ray inspection using dual-energy is required.

9.4.2 An Application

In this section, we present the use of an automated method based on multiple X-ray views to recognize certain regular objects with highly defined shapes and sizes. The method consists of two steps: 'monocular analysis', to obtain possible detections in each view of a sequence, and 'multiple-view analysis', to recognize the objects of interest using matchings in all views. The search for matching candidates is efficiently performed using a lookup table that is computed off-line. In order to illustrate the effectiveness of the proposed method, experimental results on recognizing regular objects (clips, springs, and razor blades) in pen cases are shown. In this section, we explain in further detail the proposed method. The strategy consists of two main stages: *off-line* and *on-line*.

Off-Line Stage
The first stage, performed off-line, consists of two main steps: (i) learning a model that is used for the recognition and (ii) estimation of a multiple-view geometric model that is used for data association.

Learning: In this step, we learn a classifier h to recognize parts of the objects that we are attempting to detect. It is assumed that there are $C + 1$ classes (labeled as '0' for non-object class, and '1', '2', ... 'C' for C different objects). Images are taken of representative objects of each class from different points of view. In order to model the

Table 9.5 State of art in baggage inspection

Authors	Year	Ref	$\mathbb{X}_1^*$ energies 1 2 3	$\mathbb{X}_2^*$ views 1 2 3	$\mathbb{X}_3^*$ algorithms 1 2 3
Abusaeeda et al.	2011	[2]	☒ ☒ ☐	☒ ☒ ☐	☒ ☒ ☐
Akcay and Breckon	2017	[7]	☐ ☒ ☐	☒ ☐ ☐	☐ ☐ ☒
Akcay et al.	2016	[8]	☐ ☒ ☐	☒ ☐ ☐	☐ ☐ ☒
Akcay et al.	2018	[9]	☐ ☒ ☐	☒ ☐ ☐	☐ ☐ ☒
Akcay et al.	2018	[5]	☐ ☒ ☐	☒ ☐ ☐	☐ ☐ ☒
Aydin et al.	2018	[12]	☒ ☐ ☐	☐ ☒ ☐	☐ ☒ ☐
Baştan	2015	[15]	☐ ☒ ☐	☐ ☒ ☐	☐ ☒ ☐
Baştan et al.	2011	[16]	☐ ☒ ☐	☒ ☐ ☐	☐ ☒ ☐
Chen et al.	2005	[26]	☒ ☒ ☐	☒ ☐ ☐	☒ ☒ ☐
Ding et al.	2006	[31]	☒ ☒ ☐	☒ ☐ ☐	☒ ☒ ☐
Franzel et al.	2012	[45]	☐ ☒ ☐	☐ ☒ ☐	☐ ☒ ☐
Flitton et al.	2013	[43]	☐ ☒ ☐	☐ ☐ ☒	☐ ☒ ☐
Flitton et al.	2015	[44]	☐ ☒ ☐	☐ ☐ ☒	☐ ☒ ☐
Heitz and Chechik	2010	[60]	☒ ☒ ☐	☒ ☐ ☐	☒ ☒ ☐
Liu et al.	2018	[90]	☐ ☒ ☐	☒ ☐ ☐	☐ ☐ ☒
Lu and Conners	2006	[92]	☒ ☒ ☐	☒ ☐ ☐	☒ ☒ ☐
Mansoor and Rajashankari	2012	[94]	☒ ☒ ☐	☒ ☐ ☐	☒ ☒ ☐
Mery	2015	[103]	☒ ☐ ☐	☐ ☒ ☐	☐ ☒ ☐
Mery et al.	2013	[114]	☒ ☐ ☐	☐ ☒ ☐	☐ ☒ ☐
Mery et al.	2016	[115]	☒ ☐ ☐	☒ ☐ ☐	☐ ☒ ☒
Miao et al.	2019	[116]	☐ ☒ ☐	☒ ☐ ☐	☐ ☐ ☒
Mouton and Breckon	2015	[119]	☐ ☒ ☐	☐ ☐ ☒	☐ ☒ ☐
Nercessian et al.	2008	[127]	☒ ☐ ☐	☒ ☐ ☐	☒ ☐ ☐
Riffo and Mery	2016	[143]	☒ ☐ ☐	☒ ☐ ☐	☐ ☒ ☐
Riffo and Mery	2012	[142]	☒ ☐ ☐	☐ ☒ ☐	☐ ☒ ☐
Riffo et al.	2019	[141]	☒ ☐ ☐	☐ ☐ ☒	☐ ☒ ☐
Riffo et al.	2017	[140]	☒ ☐ ☐	☐ ☒ ☐	☐ ☒ ☐
Saavedra et al.	2020	[145]	☒ ☐ ☐	☒ ☐ ☐	☐ ☐ ☒
Sangwan and Jain	2019	[146]	☒ ☐ ☐	☒ ☐ ☐	☐ ☐ ☒
Sigman and Jain	2020	[151]	☒ ☐ ☐	☒ ☐ ☐	☐ ☐ ☒
Schmidt et al.	2012	[147]	☒ ☒ ☐	☒ ☐ ☐	☒ ☒ ☐
Steitz et al.	2018	[159]	☐ ☒ ☐	☐ ☒ ☐	☐ ☐ ☒
Turcsany et al.	2013	[165]	☐ ☒ ☐	☒ ☐ ☐	☐ ☒ ☐
Uroukov and Speller	2015	[167]	☒ ☒ ☐	☒ ☐ ☐	☒ ☒ ☐
Yuanxi and Liu	2019	[183]	☒ ☐ ☐	☒ ☐ ☐	☐ ☐ ☒
Xu et al.	2018	[178]	☐ ☒ ☐	☒ ☐ ☐	☐ ☐ ☒
Zhang and Zhue	2015	[188]	☒ ☒ ☐	☒ ☐ ☐	☒ ☒ ☐
Zou et al.	2018	[194]	☒ ☐ ☐	☒ ☐ ☐	☐ ☐ ☒
		*1	Mono	Mono	Simple
		2	Dual	Multi	Medium
		3	Multi	CT	Complex

☐ not used, ☒ used

details of the objects from different poses, several keypoints per image are detected, and for each keypoint a descriptor $\mathbf{d}$ is extracted using, for example, LBP, SIFT, HOG, and SURF, among others (see Sect. 5.4). In this supervised approach, each descriptor $\mathbf{d}$ is manually labeled according to its corresponding class $c \in \{0, 1, \ldots C\}$. Given the training data $(\mathbf{d}_t, c_t)$, for $t = 1, \ldots, N$, where N is the total number of descriptors extracted in all training images, a classifier h is designed which maps $\mathbf{d}_t$ to their classification label c_t, thus, $h(\mathbf{d}_t)$ should be c_t. This classifier will be used in the on-line stage by monocular and multiple-view analysis.

Geometry: Our strategy deals with multiple monocular detections in multiple views. In this problem of data association, the aim is to find the correct correspondence among different views. For this reason, we use multiple-view geometric constraints to reduce the number of matching candidates between monocular detections. For an image sequence with n views $\mathbf{I}_1 \ldots \mathbf{I}_n$, the fundamental matrices $\{\mathbf{F}_{ij}\}$ between consecutive frames $\mathbf{I}_i$ and $\mathbf{I}_{j=i+1}$ are computed for $i = 1, \ldots n - 1$. In our approach, the fundamental matrix $\mathbf{F}_{ij}$ is calculated from projection matrices $\mathbf{P}_i$ and $\mathbf{P}_j$ that can be estimated using calibration (see Sect. 3.4) or bundle adjustment algorithms (see Sect. 9.4.3).

The geometric constraints are expressed in homogeneous coordinates. Therefore, given a point $\mathbf{m}_i = [x_i \ y_i \ 1]^T$ in image $\mathbf{I}_i$, a corresponding point $\mathbf{m}_j = [x_j \ y_j \ 1]^T$ in image $\mathbf{I}_j$ must fulfill: (i) epipolar constraint (see Sect. 3.5.1): $\mathbf{m}_j$ must lie near the epipolar line $\ell = \mathbf{F}_{ij}\mathbf{m}_i$, and (ii) location constraint: for small variations of the point of views between $\mathbf{I}_i$ and $\mathbf{I}_j$, $\mathbf{m}_j$ must lie near $\mathbf{m}_i$. Thus, a candidate $\mathbf{m}_j$ must fulfill:

$$\frac{|\mathbf{m}_j^T \mathbf{F}_{ij} \mathbf{m}_i|}{\sqrt{\ell_1^2 + \ell_2^2}} < e \text{ and } ||\mathbf{m}_i - \mathbf{m}_j|| < r. \tag{9.5}$$

In order to accelerate the search of candidates, we propose the use of a lookup table as follows: Points in images $\mathbf{I}_i$ and $\mathbf{I}_j$ are arranged in a grid format with rows and columns. For each grid point (x, y) of image $\mathbf{I}_i$, we look for the grid points of image $\mathbf{I}_j$ that fulfill (9.5), as illustrated in Fig. 9.24. Therefore, the possible corresponding points of (x, y) will be the set $\mathbf{S}_{xy} = \{(x_p, y_p)\}_{p=1}^q$, where $x_p = X(x, y, p)$, $y_p = Y(x, y, p)$ and $q = Q(x, y)$ are stored (off-line) in a lookup table. In the on-line stage, given a point $\mathbf{m}_i$ (in image $\mathbf{I}_i$), the matching candidates in image $\mathbf{I}_j$ are those that lie near to $\mathbf{S}_{xy}$, where (x, y) is the nearest grid point to $\mathbf{m}_i$. This search can be efficiently implemented using k-d tree structures [21].

In a controlled and calibrated environment, we can assume that the fundamental matrices are stable and we do not need to estimate them in each new image sequence, i.e., the lookup tables are constant. Additionally, when the relative motion of the point of view between consecutive frames is the same, the computed fundamental matrices are constant, i.e., $\mathbf{F}_{ij} = \mathbf{F}$, and we need to store only one lookup table.

On-Line Stage

The on-line stage is performed in order to recognize the objects of interest in a test image sequence of n images $\{\mathbf{I}_i\}$, for $i = 1, \ldots n$. The images are acquired by

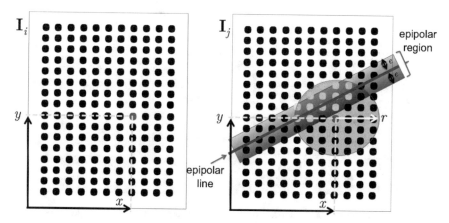

Fig. 9.24 Given the grid point illustrated as the red point at (x, y), in image $\mathbf{I}_i$, the set of possible corresponding points in image $\mathbf{I}_j$ can be those grid points (yellow points) represented by the intersection of the epipolar region (blue rectangle) and neighborhood around (x, y) (orange circle with radius r centered at red point). The use of grid points allows us to use a lookup table in order to search the matching candidates in $\mathbf{I}_j$ efficiently

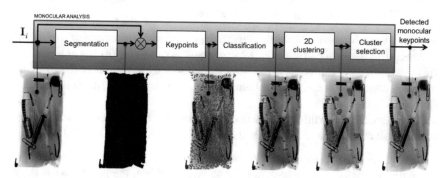

Fig. 9.25 Monocular analysis for each image of the sequence, i.e., for $i = 1, \ldots n$. In this example, the class of interest is 'razor blade'

rotation of the object being tested at β degrees (in our experiments we used $n = 4$, and $\beta = 10^0$). This stage consisted of two main steps: monocular and multiple-view analysis that will be described in further detail as follows.

Monocular Analysis: This step is performed in each image $\mathbf{I}_i$ of the test image sequence, as illustrated in Fig. 9.25 in a real case. The whole object contained in image $\mathbf{I}_i$ is segmented from the background using threshold and morphological operations. SIFT–keypoints (or other descriptors)— are only extracted in the segmented portion. The descriptor $\mathbf{d}$ of each keypoint is classified using classifier $h(\mathbf{d})$ trained in the off-line stage, and explained above. All keypoints classified as class c, where c is the class of interest, with $c \in \{1 \ldots C\}$ are selected. As we can see in Fig. 9.25 for the classification of 'razor blade', there are many keypoints misclassified. For this

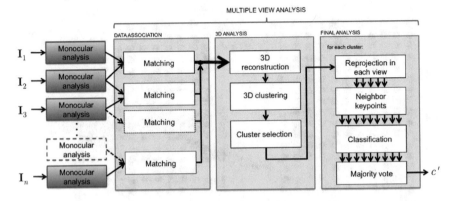

Fig. 9.26 Multiple-view analysis. An explanation of last step (final analysis) is illustrated in Fig. 9.27

reason, neighbor keypoints are clustered in the 2D space using Mean Shift algorithm [28]. Only those clusters that have a large enough number of keypoints are selected. They will be called *detected monocular keypoints*.

Multiple-View Analysis: Multiple view analysis performs the recognition of objects of interest in three steps (see Fig. 9.26): (i) data association, (ii) 3D analysis, and (iii) final analysis. The input is the detected monocular keypoints obtained by the mentioned monocular analysis explained above. The output is c', the assigned class for each detected object.

• Data Association: In this step, we find matchings for all detected monocular keypoints in all consecutive images $\mathbf{I}_i$ and $\mathbf{I}_{j=i+1}$, for $i = 1, \ldots n-1$, as follows:

+ For each detected monocular keypoint in image $\mathbf{I}_i$ (located at position (x_i, y_i) with descriptor $\mathbf{d}_i$), we seek in a dense grid of points, the nearest point (x, y) (see red point in Fig. 9.24-left) using a k-d tree structure.
+ We determine $\mathbf{S}_{xy}$, the set of matching candidates in image $\mathbf{I}_{j=i+1}$ arranged in a grid manner by reading the lookup table explained above (see yellow points in Fig. 9.24-right).
+ We look for the detected monocular keypoints in image $\mathbf{I}_j$ that are located in the neighborhood of $\mathbf{S}_{xy}$, again using a k-d tree structure. They will be called *neighbor keypoints*. When no neighbor keypoint is found, no match is established for (x_i, y_i).
+ From neighbor keypoints, we select that one (located at position (x_j, y_j) with descriptor $\mathbf{d}_j$) with minimum distance $||\mathbf{d}_i - \mathbf{d}_j||$. In order to ensure the similarity between matching points, the distance should be less than a threshold ε. If this constraint is not satisfied, again no match is established for (x_i, y_i).

• 3D analysis: From each pair of matched keypoints (x_i, y_i) in image $\mathbf{I}_i$ and (x_j, y_j) in image $\mathbf{I}_{j=i+1}$ established in the previous step, a 3D point is reconstructed using the projection matrices $\mathbf{P}_i$ and $\mathbf{P}_j$ of our geometric model (see Sect. 3.6). Similar to

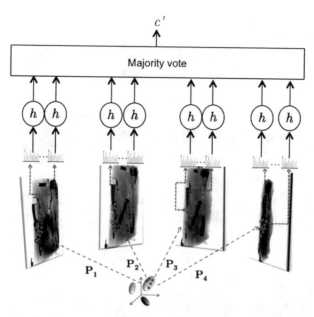

Fig. 9.27 Final analysis: using the geometric model, the reconstructed 3D points in each cluster are reprojected in each view (blue points). The keypoints that are near to the reprojected points are identified (red points). The descriptors of these keypoints (orange histograms) are classified using trained classifier h. The class c' of this cluster is determined by majority vote. In this example of $n = 4$ views, only the green cluster is represented

the monocular detection approach, neighbor 3D points are clustered in the 3D space using Mean Shift algorithm [28], and only those clusters that have a large enough number of 3D points are selected.

• Final analysis: For each selected 3D cluster, all 3D reconstructed points belonging to the cluster are re-projected onto all images of the sequence using the projection matrices of geometric model (see Fig. 9.27). The extracted descriptors of the keypoints located near these re-projected points are classified individually using classifier h. The cluster will be classified as class c' if there is a large number of keypoints individually classified as c', and this number represents a majority in the cluster.

This majority vote strategy can overcome the problem of false monocular detections when the classification of the minority fails. A cluster can be misclassified if the part that we are trying to recognize is occluded by a part of another class. In this case, there will be keypoints in the cluster assigned to both classes; however, we expect that the majority of keypoints will be assigned to the true class if there are a small number of keypoints misclassified.

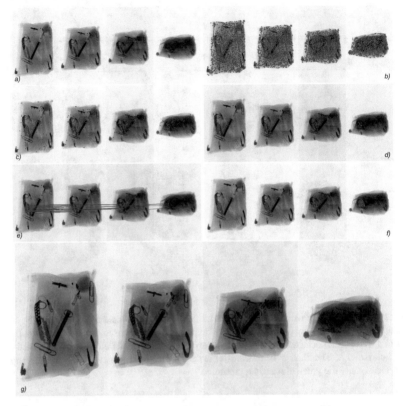

Fig. 9.28 Recognition of a razor blade using our approach. **a** original sequence, **b** keypoints, **c** classified keypoints, **d** detected monocular keypoints, **e** matched keypoints, **f** reprojected 3D points (blue) and neighbor keypoins (red), **g** final detection

Experiments and Results

In our experiments, the task was to recognize three different classes of objects that are present in a pencil case (see, for example, a sequence in Fig. 9.28a). These classes are: 'clips', 'springs', and 'razor blades'. We followed the recognition approach explained above.

In the off-line stage we used a structure from a motion algorithm in order to estimate the projection matrices of each view[4]. Additionally, in the learning phase, we used only 16 training images of each class. Due to the small intra-class variation of our classes, this number of training images was deemed sufficient. The training objects were posed at different angles. SIFT descriptors were extracted as explained in [91], and a k-Nearest Neighbor (KNN) classifier with $k = 3$ neighbors was ascertained

---

[4]We use in our experiments a fast implementation of multiple-view geometry algorithms from Balu Toolbox [97].

a   b   c   d

Fig. 9.29 Recognition using our approach in cases with some degree of overlap: **a** one spring, **b** two springs, **c** one clip, **d** one clip. Each figure shows a part of one image of the whole sequence

using the SIFT descriptors of the four classes[5]. Other descriptors (like LBP and HOG) and other classifiers (like SVM or KNN with other values of k) were also tested, although the best performance was achieved with the aforementioned configuration.

In order to illustrate step by step the on-line stage, the recognition of a razor blade is illustrated in Fig. 9.28a–d for monocular analysis and in Fig. 9.28e–g for multiple-view analysis[6]. It is worth mentioning that in monocular detection there are false alarms, however, they can be filtered out after multiple-view analysis. The reason is because false alarms cannot be tracked in the sequence or because the tracked points, when validating the corresponding points in other views of the sequence, do not belong to the class of interest. Other results with some degree of overlap, where the task was the recognition of springs and clips, are illustrated in Fig 9.29.

Testing experiments were carried out by recognizing the three mentioned classes ('clips', 'springs', and 'razor blades') in 45 different sequences of 4 views (15 sequences for each class)[7]. The size of an individual image was 1430 ×900 pixels. In these experiments there were 30 clips, 75 springs and 15 razor blades to be recognized. A summary of the results using the proposed algorithm is presented in Table 9.6, in which the performance in the recognition of each class is presented in two different parts of our algorithm: after monocular analysis (Mono) and after multiple-view analysis (Multi). These parts are illustrated in Fig. 9.28d and 9.28g respectively for a razor blade. In this table, Ground Truth (GT) is the number of existing objects to be recognized. The number of detected objects by our algorithm is D = TP + FP, including False Positives (FP) and true positives (TP). Ideally, FP = 0 and TP = GT. In our experiments, precision (PR), computed as PR=TP/D, is 71.4% and 95.7% in each part; and recall (RE), computed as RE=TP/GT, is 90.8% and 92.5% in each step. If we compare single versus multiple view detection, both precision and recall are incremented. Precision, however, is drastically incremented because our approach achieves good discrimination from false alarms.

The amount of time required in our experiments was about 15 minutes for the off-line stage and about 16s for testing each sequence on a iMac OS X 10.7.3, processor 3.06 GHz Intel Core 2 Duo, 4 GB 1067 MHz DDR3 memory. The code of the program—implemented in Matlab—is available on our website.

---

[5]We used in our experiments fast implementations of SIFT and KNN (based on k-d tree) from VLFeat Toolbox [169].

[6]We used in our experiments a fast implementation of Mean Shift from PMT Toolbox [32].

[7]The images tested in our experiments come from public GDXray database [113].

Table 9.6 Recognition performance

	Mono			Multi		
Class	TP	FP	GT	TP	FP	GT
Clip	114	127	120	26	2	30
Spring	263	30	300	71	3	75
Blade	59	18	60	14	0	15
Total	436	175	480	111	5	120
Precision		71.4%			95.7%	
Recall		90.8%			92.5%	

Conclusions

In this section, we presented a new method that can be used to recognize certain parts of interest in complex objects using multiple X-ray views. The proposed method filters out false positives resulting from monocular detection performed on single views by matching information across multiple views. This step is performed efficiently using a lookup table that is computed off-line. In order to illustrate the effectiveness of the proposed method, experimental results on recognizing regular objects—clips, springs, and razor blades—in pen cases are shown achieving around 93% accuracy in the recognition of 120 objects. We believe that it would be possible to design an automated aid in a target detection task using the proposed algorithm. In our future work, the approach will be tested in more complex scenarios recognizing objects with a larger intra-class variation.

9.4.3 An Example Using Multiple Views

In this example, we show how to detect objects in a non-calibrated image sequence as illustrated in Fig. 9.31. The approach has two parts: *structure estimation* and *parts detection*[8]. The approach follows the same strategy of method explained in Sect. 9.2.2. The results are shown in Fig. 9.32.

Structure Estimation

In case the X-ray imaging system is not calibrated, a geometric model must be estimated. The estimation of the geometric model is based on well-known structure-from-motion (SfM) methodologies. For the sake of completeness, a brief description of this model is presented here. In our work, SfM is estimated from a sequence of m images taken from a rigid object at different viewpoints. The original image sequence is stored in m images $\mathbf{J}_1, ..., \mathbf{J}_m$.

---

[8]See implementation in Matlab at https://github.com/domingomery/Xvis - function `Xtrgui`.

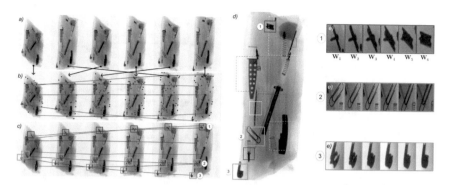

Fig. 9.30 Detection of objects in a pencil case using the proposed method: **a** Unsorted sequence with six X-ray images. The images are sorted according to their similarity (see arrows). **b** Sorted sequence, keypoints (points) and structure from motion (lines across the sequence). **c** Detection in the sequence and tracked regions. **d** Detection of parts of interest in the last image in the sequence (three of them are used in this example to illustrate the next sub-figures). **e** Tracked example regions in each view of the sequence (1: pencil sharpener, 2: clip, and 3: zipper slider body and pull-tab)

Keypoints: For each image, SIFT keypoints are extracted because they are very robust against scale, rotation, viewpoint, noise, and illumination changes [91]. Thus, not only a set of 2D image positions $\mathbf{x}$, but also descriptors $\mathbf{y}$, are obtained. Although this method is based on SIFT descriptors, there is no limitation to use other descriptors, e.g., SURF [18].

Image Sorting: If the images are not sorted, a visual vocabulary tree is constructed for fast image indexing. Thus, a new image sequence $\mathbf{I}_1, ..., \mathbf{I}_m$ is established from $\mathbf{J}_1, ..., \mathbf{J}_m$ by maximizing the total similarity defined as $\sum \text{sim}(\mathbf{I}_i, \mathbf{I}_{i+1})$, for $i = 1, ..., m-1$, where the similarity function 'sim' is computed from a normalized scalar product obtained from the visual words of the images [158]. See an example in Fig. 9.30a and 9.30b.

Matching Points: For two consecutive images, $\mathbf{I}_i$ and $\mathbf{I}_{i+1}$, SIFT keypoints are matched using the algorithm suggested by Lowe [91] that rejects too ambiguous matches. Afterwards, the Fundamental Matrix between views i and $i + 1$, $\mathbf{F}_{i,i+1}$, is estimated using RANSAC [56] to remove outliers. If keypoint k of $\mathbf{I}_i$ is matched with keypoint k' of $\mathbf{I}_{i+1}$, the match will be represented as $\mathbf{x}_{i,k} \rightarrow \mathbf{x}_{i+1,k'}$.

Structure Tracks: We look for all possible structure tracks—with one keypoint in each image of sequence—that belong to a family of the following matches:

$$\mathbf{x}_{1,k_1} \rightarrow \mathbf{x}_{2,k_2} \rightarrow \mathbf{x}_{3,k_3} \rightarrow ... \rightarrow \mathbf{x}_{m,k_m}.$$

There are many matches that are eliminated using this approach, however, having a large number of keypoints there are enough tracks to perform the bundle adjustment. We define n as the number of tracks.

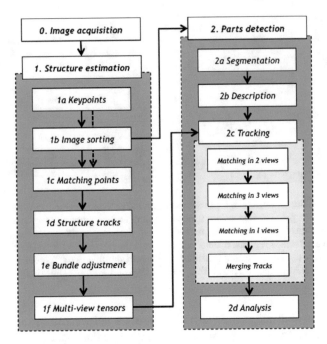

Fig. 9.31 Block diagram of the proposed approach

Bundle Adjustment: The determined tracks define n image point correspondences over m views. They are arranged as $\mathbf{x}_{i,j}$ for $i = 1, ..., m$ and $j = 1, ... n$. Bundle adjustment estimates 3D points $\hat{\mathbf{X}}_j$ and camera matrices $\mathbf{P}_i$ so that $\sum ||\mathbf{x}_{i,j} - \hat{\mathbf{x}}_{i,j}||$ is minimized, where $\hat{\mathbf{x}}_{i,j}$ is the projection of $\hat{\mathbf{X}}_j$ by $\mathbf{P}_i$. If $n \geq 4$, we can use the *factorization algorithm* [56] to perform an affine reconstruction because for our purposes the affine ambiguity of 3D space is irrelevant[9]. This method gives a fast and closed-form solution using SVD decomposition. A RANSAC approach is used to remove outliers.

Multiple-View Tensors: Bundle adjustment provides a method for computing bifocal and trifocal tensors from projection matrices $\mathbf{P}_i$ [56], that will be used in the next section.

Parts Detection
In this section, we give details of the algorithm that detects the object parts of interest. The algorithm consists of four steps: segmentation, description, tracking, and analysis as shown in Fig. 9.31.

---

[9]In this problem, the projective factorization can be used as well [56], however, our simplifying assumption is that only small depth variations occur and an affine model may be used.

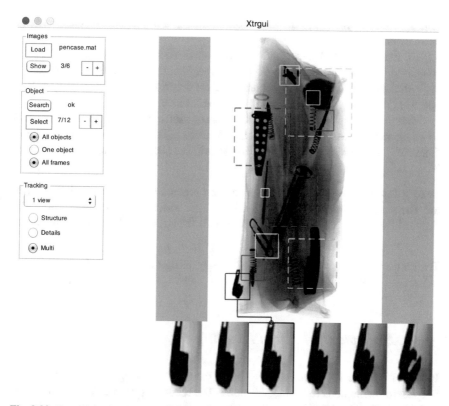

Fig. 9.32 Detection of objects in a pen case using graphic user interface `Xtrgui` [103]. In this example, we can see the zipper slider body and pull-tab in six different views

Segmentation: Potential regions of interest are segmented in each image $\mathbf{I}_i$ of the sequence. It is an *ad-hoc* procedure that depends on the application. For instance, one can be interested in detecting razor blades or pins in a bag, or flaws in a material, etc. This step ensures the detection of the object parts of interest allowing false detections. The discrimination between these two classes takes place by tracking them across multiple views (see steps 2c and 2d). In our experiments we tested three segmentation approaches.

• Spots detector: The X-ray image is filtered using a 2D median filter. The difference between original and filtered images is thresholded obtaining a binary image. A potential region r is segmented if size, shape, and contrast criteria are fulfilled. This approach was used to detect small parts (like pen tips or pins in a pencil case).

• Crossing line profile (CLP): Laplacian of Gaussian edges are computed from the X-ray image. The closed and connected contours of the edge image define region candidates. Gray level profiles along straight lines crossing each region candidate in the middle are extracted. A potential region r is segmented if the profile that contains the most similar gray levels in the extremes fulfills contrast criteria [99].

This approach was used to detect discontinuities in a homogeneous material, e.g., flaws in automotive parts.

• SIFT matching: SIFT descriptors are extracted from the X-ray image. They are compared with SIFT descriptors extracted from the image of a reference object of interest. A potential region r is segmented if the descriptors fulfill similarity criteria [49, 91]. This approach was used to detect razor blades in a bag.

Other general segmentation approaches can be used as well. For example, methods based on saliency maps [118], Haar basis features [171], histogram of oriented gradients [30], corner detectors [55], SURF descriptors [18], Maximally Stable regions [95], Local Binary Patterns [133], etc.

Description: Each segmented potential region r is characterized using a SIFT descriptor. The scale of the extracted descriptor, i.e., the width in pixels of the spatial histogram of 4×4 bins is set to $\sqrt{A_r}$, where A_r is the corresponding area of the region r.

Tracking and Analysis: The tracking and analysis algorithms were covered in detail in Sect. 9.2.2. Results are shown in Fig. 9.30.

9.4.4 Example Using Deep Learning

In Sect. 7.7.6, we illustrated already many examples in baggage inspection using deep learning methods for object detection. The reader is referred to those examples and Sect. 7.7 to see the detection methods that are proposed for baggage inspection. Here, we include additional results in Fig. 9.33 to illustrate an example using deep learning.

9.5 Natural Products

In order to ensure food safety inspection, several applications have been developed by the natural products industry. The difficulties inherent in the detection of defects and contaminants in food products have limited the use of X-ray into the packaged foods sector. However, the need for NDT has motivated a considerable research effort in this field spanning many decades [54].

9.5.1 State of the Art

The most important advances are: detection of foreign objects in packaged foods [76]; detection of fish bones in fishes [111]; identification of insect infestation in

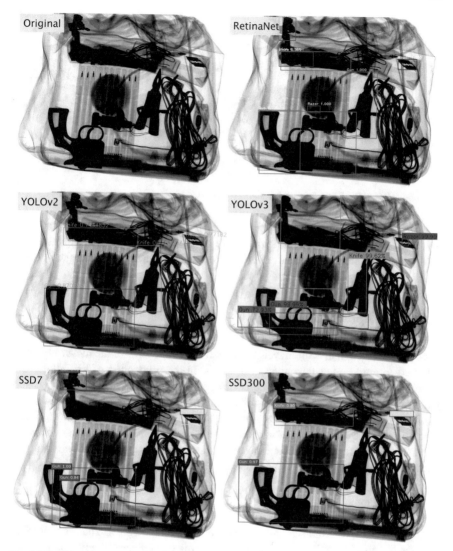

Fig. 9.33 Object detection on $\mathbb{GDX}$ray+ image B0046_0016 using detection methods RetinaNet, YOLOv2, YOLOv3, SSD7, and SSD300 explained in Sect. 7.7

citrus [65]; detection of codling moth larvae in apples [54]; fruit quality inspection like split-pits, water content distribution and internal structure [132]; and detection of larval stages of the granary weevil in wheat kernels [53]. In these applications, only single-view analysis is required. An example is illustrated in Fig. 9.34. The reader is referred to following survey papers for further analysis of the state of the art in the field: detection of contaminants [54], quality inspection of agricultural product using compute tomography [35], and inspection using X-ray fluorescence [39]. In Table

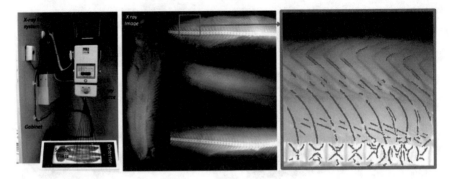

Fig. 9.34 Detection of fish bones using sliding-windows [111]

9.7, some applications are summarized. We observe that deep learning methods in this field are rarely used. This trend is sure to change in the next few years.

9.5.2 An Application

In countries where fish is often consumed, fish bones are some of the most frequently ingested foreign bodies encountered in foods. In the production of fish fillets, fish bone detection is performed by human inspection using their sense of touch and vision which can lead to misclassification. Effective detection of fish bones in the quality control process would help avoid this problem. For this reason, an X-ray machine vision approach to automatically detect fish bones in fish fillets was developed. This section describes our approach to detect fish bones automatically and the corresponding experiments with salmon and trout fillets based on [111].

Pre-Processing and Segmentation

The fish bones are only present in certain space frequencies of the spectrum: they are not too thin (minimal 0.5mm) nor too thick (maximal 2mm). The segmentation of potential fish bones is based on a band-pass filter to enhance the fish bones with respect to their surroundings as shown in Fig. 9.35. The proposed approach to detect potential fish bones has four steps:

Enhancement: The original X-ray image $\mathbf{X}$ (Fig. 9.35b) is enhanced linearly by modifying the original histogram in order to increase contrast [50]: The enhanced image $\mathbf{Y}$ is

$$\mathbf{Y} = a\mathbf{X} + b \tag{9.6}$$

Band-Pass Filtering: The enhanced image $\mathbf{Y}$ is filtered using a radial symmetric $\overline{17 \times 17}$ pixels mask $\mathbf{H}$ (Fig. 9.35a). Mask $\mathbf{H}$ was estimated from 20 X-ray images

Table 9.7 State of art on natural products

Authors	Year	Ref	$\mathbb{X}_1^*$ energies 1 2 3	$\mathbb{X}_2^*$ views 1 2 3	$\mathbb{X}_3^*$ algorithms 1 2 3
Bej et al.	2015	[19]	⊠ □ □	⊠ □ □	⊠ ⊠ □
van Deal et al.	2016	[29]	⊠ □ □	⊠ □ □	⊠ ⊠ □
van Deal et al.	2019	[168]	⊠ □ □	□ □ ⊠	⊠ ⊠ □
Douarre et al.	2016	[33]	⊠ □ □	□ □ ⊠	⊠ ⊠ □
Guelpa et al.	2015	[52]	⊠ □ □	⊠ □ □	⊠ ⊠ □
Haff and Slaughter	2004	[53]	⊠ □ □	⊠ □ □	⊠ ⊠ □
Jiang et al.	2008	[65]	⊠ □ □	⊠ □ □	⊠ ⊠ □
Karunakaran et al.	2004	[69]	⊠ □ □	⊠ □ □	⊠ ⊠ □
Kelkar et al.	2015	[70]	⊠ □ □	⊠ □ □	⊠ ⊠ □
Kotwaliwale et al.	2014	[73]	⊠ □ □	⊠ □ □	⊠ ⊠ □
Kwon et al.	2008	[76]	⊠ □ □	⊠ □ □	⊠ ⊠ □
Mathanker et al.	2011	[96]	⊠ □ □	⊠ □ □	⊠ ⊠ □
Mery et al.	2011	[111]	⊠ □ □	⊠ □ □	⊠ ⊠ □
Neethirajan et al.	2014	[125]	⊠ □ □	⊠ □ □	⊠ ⊠ □
Nielsen et al.	2014	[128]	⊠ □ □	⊠ □ □	⊠ ⊠ □
Nugraha et al.	2019	[130]	⊠ □ □	□ □ ⊠	⊠ ⊠ □
Ogawa et al.	2003	[132]	⊠ □ □	⊠ □ □	⊠ ⊠ □
Orina et al.	2018	[134]	⊠ □ □	⊠ □ □	⊠ ⊠ □
Schoeman et al.	2016	[148]	⊠ □ □	□ □ ⊠	⊠ ⊠ □
van Deal et al.	2019	[168]	⊠ □ □	□ □ ⊠	⊠ ⊠ □
Zhong et al.	2019	[191]	⊠ □ □	⊠ □ □	⊠ ⊠ □
		*1	Mono	Mono	Simple
		2	Dual	Multi	Medium
		3	Multi	CT	Complex

□ not used, ⊠ used

by minimizing the error rate as mention in [23] and applied to fish bones (all fish bones should be found and there should be no false alarms). The filtered image $\mathbf{Z}$ (Fig. 9.35c) is then the convolution of $\mathbf{Y}$ with mask $\mathbf{H}$:

$$\mathbf{Z} = \mathbf{Y} * \mathbf{H} \qquad (9.7)$$

Thresholding: Those pixels in $\mathbf{Z}$ that have gray values greater than a certain threshold θ are marked in a binary image $\mathbf{B}$. The threshold is defined to ensure that all fish bones are detected, i.e., false alarms are allowed in this step. The pixels of $\mathbf{B}$ are defined as

$$B_{ij} = \begin{cases} 1 \text{ if } Z_{ij} > \theta \\ 0 \text{ else} \end{cases} \qquad (9.8)$$

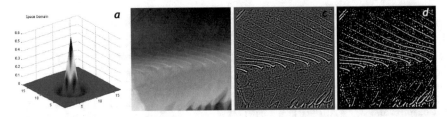

Fig. 9.35 Segmentation of potential fish bones: **a** Convolution mask **H** in space domain, **b** original X-ray image **X** of a salmon fillet, **c** filtered image **Z**, **d** potential fish bones image **P** after thresholding and removing objects deemed too small

Removal of Small Objects: All connected pixels in **B** containing fewer than A pixels are removed as shown in Fig. 9.35d. This image, called **P**, defines the potential fish bones.

Feature Extraction, Selection, And Classification
The segmented potential fish bones—contained in image **P**—are divided into small 10×10 pixels windows called *detection windows*. In a training phase, using a priori knowledge of the fish bones, the detection windows are manually labeled as one of two classes: *bones* and *no-bones*. The first class corresponds to those regions where the potential fish bones are indeed fish bones. Alternatively, the second class corresponds to false alarms. Intensity features of the enhanced X-ray image **Y** are extracted for both classes. We use enhanced image **Y**, instead of pre-processed image **X**, because after our experiments the detection performance was higher. Features extracted from each area of an X-ray image region are divided into four groups as shown in Sect. 9.3.2. In these experiments, 279 features are extracted from each detection window. Afterwards, the features are selected in order to decide on the relevant features for the two defined classes. In addition, a classifier is designed. The best results, after evaluation a 10-fold cross-validation was achieved by Sequential Forward Selection (as feature selection technique) and Support Vector Machine with RBF kernel (as classifier).

Experimental Results
First, the proposed method was tested with 20 representative salmon fillets obtained at a local fish market. The average size of these fillets was 15×10 cm^2. According to pre-processing and segmentation techniques explained above, several regions of interest were obtained where fish bones could be located. The area occupied by these regions of interest corresponds to approx. 12% of the salmon fillets as shown in Fig. 9.35. More results are presented in Fig. 9.36.

From the mentioned regions of interest 7697, detection windows of 10×10 pixels were obtained (available in series N0003 of GDXray+). Each window was labeled with '1' for class *bones* and '0' for *no-bones* (see file labels.txt in directory of N0003). From each window, 279 features were extracted. After the feature extrac-

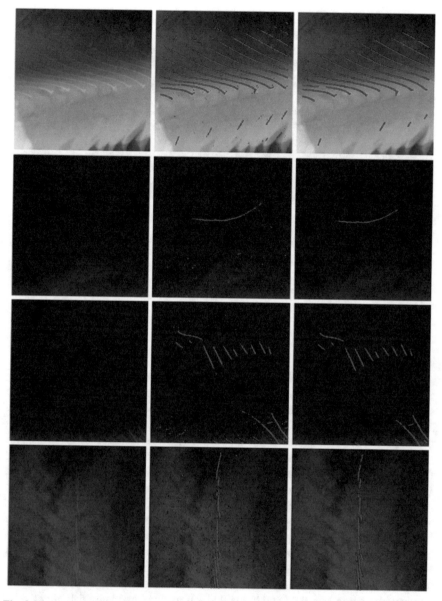

Fig. 9.36 Results obtained in four X-ray images. The columns correspond to enhanced images, classified fish bones and post processed fish bones. The first row corresponds to the example shown in Fig. 9.35

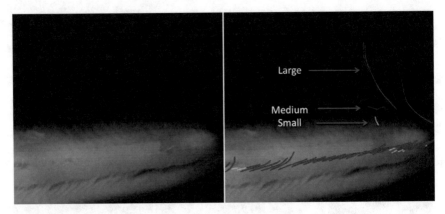

Fig. 9.37 Results obtained on 3878 samples using cross-validation with five folds. See results in Table 9.8

Table 9.8 Performance in the detection of fishbones

Fishbone	Sensibility (%)	1-Specificity (%)	Size
Large (red)	100	0	>0.64 mm × 12 mm
Medium (blue)	100	3	Between
Small (green)	93	6	<0.48 mm × 8.5 mm

tion, 75% of the samples from each class were randomly chosen to perform the feature selection. The best performance was achieved using Sequential Forward Selection. 24 features were selected. The features gave information about the spatial distribution of pixels, i.e., how coarse or fine the texture is. The selected features correspond mainly to statistical features (12) and filter banks (7), however, it is worth nothing that the two most discriminative features are LBP features (in this case LBP 48 and LBP 11). On the other hand, from the standard features there is only one feature (standard deviation of the intensity).

In order to investigate the sensibility (S_n) and 1-specificity ($1 - S_p$) of the fish bones depending on their largeness, three size groups were constructed: *large* for fish bones larger than 12mm, *small* for fish bones smaller than 8.5mm, and *medium* for fish bones between both sizes. In this experiment, 3878 fish bones were manually selected. The performance was calculated using a cross-validation with 5 folds. The results are summarized in Fig. 9.37 and Table 9.8. All medium and large fish bones were detected (with $1 - S_p = 0\%$ and 3% respectively), whereas 93% of small fish bones were correctly detected with $1 - S_p = 6\%$. This means that cross-validation yielded a detection performance of 100%, 98.5%, and 93.5% (computed using $(S_n + S_p)/2$) for large, medium, and small fish bones respectively.

Finally, in order to validate the proposed methodology, the last experiment was carried out using representative fish bones and representative trout fillets provided by a Chilean salmon industry. The size of the fish bones were between 14 and 47 mm

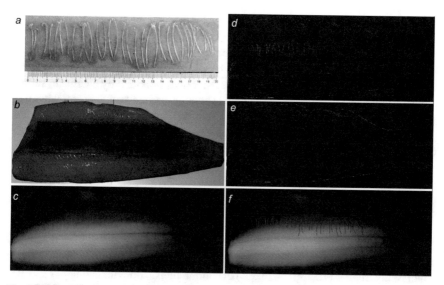

Fig. 9.38 Results obtained on a trout fillet using a fish bone strip with 33 fish bones: **a** strip, **b** strip over the fillet, **c** X-ray image, **d** segmentation, **e** classification, **f** post-processing. All fish bones were detected ($S_n = 1$), in this example there was no false alarm ($1 - S_p = 0$)

(larger than the small-size and mid-size groups considered above). The fish bones were arranged in strips that were superimposed onto trout fillets. Thus, the number of fish bones to be detected was a priori known. According to the absorption law, an X-ray image of a fillet with a fish bone inside, and an X-ray image with a fish bone laid on the fillet top are almost identical. Similar methodologies are used in industrial X-ray inspection of materials in order to simulate discontinuities [98]. The only difference could be that the position of a real fish bone (inside of a fillet) achieves a more realistic location related to the fish tissues, however, after our experience, the obtained images were found to be very similar. Fig. 9.38 shows the detection of one fish bone strip on a trout fillet. Using the same classifier trained in the last experiment, i.e., no new training was necessary, the proposed method was able to detect all fish bones with a 1% false positive rate. In this case, 15 X-ray images were tested, with 459 *bones* and 10413 *no–bones*.

Conclusion

The need for more information on the quality control of several fish types by means of quantitative methods can be satisfied using X-ray testing, a non-destructive technique that can be used to objectively measure intensity and geometric patterns in non-uniform surfaces. In addition the method can also determine other physical features such as image texture, morphological elements, and defects in order to automatically determine the quality of a fish fillet. The promising results outlined in this work show that a very high classification rate was achieved in the quality control of salmon and trout when using a large number of features combined with efficient feature selection and classification. The key idea of the proposed method was to select, from a large universe of features, only those features that were relevant for the separation of the classes. Cross-validation yielded a detection performance of 100%, 98.5%, and 93.5% for large, medium, and small fish bones respectively. The proposed method was validated on trout with representative fish bones provided by a Chilean salmon industry yielding a performance of 99%. Although the method was validated with salmon and trout fillets only, we believe that the proposed approach opens new possibilities not only in the field of automated visual inspection of salmons and trout, but also in other similar fish.

9.5.3 An Example

In order to illustrate the methodology explained in the previous section, the reader can see Example 6.13, where the whole process is presented. In this example, 200 small X-ray images (100×100 pixels) of salmon filets, 100 with fish bones and 100 without fish bones are used. The images are available in series N0002 of GDXray+.

9.6 Further Applications

There are many applications in which X-rays can be used as a NDT and E method. In this section, we mention only cargos and electronic circuits.

9.6.1 Cargo Inspection

With the ongoing development of international trade, cargo inspection becomes more and more important. X-ray testing has been used for the evaluation of the contents of cargo, trucks, containers, and passenger vehicles to detect the possible presence of many types of contraband. See an example in Fig. 9.39. Some approaches are presented in Table 9.9. There still is not much research on cargo inspection, and the complexity of this inspection task is very high. Nowadays, there are some approaches that use dual-energy, computed tomography, and deep learning. For this reason, X-ray systems are still only semi-automatic, and they require human supervision.

9.6.2 Electronic Circuits

In this industrial application of X-rays, the idea is to inspect circuit boards or integrated circuits in order to detect flaws in manufacturing, e.g., broken traces, missing

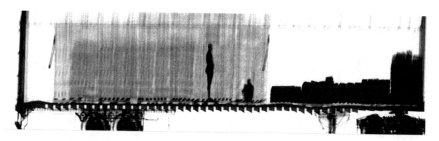

Fig. 9.39 X-ray image of a cargo. Collected by U.S. Customs and Border Protection a bureau of the United States Department of Homeland Security, via Wikimedia Commons

Table 9.9 State of art on cargo inspection

Authors	Year	Ref	X_1^* energies 1 2 3			X_2^* views 1 2 3			X_3^* algorithms 1 2 3		
Duan	2008	[36]	☒	☐	☐	☒	☒	☐	☒	☒	☐
Frosio	2011	[46]	☒	☐	☐	☒	☐	☐	☒	☒	☐
Kolkoori	2014	[71]	☒	☐	☐	☒	☐	☐	☒	☐	☐
Kolokytha et al.	2018	[72]	☒	☐	☐	☒	☐	☐	☒	☒	☐
Jaccard et al.	2016	[63]	☒	☐	☐	☒	☐	☐	☒	☒	☐
Jaccard et al.	2017	[64]	☒	☐	☐	☒	☐	☐	☒	☒	☐
Lee et al.	2018	[78]	☐	☒	☐	☒	☐	☐	☒	☒	☐
Rogers et al.	2017	[144]	☐	☒	☐	☒	☐	☐	☐	☐	☒
Zhu	2008	[193]	☒	☐	☐	☒	☐	☐	☒	☒	☐
Zhu	2010	[192]	☒	☐	☐	☒	☐	☐	☒	☒	☐
		*1	Mono			Mono			Simple		
		2	Dual			Multi			Medium		
		3	Multi			CT			Complex		

☐ not used , ☒ used

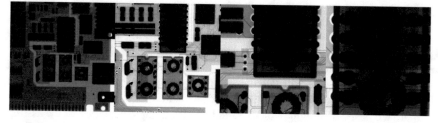

Fig. 9.40 X-ray image of a printed circuit board. By SecretDisc (Own work) via Wikimedia Commons

components, cracks, dilapidations, etc. An example is shown in Fig. 9.40. Some approaches are presented in Table 9.10. In this area, automated systems are very effective, and the inspection task is very fast and obtains a high performance.

Table 9.10 State of art on electronic circuit boards

Authors	Year	Ref	X_1^* energies 1 2 3	X_2^* views 1 2 3	X_3^* algorithms 1 2 3
Adato et al.	2016	[3]	⊞ ⊞ ⊞	⊞ □ □	⊞ ⊞ □
Alam et al.	2017	[10]	⊞ □ □	□ □ ⊞	⊞ ⊞ □
Ghosh et al.	2018	[48]	⊞ □ □	⊞ □ □	⊞ ⊞ □
Goto et al.	2019	[51]	⊞ □ □	⊞ □ □	□ □ ⊞
Favata and Shahbazmohamadi	2018	[38]	⊞ □ □	⊞ □ ⊞	⊞ ⊞ □
Lin et al.	2017	[85]	⊞ □ □	⊞ □ □	⊞ ⊞ □
Mahmood et al.	2015	[93]	⊞ □ □	⊞ □ □	⊞ ⊞ □
Uehara et al.	2013	[166]	⊞ □ □	⊞ □ □	⊞ ⊞ □
Wang et al.	2014	[175]	⊞ □ □	⊞ □ □	⊞ ⊞ □
Wu et al.	2014	[177]	⊞ □ □	⊞ □ □	⊞ ⊞ □
Zakaria et al.	2020	[184]	⊞ □ □	⊞ □ □	⊞ ⊞ □
		*1	Mono	Mono	Simple
		2	Dual	Multi	Medium
		3	Multi	CT	Complex

□ not used, ⊞ used

9.7 Summary

In this chapter, relevant applications on X-ray testing were described. We covered X-ray testing in:

- Castings: To ensure the safety of the construction of automative parts, it is necessary to check every part thoroughly using X-ray testing. We presented the state of the art, a defect detection approach based on a tracking principle, and a Python implementation of a classifier that is able to defect casting defects in single X-ray images.
- Welds: In welding processes, a mandatory inspection using X-ray testing is required in order to detect defects like porosity, inclusion, lack of fusion, lack of penetration, and cracks. We presented the state of the art, a defect detection approach based on sliding windows, and a Python implementation of a classifier that is able to detect defects using sliding windows methodology in single X-ray images.
- Baggage: In baggage screening, every piece of luggage must be inspected using X-ray testing in order to detect dangerous objects. We presented the state of the art, a recognition approach based on multiple-view analysis, and a Matlab implementation of tracking principle that is able to detect objects in the sequence X-ray images of a pen case.

- Natural products: We presented some applications of X-ray testing in natural products, such as inspection of fruit, identification of infections, and detection of fish bones. We reviewed the state of the art, a fish bones detection approach based on sliding windows, and a Python implementation of a classifier that is able to detect fish bones in cropped images with and without fish bones.
- Others: There are several industrial applications that use X-ray testing. We mentioned only cargos and electronic circuits giving some references of the state of art.

References

1. Abidi, B.R., Zheng, Y., Gribok, A.V., Abidi, M.A.: Improving weapon detection in single energy X-ray images through pseudocoloring. IEEE Trans. Syst., Man, Cybern., Part C: Appl. Rev. **36**(6), 784–796 (2006)
2. Abusaeeda, O., Evans, J., Downes, D., Chan, J.: View synthesis of KDEX imagery for 3D security X-ray imaging. In: Proceedings of the 4th International Conference on Imaging for Crime Detection and Prevention (ICDP-2011) (2011)
3. Adato, R., Uyar, A., Zangeneh, M., Zhou, B., Joshi, A., Goldberg, B., Unlu, M.S.: Rapid mapping of digital integrated circuit logic gates via multi-spectral backside imaging. arXiv preprint arXiv:1605.09306 (2016)
4. Ajmi, C., El Ferchichi, S., Laabidi, K.: New procedure for weld defect detection based-gabor filter. In: 2018 International Conference on Advanced Systems and Electric Technologies (IC_ASET), pp. 11–16. IEEE (2018)
5. Akcay, S., Atapour-Abarghouei, A., Breckon, T.P.: Ganomaly: Semi-supervised anomaly detection via adversarial training. arXiv preprint arXiv:1805.06725 (2018)
6. Akcay, S., Breckon, T.: Towards automatic threat detection: a survey of advances of deep learning within X-ray security imaging. arXiv preprint arXiv:2001.01293 (2020)
7. Akcay, S., Breckon, T.P.: An evaluation of region based object detection strategies within X-ray baggage security imagery. In: 2017 IEEE International Conference on Image Processing (ICIP), pp. 1337–1341. IEEE (2017)
8. Akçay, S., Kundegorski, M.E., Devereux, M., Breckon, T.P.: Transfer learning using convolutional neural networks for object classification within X-ray baggage security imagery. In: 2016 IEEE International Conference on Image Processing (ICIP), pp. 1057–1061. IEEE (2016)
9. Akcay, S., Kundegorski, M.E., Willcocks, C.G., Breckon, T.P.: Using deep convolutional neural network architectures for object classification and detection within X-ray baggage security imagery. IEEE Trans. Inf. Forens. Secur. **13**(9), 2203–2215 (2018)
10. Alam, M., Shen, H., Asadizanjani, N., Tehranipoor, M., Forte, D.: Impact of x-ray tomography on the reliability of integrated circuits. IEEE Trans. Device Mater. Reliability **17**(1), 59–68 (2017)
11. Anand, R., Kumar, P., et al.: Flaw detection in radiographic weldment images using morphological watershed segmentation technique. NDT & E Int. **42**(1), 2–8 (2009)
12. Aydin, I., Karakose, M., Erhan, A.: A new approach for baggage inspection by using deep convolutional neural networks. In: 2018 International Conference on Artificial Intelligence and Data Processing (IDAP), pp. 1–6. IEEE (2018)
13. Bandara, A., Kan, K., Morii, H., Koike, A., Aoki, T.: X-ray computed tomography to investigate industrial cast al-alloys. Product. Eng. **14**(2), 147–156 (2020)
14. Baniukiewicz, P.: Automated defect recognition and identification in digital radiography. J. Nondestruct. Eval. **33**(3), 327–334 (2014)

15. Baştan, M.: Multi-view object detection in dual-energy X-ray images. Mach. Vis. Appl. **26**(7–8), 1045–1060 (2015)

16. Baştan, M., Yousefi, M.R., Breuel, T.M.: Visual words on baggage X-ray images. In: Computer Analysis of Images and Patterns, pp. 360–368. Springer, Berlin (2011)

17. von Bastian, C., Schwaninger, A., Michel, S.: Do Multi-view X-ray Systems Improve X-ray Image Interpretation in Airport Security Screening?, vol. 52. GRIN Verlag (2010)

18. Bay, H., Tuytelaars, T., Van Gool, L.: Surf: speeded up robust features. In: 9th European Conference on Computer Vision (ECCV2006). Graz Austria (2006)

19. Bej, G., Akuli, A., Pal, A., Dey, T., Chaudhuri, A., Alam, S., Khandai, R., Bhattacharyya, N.: X-ray imaging and general regression neural network (GRNN) for estimation of silk content in cocoons. In: Proceedings of the 2nd International Conference on Perception and Machine Intelligence, pp. 71–76. ACM (2015)

20. Bengio, Y., Courville, A., Vincent, P.: Representation learning: A review and new perspectives. IEEE Trans. Pattern Anal. Mach. Intell. **35**(8), 1798–1828 (2013)

21. Bentley, J.: Multidimensional binary search trees used for associative searching. Commun. ACM **18**(9), 509–517 (1975)

22. Boerner, H., Strecker, H.: Automated X-ray inspection of aluminum casting. IEEE Trans. Pattern Anal. Mach. Intell. **10**(1), 79–91 (1988)

23. Canny, J.: A computational approach to edge detection. IEEE Trans. Pattern Anal. Mach. Intell. **PAMI-8**(6), 679–698 (1986)

24. Carrasco, M., Mery, D.: Automatic multiple view inspection using geometrical tracking and feature analysis in aluminum wheels. Mach. Vis. Appl. **22**(1), 157–170 (2011)

25. Chan, J., Evans, P., Wang, X.: Enhanced color coding scheme for kinetic depth effect X-ray (KDEX) imaging. In: 2010 IEEE International Carnahan Conference on Security Technology (ICCST), pp. 155–160 (2010)

26. Chen, Z., Zheng, Y., Abidi, B.R., Page, D.L., Abidi, M.A.: A combinational approach to the fusion, denoising and enhancement of dual-energy X-ray luggage images. In: Workshop of IEEE Conference on Computer Vision and Pattern Recognition (CVPR-2005) (2005)

27. Cogranne, R., Retraint, F.: Statistical detection of defects in radiographic images using an adaptive parametric model. Signal Process. **96**, 173–189 (2014)

28. Comaniciu, D., Meer, P.: Mean shift: a robust approach toward feature space analysis. IEEE Trans. Pattern Anal. Mach. Intell. **24**(5), 603–619 (2002)

29. van Dael, M., Lebotsa, S., Herremans, E., Verboven, P., Sijbers, J., Opara, U., Cronje, P., Nicolaï, B.: A segmentation and classification algorithm for online detection of internal disorders in citrus using X-ray radiographs. Postharvest Biol. Technol. **112**, 205–214 (2016)

30. Dalal, N., Triggs, B.: Histograms of oriented gradients for human detection. In: Conference on Computer Vision and Pattern Recognition (CVPR2005), vol. 1, pp. 886–893 (2005)

31. Ding, J., Li, Y., Xu, X., Wang, L.: X-ray image segmentation by attribute relational graph matching. In: 8th IEEE International Conference on Signal Processing, vol. 2 (2006)

32. Dollár, P.: Piotr's Image and Video Matlab Toolbox (PMT). http://vision.ucsd.edu/~pdollar/toolbox/doc/index.html

33. Douarre, C., Schielein, R., Frindel, C., Gerth, S., Rousseau, D.: Deep learning based root-soil segmentation from X-ray tomography images. bioRxiv p. 071662 (2016)

34. Du, W., Shen, H., Fu, J., Zhang, G., He, Q.: Approaches for improvement of the X-ray image defect detection of automobile casting aluminum parts based on deep learning. NDT & E Int. **107**, 102,144 (2019)

35. Du, Z., Hu, Y., Buttar, N.A., Mahmood, A.: X-ray computed tomography for quality inspection of agricultural products: a review. Food Sci. Nutr. **7**(10), 3146 (2019)

36. Duan, X., Cheng, J., Zhang, L., Xing, Y., Chen, Z., Zhao, Z.: X-ray cargo container inspection system with few-view projection imaging. Nuclear Instrum. Methods Phys. Res. A **598**, 439–444 (2009)

37. Faugeras, O.: Three-Dimensional Computer Vision: A Geometric Viewpoint. The MIT Press, Cambridge (1993)

38. Favata, J., Shahbazmohamadi, S.: Realistic non-destructive testing of integrated circuit bond wiring using 3-d x-ray tomography, reverse engineering, and finite element analysis. Microelectron. Reliability **83**, 91–100 (2018)
39. Feng, X., Zhang, H., Yu, P.: X-ray fluorescence application in food, feed, and agricultural science: a critical review. Critic. Rev. Food Sci. Nutrit. 1–11 (2020)
40. Ferguson, M., Ak, R., Lee, Y.T.T., Law, K.H.: Automatic localization of casting defects with convolutional neural networks. In: 2017 IEEE International Conference on Big Data (Big Data), pp. 1726–1735. IEEE (2017)
41. Ferguson, M.K., Ronay, A., Lee, Y.T.T., Law, K.H.: Detection and segmentation of manufacturing defects with convolutional neural networks and transfer learning. Smart Sustain Manufact. Syst. **2** (2018)
42. Filbert, D., Klatte, R., Heinrich, W., Purschke, M.: Computer aided inspection of castings. In: IEEE-IAS Annual Meeting, pp. 1087–1095. Atlanta, USA (1987)
43. Flitton, G., Breckon, T.P., Megherbi, N.: A comparison of 3d interest point descriptors with application to airport baggage object detection in complex ct imagery. Pattern Recognit. **46**(9), 2420–2436 (2013)
44. Flitton, G., Mouton, A., Breckon, T.P.: Object classification in 3d baggage security computed tomography imagery using visual codebooks. Pattern Recognit. **48**(8), 2489–2499 (2015)
45. Franzel, T., Schmidt, U., Roth, S.: Object detection in multi-view X-Ray images. Pattern Recognit. 144–154 (2012)
46. Frosio, I., Borghese, N., Lissandrello, F., Venturino, G., Rotondo, G.: Optimized acquisition geometry for X-ray inspection. In: 2011 IEEE Instrumentation and Measurement Technology Conference (I2MTC), pp. 1–6 (2011)
47. Gao, W., Hu, Y.H.: Real-time X-ray radiography for defect detection in submerged arc welding and segmentation using sparse signal representation. Insight-Non-Destruct. Test. Condit. Monitor. **56**(6), 299–307 (2014)
48. Ghosh, P., Forte, D., Woodard, D.L., Chakraborty, R.S.: Automated detection of pin defects on counterfeit microelectronics. In: ISTFA 2018: Proceedings from the 44th International Symposium for Testing and Failure Analysis, p. 57. ASM International (2018)
49. Gobi, A.F.: Towards generalized benthic species recognition and quantification using computer vision. In: 4th Pacific-Rim Symposium on Image and Video Technology (PSIVT2010), Singapore, Nov. 14–17, 2010, pp. 94–100 (2010)
50. Gonzalez, R., Woods, R.: Digital Image Processing, 3rd edn. Prentice Hall, Pearson (2008)
51. Goto, K., Kato, K., Nakatsuka, S., Saito, T., Aizawa, H.: Anomaly detection of solder joint on print circuit board by using adversarial autoencoder. In: Fourteenth International Conference on Quality Control by Artificial Vision, vol. 11172, p. 111720T. International Society for Optics and Photonics (2019)
52. Guelpa, A., du Plessis, A., Kidd, M., Manley, M.: Non-destructive estimation of maize (zea mays l.) kernel hardness by means of an X-ray micro-computed tomography (μct) density calibration. Food Bioprocess Technol. **8**(7), 1419–1429 (2015)
53. Haff, R., Slaughter, D.: Real-time X-ray inspection of wheat for infestation by the granary weevil, sitophilus granarius (l.). Trans. Am. Soc. Agricul. Eng. **47**, 531–537 (2004)
54. Haff, R., Toyofuku, N.: X-ray detection of defects and contaminants in the food industry. Sens. Instrum. Food Quality Safety **2**(4), 262–273 (2008). https://doi.org/10.1007/s11694-008-9059-8
55. Harris, C., Stephens, M.: A combined corner and edge detector. In: Proceedings of the 4th Alvey Vision Conferences, pp. 147–152 (1988)
56. Hartley, R.I., Zisserman, A.: Multiple View Geometry in Computer Vision, 2nd edn. Cambridge University Press, Cambridge (2003)
57. Hassan, J., Awan, A.M., Jalil, A.: Welding defect detection and classification using geometric features. In: 2012 10th International Conference on Frontiers of Information Technology, pp. 139–144. IEEE (2012)
58. Hecker, H.: Ein neues Verfahren zur robusten Röntgenbildauswertung in der automatischen Gußteilprüfung. Ph.D. thesis, vom Fachbereich Elektrotechnik, Technische Universität Berlin (1995)

59. Heinrich, W.: Automated inspection of castings using X-ray testing. Ph.D. thesis, Institute for Measurement and Automation, Faculty of Electrical Engineering, Technical University of Berlin (1988). (in German)

60. Heitz, G., Chechik, G.: Object separation in X-ray image sets. In: IEEE Conference on Computer Vision and Pattern Recognition (CVPR-2010), pp. 2093–2100 (2010)

61. Hou, W., Wei, Y., Guo, J., Jin, Y., et al.: Automatic detection of welding defects using deep neural network. In: Journal of Physics: Conference Series, vol. 933, p. 012006. IOP Publishing (2018)

62. Hou, W., Wei, Y., Jin, Y., Zhu, C.: Deep features based on a dcnn model for classifying imbalanced weld flaw types. Measurement **131**, 482–489 (2019)

63. Jaccard, N., Rogers, T.W., Morton, E.J., Griffin, L.D.: Tackling the X-ray cargo inspection challenge using machine learning. In: Anomaly Detection and Imaging with X-Rays (ADIX), vol. 9847, p. 98470N. International Society for Optics and Photonics (2016)

64. Jaccard, N., Rogers, T.W., Morton, E.J., Griffin, L.D.: Detection of concealed cars in complex cargo X-ray imagery using deep learning. J. X-ray Sci. Technol. **25**(3), 323–339 (2017)

65. Jiang, J., Chang, H., Wu, K., Ouyang, C., Yang, M., Yang, E., Chen, T., Lin, T.: An adaptive image segmentation algorithm for X-ray quarantine inspection of selected fruits. Comput. Electron. Agricul. **60**, 190–200 (2008)

66. Jin, C., Kong, X., Chang, J., Cheng, H., Liu, X.: Internal crack detection of castings: a study based on relief algorithm and adaboost-svm. Int. J. Adv. Manufact. Technol. 1–10 (2020)

67. Kaftandjian, V., Dupuis, O., Babot, D., Zhu, Y.M.: Uncertainty modelling using dempster-shafer theory for improving detection of weld defects. Pattern Recognit. Lett. **24**(1), 547–564 (2003)

68. Kamalakannan, A., Rajamanickam, G.: Spatial smoothing based segmentation method for internal defect detection in X-ray images of casting components. In: 2017 Trends in Industrial Measurement and Automation (TIMA), pp. 1–6. IEEE (2017)

69. Karunakaran, C., Jayas, D., White, N.: Identification of wheat kernels damaged by the red flour beetle using X-ray images. Biosyst. Eng. **87**(3), 267–274 (2004)

70. Kelkar, S., Boushey, C.J., Okos, M.: A method to determine the density of foods using X-ray imaging. J. Food Eng. (2015)

71. Kolkoori, S., Wrobel, N., Deresch, A., Redmer, B., Ewert, U.: Dual high-energy X-ray digital radiography for material discrimination in cargo containers. In: 11th European Conference on Non-Destructive Testing (ECNDT 2014), October 6–10, 2014. Prague, Czech Republic (2014)

72. Kolokytha, S., Flisch, A., Lüthi, T., Plamondon, M., Visser, W., Schwaninger, A., Hardmeier, D., Costin, M., Vienne, C., Sukowski, F.: Creating a reference database of cargo inspection X-ray images using high energy radiographs of cargo mock-ups. Multimedia Tools Appl. **77**(8), 9379–9391 (2018)

73. Kotwaliwale, N., Singh, K., Kalne, A., Jha, S.N., Seth, N., Kar, A.: X-ray imaging methods for internal quality evaluation of agricultural produce. J. food Sci. Technol. **51**(1), 1–15 (2014)

74. Kumar, J., Anand, R., Srivastava, S.: Flaws classification using ann for radiographic weld images. In: 2014 International Conference on Signal Processing and Integrated Networks (SPIN), pp. 145–150 (2014)

75. Kumar, J., Anand, R., Srivastava, S.: Multi - class welding flaws classification using texture feature for radiographic images. In: 2014 International Conference on Advances in Electrical Engineering (ICAEE), pp. 1–4 (2014)

76. Kwon, J., Lee, J., Kim, W.: Real-time detection of foreign objects using X-ray imaging for dry food manufacturing line. In: Proceedings of IEEE International Symposium on Consumer Electronics (ISCE 2008), pp. 1–4 (2008)

77. LeCun, Y., Bengio, Y., Hinton, G.: Deep learning. Nature **521**(7553), 436–444 (2015)

78. Lee, D., Lee, J., Min, J., Lee, B., Lee, B., Oh, K., Kim, J., Cho, S.: Efficient material decomposition method for dual-energy X-ray cargo inspection system. Nuclear Instrum. Methods Phys. Res. Sect. A: Accelerat., Spectrom., Detect. Assoc. Equip. **884**, 105–112 (2018)

79. Li, J., Oberdorfer, B., Schumacher, P.: Determining casting defects in thixomolding mg casting part by computed tomography. In: Shape Casting, pp. 99–103. Springer, Berlin (2019)

80. Li, W., Li, K., Huang, Y., Deng, X.: A new trend peak algorithm with X-ray image for wheel hubs detection and recognition. In: Computational Intelligence and Intelligent Systems, pp. 23–31. Springer, Berlin (2015)

81. Li, X., Tso, S.K., Guan, X.P., Huang, Q.: Improving automatic detection of defects in castings by applying wavelet technique. IEEE Trans. Indust. Electron. **53**(6), 1927–1934 (2006)

82. Liao, T.: Classification of welding flaw types with fuzzy expert systems. Fuzzy Sets Syst. **108**, 145–158 (2003)

83. Liao, T.: Classification of weld flaws with imbalanced class data. Expert Systems with Applications **35**(3), 1041–1052 (2008)

84. Liao, T.W.: Improving the accuracy of computer-aided radiographic weld inspection by feature selection. NDT&E Int. **42**, 229–239 (2009)

85. Lin, C.S., Chan, B.E., Huang, Y.C., Chen, H.T., Lin, Y.C.: X-ray imaging inspection system for blind holes in the intermediate layer of printed circuit boards with neural network identification. J. Test. Eval. **45**(3), 1005–1015 (2017)

86. Lin, J., Yao, Y., Ma, L., Wang, Y.: Detection of a casting defect tracked by deep convolution neural network. Int. J. Adv. Manufact. Technol. **97**(1–4), 573–581 (2018)

87. Lindgren, E.: Detection, 3-D positioning, and sizing of small pore defects using digital radiography and tracking. EURASIP J. Adv. Signal Process. **2014**(1), 1–17 (2014)

88. Liu, B., Zhang, X., Gao, Z., Chen, L.: Weld defect images classification with vgg16-based neural network. In: International Forum on Digital TV and Wireless Multimedia Communications, pp. 215–223. Springer (2017)

89. Liu, D., Wang, Z.: A united classification system of X-ray image based on fuzzy rule and neural networks. In: 3rd International Conference on Intelligent System and Knowledge Engineering, 2008. ISKE 2008, vol. 1, pp. 717–722 (2008)

90. Liu, J., Leng, X., Liu, Y.: Deep convolutional neural network based object detector for x-ray baggage security imagery. In: 2019 IEEE 31st International Conference on Tools with Artificial Intelligence (ICTAI), pp. 1757–1761. IEEE (2019)

91. Lowe, D.: Distinctive image features from scale-invariant keypoints. Int. J. Comput. Vis. **60**(2), 91–110 (2004)

92. Lu, Q., Conners, R.: Using image processing methods to improve the explosive detection accuracy. IEEE Trans. Appl. Rev., Part C: Syst., Man, Cybern. **36**(6), 750–760 (2006)

93. Mahmood, K., Carmona, P.L., Shahbazmohamadi, S., Pla, F., Javidi, B.: Real-time automated counterfeit integrated circuit detection using x-ray microscopy. Appl. Opt. **54**(13), D25–D32 (2015)

94. Mansoor, M., Rajashankari, R.: Detection of concealed weapons in X-ray images using fuzzy K-NN. Int. J. Comput. Sci., Eng. Inf. Technol. **2**(2) (2012)

95. Matas, J., Chum, O., Urban, M., Pajdla, T.: Robust wide-baseline stereo from maximally stable extremal regions. Image Vis. Comput. **22**(10), 761–767 (2004)

96. Mathanker, S., Weckler, P., Bowser, T., Wang, N., Maness, N.: Adaboost classifiers for pecan defect classification. Comput. Electron. Agricul. **77**(1), 60–68 (2011)

97. Mery, D.: BALU: A toolbox Matlab for computer vision, pattern recognition and image processing. http://dmery.ing.puc.cl/index.php/balu

98. Mery, D.: Flaw simulation in castings inspection by radioscopy. Insight **43**(10), 664–668 (2001)

99. Mery, D.: Crossing line profile: a new approach to detecting defects in aluminium castings. In: Proceedings of the Scandinavian Conference on Image Analysis (SCIA 2003). Lecture Notes in Computer Science, vol. 2749, pp. 725–732 (2003)

100. Mery, D.: Automated radioscopic testing of aluminum die castings. Mater. Eval. **64**(2), 135–143 (2006)

101. Mery, D.: Automated detection in complex objects using a tracking algorithm in multiple X-ray views. In: Proceedings of the 8th IEEE Workshop on Object Tracking and Classification Beyond the Visible Spectrum (OTCBVS 2011), in Conjunction with CVPR 2011, Colorado Springs, pp. 41–48 (2011)

102. Mery, D.: Automated detection of welding defects without segmentation. Mater. Eval. **69**(6), 657–663 (2011)
103. Mery, D.: Inspection of complex objects using multiple-X-ray views. IEEE/ASME Trans. Mechatron. **20**(1), 338–347 (2015)
104. Mery, D.: Aluminum casting inspection using deep learning: a method based on convolutional neural networks. J. Nondestruct. Eval. **39**(1), 12 (2020)
105. Mery, D., Arteta, C.: Automatic defect recognition in X-ray testing using computer vision. In: 2017 IEEE Winter Conference on Applications of Computer Vision (WACV), pp. 1026–1035. IEEE (2017)
106. Mery, D., Berti, M.A.: Automatic detection of welding defects using texture features. Insight-Non-Destruct. Test. Condit. Monitor. **45**(10), 676–681 (2003)
107. Mery, D., Filbert, D.: Automated flaw detection in aluminum castings based on the tracking of potential defects in a radioscopic image sequence. IEEE Trans. Robot. Autom. **18**(6), 890–901 (2002)
108. Mery, D., Filbert, D.: Classification of potential defects in automated inspection of aluminium castings using statistical pattern recognition. In: 8th European Conference on Non-Destructive Testing (ECNDT 2002), pp. 1–10. Barcelona (2002)
109. Mery, D., Filbert, D., Jaeger, T.: Image processing for fault detection in aluminum castings. In: MacKenzie, D., Totten, G. (eds.) Anal. Charact. Alumin. Alloys. Marcel Dekker, New York (2003)
110. Mery, D., Filbert, D., Parspour, N.: Improvement in automated aluminum casting inspection by finding correspondence of potential flaws in multiple radioscopic images. In: Proceedings of the 15th World Conference on Non-Destructive Testing (WCNDT–2000). Rome (2000)
111. Mery, D., Lillo, I., Riffo, V., Soto, A., Cipriano, A., Aguilera, J.: Automated fish bone detection using X-ray testing. J. Food Eng. **2011**(105), 485–492 (2011)
112. Mery, D., Riffo, V., Mondragon, G., Zuccar, I.: Detection of regular objects in baggages using multiple X-ray views. Insight **55**(1), 16–21 (2013)
113. Mery, D., Riffo, V., Zscherpel, U., Mondragón, G., Lillo, I., Zuccar, I., Lobel, H., Carrasco, M.: GDXray: the database of X-ray images for nondestructive testing. J. Nondestruct. Eval. **34**(4), 1–12 (2015)
114. Mery, D., Riffo, V., Zuccar, I., Pieringer, C.: Automated X-ray object recognition using an efficient search algorithm in multiple views. In: Proceedings of the 9th IEEE CVPR Workshop on Perception Beyond the Visible Spectrum, Portland (2013)
115. Mery, D., Svec, E., Arias, M., Riffo, V., Saavedra, J.M., Banerjee, S.: Modern computer vision techniques for X-ray testing in baggage inspection. IEEE Trans. Syst., Man, Cybern.: Syst. **47**(4), 682–692 (2016)
116. Miao, C., Xie, L., Wan, F., Su, C., Liu, H., Jiao, J., Ye, Q.: Sixray: a large-scale security inspection X-ray benchmark for prohibited item discovery in overlapping images. In: Proceedings of the IEEE Conference on Computer Vision and Pattern Recognition, pp. 2119–2128 (2019)
117. Michel, S., Koller, S., de Ruiter, J., Moerland, R., Hogervorst, M., Schwaninger, A.: Computer-based training increases efficiency in X-Ray image interpretation by aviation security screeners. In: 2007 41st Annual IEEE International Carnahan Conference on Security Technology, pp. 201–206 (2007)
118. Montabone, S., Soto, A.: Human detection using a mobile platform and novel features derived from a visual saliency mechanism. Image Vis. Comput. **28**(3), 391–402 (2010)
119. Mouton, A., Breckon, T.P.: Materials-based 3d segmentation of unknown objects from dual-energy computed tomography imagery in baggage security screening. Pattern Recognit. **48**(6), 1961–1978 (2015)
120. Mu, W., Gao, J., Jiang, H., Wang, Z., Chen, F., Dang, C.: Automatic classification approach to weld defects based on pca and svm. Insight-Non-Destruct. Test. Condit. Monitor. **55**(10), 535–539 (2013)
121. Muniategui, A., del Barrio, J.A., Vinuesa, X.A., Masenlle, M., de la Yedra, A.G., Moreno, R.: One dimensional fourier transform on deep learning for industrial welding quality control. In: International Work-Conference on Artificial Neural Networks, pp. 174–185. Springer (2019)

122. Muravyov, S., Pogadaeva, E.Y.: Computer-aided recognition of defects in welded joints during visual inspections based on geometric attributes. Russian J. Nondestruct. Test. **56**, 259–267 (2020)

123. Murphy, E.: A rising war on terrorists. Spectrum, IEEE **26**(11), 33–36 (1989)

124. Murray, N., Riordan, K.: Evaluation of automatic explosive detection systems. In: 29th Annual 1995 International Carnahan Conference on Security Technology, 1995. Proceedings. Institute of Electrical and Electronics Engineers, pp. 175–179 (1995). https://doi.org/10.1109/CCST.1995.524908

125. Neethirajan, S., Karunakaran, C., Symons, S., Jayas, D.: Classification of vitreousness in durum wheat using soft X-rays and transmitted light images. Comput. Electron. Agricul. **53**(1), 71–78 (2006)

126. Nercessian, S., Panetta, K., Agaian, S.: Automatic detection of potential threat objects in X-ray luggage scan images. In: 2008 IEEE Conference on Technologies for Homeland Security, pp. 504–509 (2008). 10.1109/THS.2008.4534504

127. Nercessian, S., Panetta, K., Agaian, S.: Automatic detection of potential threat objects in X-ray luggage scan images. In: 2008 IEEE Conference on Technologies for Homeland Security, pp. 504–509 (2008)

128. Nielsen, M.S., Christensen, L.B., Feidenhans, R.: Frozen and defrosted fruit revealed with X-ray dark-field radiography. Food Control **39**, 222–226 (2014)

129. Noble, A., Gupta, R., Mundy, J., Schmitz, A., Hartley, R.: High precision X-ray stereo for automated 3D CAD-based inspection. IEEE Trans. Robot. Autom. **14**(2), 292–302 (1998)

130. Nugraha, B., Verboven, P., Janssen, S., Wang, Z., Nicolaï, B.M.: Non-destructive porosity mapping of fruit and vegetables using X-ray ct. Postharvest Biol. Technol. **150**, 80–88 (2019)

131. Oertel, C., Bock, P.: Identification of objects-of-interest in X-Ray images. In: Applied Imagery and Pattern Recognition Workshop, 2006. AIPR 2006. 35th IEEE, p. 17 (2006)

132. Ogawa, Y., Kondo, N., Shibusawa, S.: Inside quality evaluation of fruit by X-ray image. In: 2003 IEEE/ASME International Conference on Advanced Intelligent Mechatronics, 2003. AIM 2003. Proceedings. vol. 2, pp. 1360–1365 (2003)

133. Ojala, T., Pietikainen, M., Maenpaa, T.: Multiresolution gray-scale and rotation invariant texture classification with local binary patterns. IEEE Trans. Pattern Anal. Mach. Intell. **24**(7), 971–987 (2002)

134. Orina, I., Manley, M., Kucheryavskiy, S., Williams, P.J.: Application of image texture analysis for evaluation of X-ray images of fungal-infected maize kernels. Food Anal. Methods **11**(10), 2799–2815 (2018)

135. Pan, H., Pang, Z., Wang, Y., Wang, Y., Chen, L.: A new image recognition and classification method combining transfer learning algorithm and mobilenet model for welding defects. IEEE Access (2020)

136. Pieringer, C., Mery, D.: Flaw detection in aluminium die castings using simultaneous combination of multiple views. Insight **52**(10), 548–552 (2010)

137. Pizarro, L., Mery, D., Delpiano, R., Carrasco, M.: Robust automated multiple view inspection. Pattern Anal. Appl. **11**(1), 21–32 (2008)

138. Ramírez, F., Allende, H.: Detection of flaws in aluminium castings: a comparative study between generative and discriminant approaches. Insight-Non-Destruct. Test. Condit. Monitor. **55**(7), 366–371 (2013)

139. Ren, J., Ren, R., Green, M., Huang, X.: Defect detection from X-ray images using a three-stage deep learning algorithm. In: 2019 IEEE Canadian Conference of Electrical and Computer Engineering (CCECE), pp. 1–4. IEEE (2019)

140. Riffo, V., Flores, S., Mery, D.: Threat objects detection in X-ray images using an active vision approach. J. Nondestruct. Eval. **36**(3), 44 (2017)

141. Riffo, V., Godoy, I., Mery, D.: Handgun detection in single-spectrum multiple X-ray views based on 3d object recognition. J. Nondestruct. Eval. **38**(3), 66 (2019)

142. Riffo, V., Mery, D.: Active X-ray testing of complex objects. Insight **54**(1), 28–35 (2012)

143. Riffo, V., Mery, D.: Automated detection of threat objects using adapted implicit shape model. IEEE Trans. Syst., Man, Cybern.: Syst. **46**(4), 472–482 (2016)

144. Rogers, T.W., Jaccard, N., Griffin, L.D.: A deep learning framework for the automated inspection of complex dual-energy X-ray cargo imagery. In: Anomaly Detection and Imaging with X-Rays (ADIX) II, vol. 10187, p. 101870L. International Society for Optics and Photonics (2017)

145. Saavedra, D., Banerjee, S., Mery, D.: Detection of threat objects in baggage inspection with X-ray images using deep learning. Neural Comput. Appl. pp. 1–17. Springer (2020)

146. Sangwan, D., Jain, D.K.: An evaluation of deep learning based object detection strategies for threat object detection in baggage security imagery. Pattern Recognit. Lett. (2019)

147. Schmidt-Hackenberg, L., Yousefi, M.R., Breuel, T.M.: Visual cortex inspired features for object detection in X-ray images. In: 2012 21st International Conference on Pattern Recognition (ICPR), pp. 2573–2576. IEEE (2012)

148. Schoeman, L., Williams, P., du Plessis, A., Manley, M.: X-ray micro-computed tomography (μct) for non-destructive characterisation of food microstructure. Trends Food Sci. Technol. **47**, 10–24 (2016)

149. Shao, J., Du, D., Chang, B., Shi, H.: Automatic weld defect detection based on potential defect tracking in real-time radiographic image sequence. NDT & E Int. **46**, 14–21 (2012)

150. Shi, D.H., Gang, T., Yang, S.Y., Yuan, Y.: Research on segmentation and distribution features of small defects in precision weldments with complex structure. NDT & E Int. **40**, 397–404 (2007)

151. Sigman, J.B., Spell, G.P., Liang, K.J., Carin, L.: Background adaptive faster R-CNN for semi-supervised convolutional object detection of threats in X-ray images. In: Anomaly Detection and Imaging with X-Rays (ADIX) V, vol. 11404, p. 1140404. International Society for Optics and Photonics (2020)

152. da Silva, R., Mery, D.: State-of-the-art of weld seam inspection using X-ray testing: Part I - image processing. Mater. Eval. **65**(6), 643–647 (2007)

153. da Silva, R., Mery, D.: State-of-the-art of weld seam inspection using X-ray testing: Part II - pattern recognition. Mater. Eval. **65**(9), 833–838 (2007)

154. da Silva, R.R., Calôba, L.P., Siqueira, M.H., Rebello, J.M.: Pattern recognition of weld defects detected by radiographic test. Ndt & E Int. **37**(6), 461–470 (2004)

155. Simonyan, K., Zisserman, A.: Very deep convolutional networks for large-scale image recognition. CoRR arXiv:abs/1409.1556 (2014)

156. Singh, M., Singh, S.: Optimizing image enhancement for screening luggage at airports. In: Proceedings of the 2005 IEEE International Conference on Computational Intelligence for Homeland Security and Personal Safety, 2005. CIHSPS 2005, pp. 131 –136 (2005). https://doi.org/10.1109/CIHSPS.2005.1500627

157. Singh, S., Singh, M.: Explosives detection systems (eds) for aviation security. Signal Process. **83**(1), 31–55 (2003)

158. Sivic, J., Zisserman, A.: Efficient visual search of videos cast as text retrieval. IEEE Trans. Pattern Anal. Mach. Intell. **31**(4), 591–605 (2009)

159. Steitz, J.M.O., Saeedan, F., Roth, S.: Multi-view X-ray R-CNN. arXiv preprint arXiv:1810.02344 (2018)

160. Strecker, H.: Automatic detection of explosives in airline baggage using elastic X-ray scatter. Medicamundi **42**, 30–33 (1998)

161. Suyama, F.M., Delgado, M.R., da Silva, R.D., Centeno, T.M.: Deep neural networks based approach for welded joint detection of oil pipelines in radiographic images with double wall double image exposure. NDT & E Int. **105**, 46–55 (2019)

162. Tang, Y., Zhang, X., Li, X., Guan, X.: Application of a new image segmentation method to detection of defects in castings. Int. J. Adv. Manufact. Technol. **43**(5–6), 431–439 (2009)

163. Tang, Z., Tian, E., Wang, Y., Wang, L., Yang, T.: Non-destructive defect detection in castings by using spatial attention bilinear convolutional neural network. IEEE Trans. Indust. Inf. 1 (2020)

164. Tong, T., Cai, Y., Sun, D.: Defects detection of weld image based on mathematical morphology and thresholding segmentation. In: 2012 8th International Conference on Wireless Communications, Networking and Mobile Computing, pp. 1–4. IEEE (2012)

165. Turcsany, D., Mouton, A., Breckon, T.P.: Improving feature-based object recognition for X-ray baggage security screening using primed visualwords. In: IEEE International Conference on Industrial Technology (ICIT), pp. 1140–1145 (2013)

166. Uehara, M., Yashiro, W., Momose, A.: Effectiveness of X-ray grating interferometry for non-destructive inspection of packaged devices. J. Appl. Phys. **114**(13), 134,901 (2013)

167. Uroukov, I., Speller, R.: A preliminary approach to intelligent X-ray imaging for baggage inspection at airports. Signal Process. Res. **4**, 1–11 (2015)

168. Van Dael, M., Verboven, P., Zanella, A., Sijbers, J., Nicolai, B.: Combination of shape and X-ray inspection for apple internal quality control: in silico analysis of the methodology based on X-ray computed tomography. Postharvest Biol. Technol. **148**, 218–227 (2019)

169. Vedaldi, A., Fulkerson, B.: VLFeat: An open and portable library of computer vision algorithms. In: Proceedings of the International Conference on Multimedia, pp. 1469–1472. ACM (2010)

170. Vilar, R., Zapata, J., Ruiz, R.: An automatic system of classification of weld defects in radiographic images. NDT & E Int. (2009)

171. Viola, P., Jones, M.: Robust real-time object detection. Int. J. Comput. Vis. **57**(2), 137–154 (2004)

172. Wales, A., Halbherr, T., Schwaninger, A.: Using speed measures to predict performance in X-ray luggage screening tasks. In: 43rd Annual 2009 International Carnahan Conference on Security Technology, 2009, pp. 212–215 (2009)

173. Wang, Y., Shi, F., Tong, X.: A welding defect identification approach in X-ray images based on deep convolutional neural networks. In: International Conference on Intelligent Computing, pp. 53–64. Springer (2019)

174. Wang, Y., Sun, Y., Lv, P., Wang, H.: Detection of line weld defects based on multiple thresholds and support vector machine. NDT & E Int. **41**(7), 517–524 (2008)

175. Wang, Y., Wang, M., Zhang, Z.: Microfocus X-ray printed circuit board inspection system. Optik-Int. J. Light and Electron Opt. **125**(17), 4929–4931 (2014)

176. Wells, K., Bradley, D.: A review of X-ray explosives detection techniques for checked baggage. Appl. Radiat. Isotopes (2012)

177. Wu, J.h., Yan, X.y., Wang, G.: High-resolution pcb board defect detection system based on non-destructive detection. Instrum. Tech. Sens. **6**, 028 (2013)

178. Xu, C., Han, N., Li, H.: A dangerous goods detection approach based on yolov3. In: Proceedings of the 2018 2Nd International Conference on Computer Science and Artificial Intelligence, CSAI '18, pp. 600–603. ACM, New York (2018). https://doi.org/10.1145/3297156.3297199. http://doi.acm.org/10.1145/3297156.3297199

179. Yahaghi, E., Mirzapour, M., Movafeghi, A.: Enhancing flaw detection in aluminum castings by two different mixed noise removal methods. Phys. Script. **95**(7), 075,302 (2020)

180. Yang, J., Zhao, Z., Zhang, H., Shi, Y.: Data augmentation for X-ray prohibited item images using generative adversarial networks. IEEE Access **7**, 28894–28902 (2019)

181. Yirong, Z., Dong, D., Baohua, C., Linhong, J., Jiluan, P.: Automatic weld defect detection method based on kalman filtering for real-time radiographic inspection of spiral pipe. NDT & E Int. (2015)

182. Yongwei, Y., Liuqing, D., Cuilan, Z., Jianheng, Z.: Automatic localization method of small casting defect based on deep learning feature. Chinese J. Sci. Instrum. **2016**(6), 21 (2016)

183. Yuanxi, W., Liu, X.: Dangerous goods detection based on transfer learning in X-ray images. Neural Comput. Appl. (2019). https://doi.org/10.1007/s00521-019-04360-0

184. Zakaria, S., Amir, A., Yaakob, N., Nazemi, S.: Automated detection of printed circuit boards (pcb) defects by using machine learning in electronic manufacturing: Current approaches. MS&E **767**(1), 012,064 (2020)

185. Zapata, J., Vilar, R., Ruiz, R.: Automatic inspection system of welding radiographic images based on ann under a regularisation process. J. Nondestruct. Eval. **31**(1), 34–45 (2012)

186. Zentai, G.: X-ray imaging for homeland security. In: IEEE International Workshop on Imaging Systems and Techniques (IST 2008) pp. 1–6 (2008)

187. Zhang, J., Guo, Z., Jiao, T., Wang, M.: Defect detection of aluminum alloy wheels in radiography images using adaptive threshold and morphological reconstruction. Appl. Sci. **8**(12), 2365 (2018)
188. Zhang, N., Zhu, J.: A study of X-ray machine image local semantic features extraction model based on bag-of-words for airport security. Int. J. Smart Sens. Intell. Syst. **1**, 45–64 (2015)
189. Zhao, X., He, Z., Zhang, S.: Defect detection of castings in radiography images using a robust statistical feature. JOSA A **31**(1), 196–205 (2014)
190. Zhao, X., He, Z., Zhang, S., Liang, D.: A sparse-representation-based robust inspection system for hidden defects classification in casting components. Neurocomputing **153**, 1–10 (2015)
191. Zhong, J., Zhang, F., Lu, Z., Liu, Y., Wang, X.: High-speed display-delayed planar X-ray inspection system for the fast detection of small fishbones. J. Food Process Eng. **42**(3), e13,010 (2019)
192. Zhu, Z., Hu, Y.C., Zhao, L.: Gamma/X-ray linear pushbroom stereo for 3D cargo inspection. Mach. Vis. Appl. **21**(4), 413–425 (2010)
193. Zhu, Z., Zhao, L., Lei, J.: 3D measurements in cargo inspection with a gamma-ray linear pushbroom stereo system. In: Proceedings of the 2005 IEEE Computer Society Conference on Computer Vision and Pattern Recognition (CVPR-05) (2005)
194. Zou, L., Yusuke, T., Hitoshi, I.: Dangerous objects detection of X-ray images using convolution neural network. In: International Conference on Security with Intelligent Computing and Big-Data Services, pp. 714–728. Springer (2018)

Appendix A
$\mathbb{GDX}$ray+ Database

In this Appendix we show the details of each series of $\mathbb{GDX}$ray+. The database consists of 23,189 X-ray images. The images are organized in a public database called $\mathbb{GDX}$ray+ that can be used free of charge,[1] for research and educational purposes only. The database includes five groups of X-ray images: castings, welds, baggages, natural objects, and settings. Each group has several series, and each series several X-ray images. The most of the series are annotated or labeled. In those cases, the coordinates of the bounding boxes of the objects of interest or the labels of the images are available in standard text files. The size of $\mathbb{GDX}$ray+ is 4.5 GB and it can be downloaded from our website.

The details of each series are summarized in following tables: Table A.1 for natural objects, Table A.2 for castings, Table A.3 for baggages, Table A.4 for welds, and Table A.5 for setting X-ray images. See more about $\mathbb{GDX}$ray+ in Chap. 2.

---

[1] Available on https://domingomery.ing.puc.cl/material/.

© Springer Nature Switzerland AG 2021
D. Mery and C. Pieringer, *Computer Vision for X-Ray Testing*,
https://doi.org/10.1007/978-3-030-56769-9

437

Table A.1 Description of group 'Nature' of GDXray+

Series	Images	kpixels	Description	Additional
N0001	13	5935.1	Apples	
N0002	200	10.0	Cropped images of 100 × 100 pixels for fish bone detection	Labels
N0003	7,697	0.1	Cropped images of 10 × 10 pixels for fish bone detection	Labels
N0004	20	143.3	Static noisy images of a wood piece	
N0005	9	4076.7	Apples	Annotations for apples
N0006	27	5935.1	Cherries	Annotations for cherries
N0007	8	5935.1	Cherries	Annotations for cherries
N0008	3	5935.1	Kiwis	Annotations for cherries
N0009	39	585.0	Wood pieces	
N0010	99	83.6	Wood pieces	
N0011	163	5935.1	Salmon filets	
N0012	6	5935.1	Selected 6 images of N0011	Annotation for fish bones. See N0013
N0013	6	5935.1	Binary ideal segmentation of N0012	Original images in N0012

Table A.2 Description of group 'Castings' of GDXray+

Series	Images	kpixels	Description	Additional
C0001	72	439.3	Wheel: Rotation each 5 degrees	Annotations for defects, calibration
C0002	90	44.5	Crops of C0001 with and without defects	Annotations for defects
C0003	37	439.3	Wheel with slow rotation	
C0004	37	439.3	Wheel with medium rotation. No defects	
C0005	37	439.3	Wheel with fast rotation. No defects	
C0006	37	439.3	Wheel with medium rotation. No defects	
C0007	37	439.3	Wheel with medium rotation	
C0008	37	439.3	Wheel with medium rotation. Large defect	Annotations for defects
C0009	37	439.3	Wheel with medium rotation	
C0010	37	439.3	Wheel with medium rotation. Large defect	Annotations for defects
C0011	37	439.3	Wheel with medium rotation	
C0012	37	439.3	Wheel with medium rotation. No defect	
C0013	37	439.3	Wheel with medium rotation. No defect	
C0014	37	439.3	Wheel with medium rotation. Defect at axis	
C0015	37	439.3	Wheel with medium rotation. Defect at axis	Annotations for defects
C0016	37	439.3	Wheel with medium rotation. Defect at edge	
C0017	37	439.3	Wheel with medium rotation. Defect at edge	
C0018	37	439.3	Wheel with no defect	
C0019	37	439.3	Wheel with hidden defect	Annotations for defects
C0020	37	439.3	Wheel with possible defect at lateral side	
C0021	37	439.3	Wheel with many small drilled defects	Annotations for defects
C0022	37	439.3	Wheel with letters at lateral side	
C0023	37	439.3	Wheel with no defect	
C0024	37	439.3	Wheel with defects at the lateral side	Annotations for defects

(continued)

Table A.2 (continued)

Series	Images	kpixels	Description	Additional
C0025	37	439.3	Wheel with defects like a regular structure	
C0026	37	439.3	Wheel with large hidden defects	Annotations for defects
C0027	37	439.3	Wheel with no defect	
C0028	37	439.3	Wheel with no defect	
C0029	37	439.3	Wheel with no defect (lateral side)	Annotations for regular structure
C0030	37	439.3	Wheel with defect in its axis	Annotations for defects
C0031	37	439.3	Wheel with several defects	Annotations for defects
C0032	37	439.3	Wheel with several defects	Annotations for defects
C0033	37	439.3	Wheel with several defects	Annotations for defects
C0034	37	439.3	Wheel with defects. No motion	Annotations for defects
C0035	37	439.3	Wheel with large defect	Annotations for defects
C0036	37	439.3	Wheel with letters at lateral side	Annotations for letters
C0037	37	439.3	Wheel with large defect on an edge	Annotations for defects
C0038	37	439.3	Wheel with hidden defect	Annotations for defects
C0039	37	439.3	Wheel with hidden defect	Annotations for defects
C0040	37	439.3	Wheel with hidden defect	Annotations for defects
C0041	37	439.3	Wheel with defects. No motion	Annotations for defects
C0042	37	439.3	Wheel with several defects in motion	Annotations for defects
C0043	37	439.3	Wheel with large defect at axis	Annotations for defects
C0044	66	65.5	Wheel with small drilled defects	
C0045	66	65.5	Wheel with small drilled defects	Annotations for defects
C0046	65	65.5	Wheel with small drilled defects	
C0047	72	65.5	Wheel with small drilled defects	Annotations for defects
C0048	71	65.5	Wheel with small drilled defects	

(continued)

Table A.2 (continued)

Series	Images	kpixels	Description	Additional
C0049	63	65.5	Wheel with small drilled defects	
C0050	54	65.5	Wheel with small drilled defects	
C0051	77	65.5	Wheel with small drilled defects	Annotations for defects
C0052	17	440.8	Knuckle with small defects in motion	
C0053	31	440.8	Knuckle with small defects in motion	
C0054	31	440.8	Knuckle with low contrast defects in motion	Annotations for defects
C0055	28	440.8	Sink strainer	Annotations for holes
C0056	10	440.8	Sink strainer high speed	
C0057	31	440.8	Knuckle with low contrast defects in motion	Annotations for defects
C0058	56	440.8	Knuckle with small defects in motion	
C0059	43	440.8	Knuckle with small defects in motion	
C0060	14	440.8	Knuckle with small defects in motion	Annotations for defects
C0061	31	440.8	Knuckle with small defects in motion	
C0062	10	440.8	Knuckle with small defects in motion	Annotations for defects
C0063	11	440.8	Knuckle with small defects in motion	
C0064	56	440.8	Knuckle with small defects in motion	
C0065	10	440.8	Knuckle with small defects in motion	Annotations for defects
C0066	52	440.8	Knuckle with small defects in motion	
C0067	83	440.8	Knuckle with small defects in motion	

(continued)

Table A.2 (continued)

Series	Images	kpixels	Description	Additional
C0068	72	439.3	Copy of series C0001 with different sizes	
C0069	2	1000.0	Unlabeled casting object	
C0070	10	1000.0	Unlabeled casting object	
C0071	2	1000.0	Unlabeled casting object	
C0072	15	1000.0	Unlabeled casting object	
C0073	12	1000.0	Unlabeled casting object	
C0074	5	1000.0	Unlabeled casting object	
C0075	14	1000.0	Unlabeled casting object	
C0076	20	1000.0	Unlabeled casting object	
C0077	5	1000.0	Unlabeled casting object	
C0078	7	1000.0	Unlabeled casting object	
C0079	10	1000.0	Unlabeled casting object	
C0080	32	1048.6	Unlabeled casting object	
C0081	5	1000.0	Unlabeled casting object	
C0082	12	1000.0	Unlabeled casting object	
C0083	20	1000.0	Unlabeled casting object	

Table A.3 Description of group 'Baggages' of GDXray+

Series	Images	kpixels	Description	Additional
B0001	14	5935.1	Pen case with several objects	Annotations for razor blades
B0002	9	1287.0	Pen case with several objects	Annotations for razor blades
B0003	10	1287.0	Pen case with several objects	Annotations for clips
B0004	9	722.5	Pen case with occluded razor blade	Annotations for razor blades
B0005	10	722.5	Pen case with several objects	Annotations for pins
B0006	10	722.5	Pen case with occluded razor blade	Annotations for razor blades
B0007	20	129.6	Razor blade for training purposes	
B0008	361	745.8	Rotation of a knife in 1^0	
B0009	4	276.6	Backpack with handgun	Annotations for handguns
B0010	11	276.6	Backpack with handgun	Annotations for handguns
B0011	10	276.6	Backpack with handgun	Annotations for handguns
B0012	4	276.6	Backpack with handgun	Annotations for handguns
B0013	10	276.6	Backpack with handgun and knife	Annotations for knives
B0014	5	276.6	Backpack with handgun and camera	Annotations for handguns
B0015	5	276.6	Backpack with handgun	Annotations for handguns
B0016	4	276.6	Backpack with self-occluded handgun	Annotations for handguns
B0017	5	276.6	Backpack with occluded handgun	Annotations for handguns
B0018	4	276.6	Backpack with handgun and laptop	Annotations for handguns
B0019	6	276.6	Backpack with handgun and laptop	Annotations for handguns
B0020	4	276.6	Backpack with handgun	Annotations for handguns
B0021	4	276.6	Backpack with handgun	Annotations for handguns
B0022	6	276.6	Backpack with handgun	Annotations for handguns
B0023	6	276.6	Backpack with handgun	Annotations for handguns
B0024	5	276.6	Backpack with handgun	Annotations for handguns

(continued)

Table A.3 (continued)

Series	Images	kpixels	Description	Additional
B0025	4	276.6	Backpack with handgun	Annotations for handguns
B0026	5	276.6	Backpack with handgun	Annotations for handguns
B0027	5	276.6	Backpack with handgun	Annotations for handguns
B0028	5	276.6	Backpack with handgun and laptop	Annotations for handguns
B0029	7	276.6	Backpack with handgun and laptop	Annotations for handguns
B0030	7	276.6	Backpack with handgun	Annotations for handguns
B0031	4	276.6	Backpack with handgun and laptop	Annotations for handguns
B0032	4	276.6	Backpack with handgun	Annotations for handguns
B0033	5	276.6	Backpack with occluded handgun	Annotations for handguns
B0034	6	276.6	Backpack with handgun and laptop	Annotations for handguns
B0035	4	276.6	Backpack with handgun and laptop	Annotations for handguns
B0036	11	276.6	Backpack with self-occluded handgun	Annotations for handguns
B0037	11	276.6	Backpack with handgun	Annotations for handguns
B0038	11	276.6	Backpack with handgun	Annotations for handguns
B0039	9	276.6	Backpack with handgun	Annotations for handguns
B0040	12	276.6	Backpack with handgun	Annotations for handguns
B0041	10	276.6	Backpack with handgun	Annotations for handguns
B0042	19	276.6	Backpack with handgun and knives	Annotations for handguns
B0043	9	276.6	Backpack with handgun and camera	Annotations for handguns
B0044	178	5935.1	Backpack with handgun	Calibration parameters
B0045	90	1287.0	Pen case in 90 positions	Annotations for razor blades
B0046	200	5844.0	Backpack with handgun	Annotations for handguns
B0047	200	5896.9	Backpack with shuriken	Annotations for shuriken

(continued)

Table A.3 (continued)

Series	Images	kpixels	Description	Additional
B0048	200	5412.0	Backpack with razor blade	Annotations for razor blade
B0049	200	759.5	Handguns for training purposes	
B0050	100	741.3	Shuriken for training purposes	
B0051	100	165.6	Razor blades for training purposes	
B0052	144	741.3	Shuriken with 8 points for training purposes	
B0053	144	741.3	Shuriken with 7 points for training purposes	
B0054	144	741.3	Shuriken with 6 points for training purposes	
B0055	800	16.9	200 4-image sequences of single objects	Labels
B0056	1200	18.1	200 6-image sequences of single objects	Labels
B0057	1600	18.0	200 8-image sequences of single objects	Labels
B0058	64	196.6	Crops of clips, springs, razor blades and others	Labels. See B0059
B0059	64	196.6	Binary ideal segmentation of images of B0058	Labels. Original images in B0058
B0060	2	5935.1	Images for dual-energy experiments	Annotations for shuriken
B0061	21	3656.8	Razor blade in a can	
B0062	22	3656.8	Razor blade in a wallet	
B0063	19	3656.8	Razor blade in a CD case	Annotations for razor blades
B0064	19	3656.8	Razor blade in a pen case	
B0065	21	3656.8	Razor blade in a pen case	Annotations for razor blades
B0066	22	3656.8	Razor blade in a pen case	
B0067	17	2856.1	Razor blade in a wallet	Annotations for razor blades
B0068	20	2856.1	Razor blade in a large wallet	
B0069	25	2856.1	Razor blade in a pen case	
B0070	21	2856.1	Razor blade in a pen case	Annotations for razor blades
B0071	22	2856.1	Razor blade in a pen case	

(continued)

Table A.3 (continued)

Series	Images	kpixels	Description	Additional
B0072	22	2856.1	Razor blade in a small pen case	
B0073	20	2856.1	Razor blade in a can	Annotations for razor blades
B0074	37	2856.1	Rotation of a door key in 10^0	
B0075	576	5935.1	Knife in 576 positions	
B0076	576	1581.8	Knife in 576 positions	
B0077	576	1582.6	Knife in 576 positions	
B0078	500	443.9	Non-baggage X-ray images	
B0079	150	1205.6	Cropped handguns	
B0080	150	778.2	Cropped shuriken	
B0081	150	144.7	Cropped razor blades	
B0082	600	133.4	Cropped non-threat objects	
B0083	48	5935.1	Backpacks with no threat objects (original name: BX-100)	
B0084	67	59351.	Dual-energy X-ray images with fruits and knifes	
B0085	48	5935.1	Backpacks with no threat objects (original name: BX-55)	
B0086	1000	4.1	GAN generated patches (64×64 pixels) of background	

Table A.4 Description of group 'Welds' of GDXray+

Series	Images	kpixels	Description	Additional
W0001	10	3323.8	Selection of 10 images of W0003	Annotations for defects. See W0002
W0002	10	3323.8	Binary ideal segmentation of images of W0001	
W0003	68	6693.8	Radiographs from a round robin test performed by BAM	Excel file with real-values
W0004	68	3323.8	Same images of W0002 but with 0 and 255 values	

Table A.5 Description of group 'Settings' of GDXray+

Series	Images	kpixels	Description	Additional
S0001	18	5935.1	Checkerboard captured by flat panel	Calibration parameters
S0002	1	427.9	Regular grid captured by image intensifier	Coordinates of calibration points
S0003	36	440.8	Circular pattern in different positions	Manipulator coordinates, 3D coordinates
S0004	23	440.8	Circular pattern in different positions	Manipulator coordinates, 3D coordinates
S0005	27	440.8	Circular pattern in different positions	Manipulator coordinates, 3D coordinates
S0006	17	440.8	Circular pattern in different positions	Manipulator coordinates, 3D coordinates
S0007	29	440.8	Circular pattern in different positions	Coordinates of calibration points (2D & 3D)
S0008	18	5935.1	Checkerboard of series S0001 with corners	Calibration parameters

Index

© Springer Nature Switzerland AG 2021
D. Mery and C. Pieringer, *Computer Vision for X-Ray Testing*,
https://doi.org/10.1007/978-3-030-56769-9